Thermoelectric Materials 2000—
The Next Generation Materials for
Small-Scale Refrigeration and
Power Generation Applications

MATERIALS RESEARCH SOCIETY
SYMPOSIUM PROCEEDINGS VOLUME 626

Thermoelectric Materials 2000—The Next Generation Materials for Small-Scale Refrigeration and Power Generation Applications

Symposium held April 24–27, 2000, San Francisco, California, U.S.A.

EDITORS:

Terry M. Tritt
Clemson University
Clemson, South Carolina, U.S.A.

George S. Nolas
Marlow Industries
Dallas, Texas, U.S.A.

Gerald D. Mahan
University of Tennessee
Knoxville, Tennessee, U.S.A.

David Mandrus
Oak Ridge National Laboratory
Oak Ridge, Tennessee, U.S.A.

Mercouri G. Kanatzidis
Michigan State University
East Lansing, Michigan, U.S.A.

Materials Research Society
Warrendale, Pennsylvania

CAMBRIDGE UNIVERSITY PRESS
Cambridge, New York, Melbourne, Madrid, Cape Town,
Singapore, São Paulo, Delhi, Mexico City

Cambridge University Press
32 Avenue of the Americas, New York NY 10013-2473, USA

Published in the United States of America by Cambridge University Press, New York

www.cambridge.org
Information on this title: www.cambridge.org/9781107413016

Materials Research Society
506 Keystone Drive, Warrendale, PA 15086
http://www.mrs.org

© Materials Research Society 2001

This publication is in copyright. Subject to statutory exception
and to the provisions of relevant collective licensing agreements,
no reproduction of any part may take place without the written
permission of Cambridge University Press.

This publication has been registered with Copyright Clearance Center, Inc.
For further information please contact the Copyright Clearance Center,
Salem, Massachusetts.

First published 2001
First paperback edition 2013

Single article reprints from this publication are available through
University Microfilms Inc., 300 North Zeeb Road, Ann Arbor, MI 48106

CODEN: MRSPDH

ISBN 978-1-107-41301-6 Paperback

Cambridge University Press has no responsibility for the persistence or
accuracy of URLs for external or third-party internet websites referred to in
this publication, and does not guarantee that any content on such websites is,
or will remain, accurate or appropriate.

CONTENTS

Preface xiii

Materials Research Symposium Proceedings xv

SKUTTERUDITES

*The Synthesis of Metastable Skutterudites and
Crystalline Superlattices Z1.1
    Heike Sellinschegg, Joshua R. Williams, Gene Yoon,
    David C. Johnson, Michael Kaeser, Terry M. Tritt,
    George S. Nolas, and E. Nelson

How Cerium Filling Fraction Influences Thermal Factors and
Magnetism in $Ce_yFe_{4-x}Ni_xSb_{12}$ Z1.2
    L. Chapon, D. Ravot, J.C. Tedenac, and F. Bouree-Vigneron

Thermoelectric Properties of Some Cobalt Phosphide-Arsenide
Compounds Z1.4
    Anucha Watcharapasorn, Robert C. DeMattei, Robert S. Feigelson,
    Thierry Caillat, Alexander Borshchevsky, G. Jeffrey Snyder,
    and Jean-Pierre Fleurial

SUPERLATTICE

*Epitaxial Growth and Thermoelectric Properties of Bi_2Te_3
Based Low Dimensional Structures Z2.1
    Joachim Nurnus, Harald Beyer, Armin Lambrecht,
    and Harald Böttner

Synthesis and Physical Properties of Skutterudite Superlattices Z2.3
    Joshua R. Williams, David C. Johnson, Michael Kaeser,
    Terry M. Tritt, George S. Nolas, and E. Nelson

Artificially Atomic-Scale Ordered Superlattice Alloys For
Thermoelectric Applications Z2.4
    Sunglae Cho, Yunki Kim, Antonio DiVenere, George K.L. Wong,
    Arthur J. Freeman, John B. Ketterson, Linda J. Olafsen, Igor Vurgaftman,
    Jerry R. Meyer, Craig A. Hoffman, and Gang Chen

---

*Invited Paper

Thermoelectric Properties of PbSr(Se,Te)-Based Low
Dimensional Structures Z2.5
  Harald Beyer, Joachim Nurnus, Harald Böttner,
  Armin Lambrecht, Lothar Schmitt, and Friedemann Völklein

NEW MATERIALS I

Thermoelectric Figure of Merit, ZT, of Single Crystal
Pentatellurides ($MTe_{5-x}Se_x$: M = Hf, Zr and x = 0, 0.25) Z3.1
  Roy T. Littleton IV, Terry M. Tritt, Bartosz Zawilski,
  Joseph W. Kolis, Douglas R. Ketchum, and M. Brooks Derrick

*Thermoelectric Properties of Selenide Spinels Z3.3
  G. Jeffrey Snyder, Thierry Caillat, and J-P. Fleurial

Thermoelectric Properties of Tl_9BiTe_6/Tl_9BiSe_6 Solid Solutions Z3.4
  Bernd Wölfing, Christian Kloc, and Ernst Bucher

Investigations of Solid Solutions of $CsBi_4Te_6$ Z3.5
  Duck-Young Chung, Timothy P. Hogan, Nishant Ghelani,
  Paul W. Brazis, Melissa A. Lane, Carl R. Kannewurf,
  and Mercouri G. Kanatzidis

QUANTUM WIRES AND DOTS

*Carrier Pocket Engineering for the Design of Low Dimensional
Thermoelectrics With High $Z_{3D}T$ Z4.3
  Takaaki Koga, Stephen B. Cronin, and Mildred S. Dresselhaus

HALF-HEUSLER ALLOYS AND QUASICRYSTALS

Effects of the Addition of Rhenium on the Thermoelectric
Properties of the AlPdMn Quasicrystalline System Z5.1
  Amy L. Pope, Robert Gagnon, Robert Schneidmiller,
  Paola N. Alboni, Roy T. Littleton IV, Bartosz Zawilski,
  Donny Winkler, Terry M. Tritt, John Strom-Olsen,
  Joseph Kolis, and Stephan Legault

*Invited Paper

Effect of Substitutional Doping on the Thermal Conductivity
of Ti-Based Half-Heusler Compounds Z5.2
    Sriparna Bhattacharya, Vijayabharathi Ponnambalam, Amy L. Pope,
    Y. Xia, S. Joseph Poon, Roy T. Littleton IV, and Terry M. Tritt

High Temperature Thermal Conductivity Measurements of
Quasicrystalline $Al_{70.8}Pd_{20.9}Mn_{8.3}$ Z5.4
    Philip S. Davis, Peter A. Barnes, Cronin B. Vining, Amy L. Pope,
    Robert Schneidmiller, Terry M. Tritt, and Joseph Kolis

TE THEORY

Theoretical Evaluation of the Thermal Conductivity in
Framework (Clathrate) Semiconductors Z6.1
    Jianjun Dong, Otto F. Sankey, Charles W. Myles,
    Ganesh K. Ramachandran, Paul F. McMillan, and Jan Gryko

Electronic Structure of $CsBi_4Te_6$ Z6.2
    Paul Larson, Subhendra D. Mahanti, Duck-Young Chung,
    and Mercouri G. Kanatzidis

*Where Should We Look for High ZT Materials: Suggestions
From Theory Z6.3
    Marco Fornari, David J. Singh, Igor I. Mazin,
    and Joseph L. Feldman

Enhancement of Power Factor in a Thermoelectric Composite
With a Periodic Microstructure Z6.5
    Leonid G. Fel, Yakov M. Strelniker, and David J. Bergman

NEW MATERIALS II

*Connections Between Crystallographic Data and New
Thermoelectric Compounds Z7.1
    Brian C. Sales, Bryan C. Chakoumakos, and David Mandrus

Investigation of the Thermal Conductivity of the
Pentatellurides ($Hf_{1-x}Zr_xTe_5$) Using the Parallel Thermal
Conductance Technique Z7.3
    Bartosz M. Zawilski, Roy T. Littleton IV, Terry M. Tritt,
    Douglas R. Ketchum, and Joseph W. Kolis

*Invited Paper

Compositional and Structural Modifications in Ternary
Bismuth Chalcogenides and Their Thermoelectric Properties Z7.4
    Duck-Young Chung, Melissa A. Lane, John R. Ireland,
    Paul W. Brazis, Carl R. Kannewurf, and Mercouri G. Kanatzidis

Doping Studies of n-Type $CsBi_4Te_6$ Thermoelectric Materials Z7.5
    Melissa A. Lane, John R. Ireland, Paul W. Brazis, Theodora Kyratsi,
    Duck-Young Chung, Mercouri G. Kanatzidis, and Carl R. Kannewurf

Exploring Complex Chalcogenides for Thermoelectric
Applications Z7.6
    Ying C. Wang and Francis J. DiSalvo

POSTER SESSION

Semiconductors With Tetrahedral Anions as Potential
Thermoelectric Materials Z8.2
    Thomas P. Braun, Christopher B. Hoffman, and Francis J. DiSalvo

Lattice Dynamics Study of Anisotropic Heat Conduction in
Superlattices..Z8.3
    B. Yang and G. Chen

Structure and Thermoelectric Properties of New Quaternary Tin
and Lead Bismuth Selenides, $K_{1+x}M_{4-2x}Bi_{7+x}Se_{15}$ (M = Sn, Pb)
and $K_{1-x}Sn_{5-x}Bi_{11+x}Se_{22}$ Z8.4
    Antje Mrotzek, Kyoung-Shin Choi, Duck-Young Chung,
    Melissa A. Lane, John R. Ireland, Paul W. Brazis, Timothy P. Hogan,
    Carl R. Kannewurf, and Mercouri G. Kanatzidis

Processing, Characterization, and Measurement of the Seebeck
Coefficient of Bismuth Microwire Array Composites............................Z8.5
    T.E. Huber and P. Constant

Characterization of New Materials in a Four-Sample
Thermoelectric Measurement System Z8.6
    Nishant A. Ghelani, Sim Y. Loo, Duck-Young Chung,
    Sandrine Sportouch, Stephan de Nardi, Mercouri G. Kanatzidis,
    Timothy P. Hogan, and George S. Nolas

**Crystal Growth of Ternary and Quaternary Alkali Metal
Bismuth Chalcogenides Using Bridgman Technique** Z8.8
    Theodora Kyratsi, Duck-Young Chung, Kyoung-Shin Choi,
    Jeffrey S. Dyck, Wei Chen, Ctirad Uher, and Mercouri G. Kanatzidis

Thermoelectric Properties of Doped Iron Disilicide Z8.10
    Jun-ichi Tani and Hiroyasu Kido

**Transport Properties of the Doped Thermoelectric Material
$K_2Bi_{8-x}Sb_xSe_{13}$** Z8.11
    Paul W. Brazis, John R. Ireland, Melissa A. Lane, Theodora Kyratsi,
    Duck-Young Chung, Mercouri G. Kanatzidis, and Carl R. Kannewurf

**Structural Properties of Strain Symmetrized
Silicon/Germanium (111) Superlattices** Z8.13
    Christoph A. Kleint, Armin Heinrich, Thomas Muehl,
    Joachim Schumann, and Michael Hecker

**Electric and Thermoelectric Properties of Quantum Wires
Based on Bismuth Semimetal and Its Alloys** Z8.16
    Albina A. Nikolaeva, Pavel P. Bodiul, Dmitrii V. Gitsu,
    and Gheorgii Para

**High-Z Lanthanum-Cerium Hexaborate Thin Films for
Low-Temperature Applications** Z8.21
    Armen Kuzanyan, George Badalyan, Sergey Harutyunyan,
    Ashot Gyulamiryan, Violetta Vartanyan, Silvia Petrosyan,
    Nicholas Giordano, Todd Jacobs, Kent Wood, Gilbert Fritz,
    Syed B. Qadri, James Horwitz, Huey-Dau Wu,
    Deborah Van Vechten, and Armen Gulian

THERMIONICS

Thermal Conductivity of Bi/Sb Superlattice Z9.1
    David W. Song, G. Chen, Sunglae Cho, Yunki Kim,
    and John B. Ketterson

**Upper Limitation to the Performance of Single-Barrier
Thermionic Emission Cooling** Z9.4
    Marc D. Ulrich, Peter A. Barnes, and Cronin B. Vining

Umklapp Scattering and Heat Conductivity of Superlattices Z9.5
    Mikhail V. Simkin and Gerald D. Mahan

SKUTTERUDITES II

Partially-Filled Skutterudites: Optimizing the Thermoelectric Properties Z10.1
George S. Nolas, Michael Kaeser, Roy T. Littleton IV,
Terry M. Tritt, Heike Sellinschegg, David C. Johnson,
and Elizabeth Nelson

Bulk Synthesis of Completely and Partially Sn Filled $CoSb_3$ Using the Multilayer Repeat Method Z10.2
Heike Sellinschegg, David C. Johnson, Michael Kaeser,
Terry M. Tritt, George S. Nolas, and Elizabeth Nelson

*__The Influence of Ni on the Transport Properties of $CoSb_3$__ Z10.3
Ctirad Uher, Jeffrey S. Dyck, Wei Chen, Gregory P. Meisner,
and Jihui Yang

Structural Defects in a Partially-Filled Skutterudite Z10.4
Jennifer S. Harper and Ronald Gronsky

DEVICES, MEASUREMENTS, AND APPLICATIONS

Optimization of Bismuth Nanowire Arrays by Electrochemical Deposition Z11.1
John H. Barkyoumb, Jack L. Price, Noel A. Guardala,
Norris Lindsey, David L. Demske, Jagadish Sharma,
Hyoung Ho Kang, and Lourdes Salamanca-Riba

Evaluation of a Thermoelectric Device Utilizing Porous Medium Z11.2
Hideyuki Yasuda, Itsuo Ohnaka, Yoichi Inada,
and Kimitaka Nomura

Electrochemical Deposition of $(Bi,Sb)_2Te_3$ for Thermoelectric Microdevices Z11.3
Jean-Pierre Fleurial, Jennifer A. Herman, G. Jeffrey Snyder,
Margaret A. Ryan, Alexander Borshchevsky, and Chen-Kuo Huang

Transient Thermoelectric Cooling of Thin Film Devices Z11.4
A. Ravi Kumar, Rong Gui Yang, Gang Chen,
and Jean-Pierre Fleurial

---

*Invited Paper

P-Type (SiGe) Si Superlattice Cooler Z11.5
    Xiaofeng Fan, Gehong Zeng, Edward Croke, Gerry Robinson,
    Christopher J. LaBounty, Channing C. Ahn, Ali Shakouri,
    and John E. Bowers

Progress in the Development of Segmented Thermoelectric Unicouples at the Jet Propulsion Laboratory Z11.6
    Thierry Caillat, Jean-Pierre Fleurial, G. Jeffrey Snyder,
    and Alexander Borshchevsky

CLATHRATES

Thermal Conductivity of Type-I and -II Clathrate Compounds Z13.1
    George S. Nolas, Joshua L. Cohn, Michael Kaeser, and Terry M. Tritt

Framework Stoichiometry and Electrical Conductivity of Si-Ge Based Structure-I Clathrates Z13.2
    Ganesh K. Ramachandran, Paul F. McMillan, Jianjun Dong,
    Jan Gryko, and Otto F. Sankey

*****Ultrasound Studies of Clathrate Thermoelectrics** Z13.3
    Veerle M. Keppens, Brian C. Sales, David Mandrus,
    Bryan C. Chakoumakos, and Christiane Laermans

Synthesis and Characterization of Large Single Crystals of Silicon and Germanium Clathrate-II Compounds and a New Tin Compound With Clathrate Layers Z13.5
    Svilen Bobev and Slavi C. Sevov

THIN FILMS TE

Electrodeposition of Bi_2Te_3 Nanowire Composites Z14.1
    Amy L. Prieto, Melissa S. Sander, Angelica M. Stacy,
    Ronald Gronsky, and Timothy Sands

Thermopower of Bi Nanowire Array Composites Z14.2
    Tito E. Huber, Michael J. Graf, Colby A. Foss, Jr.,
    and Pierre Constant

---

*Invited Paper

Experimental Investigation of Thin Film InGaAsP Coolers Z14.4
Christopher J. LaBounty, Ali Shakouri, Gerry Robinson,
Luis Esparza, Patrick Abraham, and John E. Bowers

Author Index

Subject Index

PREFACE

Symposium Z, "Thermoelectric Materials 2000—The Next Generation Materials for Small-Scale Refrigeration and Power Generation Applications," held April 24–27 at the 2000 MRS Spring Meeting in San Francisco, California, was the fourth in a series of MRS symposia which are specifically related to research in new thermoelectric materials [see MRS Symposium Proceedings Vol. 234 (1991), Vol. 478 (1997) and Vol. 545 (1999).] At this Meeting there were over 60 contributed oral presentations, 11 invited talks and 22 poster presentations. Some of the highlights of this Meeting were results presented by Ted Harmon of MIT Lincoln Labs on quantum dot nanostructures of PbTe materials. These materials appear to exhibit a substantial improvement over their bulk counterparts with a $ZT \approx 2$. The Kanatzidis group at Michigan State University presented an update on the $CsBi_4Te_6$ material since their announcement in *Science* Vol. 287, pp.1024-7 (2000). In addition, George Nolas of Marlow Industries and Brian Sales of Oak Ridge National Laboratory presented results on a Yb partially filled skutterudite with $ZT > 1$ at 300 °C. There were also many interesting results in the clathrates, skutterudites, quasicrystals and novel chalcogenide materials systems as well as in III-V compounds for thermionic refrigeration. As in past symposia, both bulk and thin film thermoelectric materials research was well represented. Thermoelectric materials are used in a wide variety of applications related to small-scale solid state refrigeration or power generation.

Over the past thirty years, alloys based on the Bi_2Te_3 compounds (refrigeration) and $Si_{1-x}Ge_x$ compounds (power generation) have been extensively studied and optimized for their use as thermoelectric materials. Thermoelectric cooling is an environmentally "friendly" method of small-scale cooling in specific applications such as cooling computer chips, small beverage coolers, cooling laser diodes and infrared detectors. Another very important application of thermoelectric materials is in power generation for deep space probes such as in the Voyager and Cassini missions. Despite the extensive investigation of these traditional thermoelectric materials, there is still substantial room for improvement, and thus, entirely new classes of compounds will have to be investigated. Therefore, the focus of this symposium centers around the development of the next generation materials for small-scale refrigeration and power generation applications.

The essence of a good thermoelectric is given by the determination of the material's dimensionless figure of merit, $ZT = (\alpha^2 \sigma / \lambda)T$, where α is the Seebeck coefficient, σ is the electrical conductivity, and λ is the total thermal conductivity. The thermal conductivity consists of two parts, the electronic and the lattice thermal conductivity. Many of the papers presented in this proceedings revolve around either maximizing the numerator of ZT called the power factor, $PF = \alpha^2 \sigma T$, or by minimizing the lattice thermal conductivity. As previously described by Glen Slack, a promising thermoelectric material should possess the thermal properties of a glass and the electronic properties of a crystal, i.e. a phonon-glass and electron-crystal (PGEC). This theme is quite prevalent in the many papers presented in this volume. The best thermoelectric materials have a value of $ZT \approx 1$, which has been an upper limit for more than 30 years, yet no theoretical or thermodynamic reason exists for why it cannot be larger. We believe that the future advances in thermoelectric applications will come through research in new materials, and that is why we have focused the symposium on research in new materials, instead of further optimization of established materials. There are currently many new methods of materials synthesis and much more rapid characterization of thermoelectric materials than were available 20 to 30 years ago. Many new researchers and new ideas are appearing in this field, which gives us much anticipation about future advances. It is the hope of the organizers of this symposium that these proceedings will provide a benchmark for the current state in the field of new thermoelectric materials at this time.

At the end of the symposium, there was a special session where program managers Wendy Fuller-Mora (NSF), Valarie Browning (DARPA), Jack Rowe (ARO), and Jerry Smith (DOE) discussed potential opportunities for thermoelectric research. This was followed by a short question and answer period. One distinction at this symposium was the large number of graduate student presentations, numbering over twenty. The symposium organizers were able to give three graduate student presentation awards and a poster award. The student awards for best papers and presentations were: Ganesh Ramachandran, Arizona State University (who also won one of the general MRS best paper awards), Marc Ulrich, Auburn University, and Melissa Lane, Northwestern University. The best student poster award went to Nishant Ghelani of Michigan State University. These awards were enabled by the support of High-Z Corporation, Marlow Industries, MMR Technologies and MRS. The organizers appreciate and acknowledge the support of these sponsors. Much of the research and the results which are presented in this volume were supported by DARPA, the Office of Naval Research and the Army Research Office. Their financial support is greatly appreciated and acknowledged. We also acknowledge the assistance of Marian Littleton and Lori McGowan of Clemson University in all phases of assistance concerning the symposium organization and subsequent proceedings. Their diligence and hard work both before and after the symposium allowed for the timely progress of the manuscripts and proceedings in preparation for publication of this volume.

Terry M. Tritt
George S. Nolas
Gerald D. Mahan
David Mandrus
Mercouri G. Kanatzidis

September 2000

MATERIALS RESEARCH SOCIETY SYMPOSIUM PROCEEDINGS

Volume 578— Multiscale Phenomena in Materials—Experiments and Modeling, I.M. Robertson, D.H. Lassila, R. Phillips, B. Devincre, 2000, ISBN: 1-55899-486-6
Volume 579— The Optical Properties of Materials, J.R. Chelikowsky, S.G. Louie, G. Martinez, E.L. Shirley, 2000, ISBN: 1-55899-487-4
Volume 580— Nucleation and Growth Processes in Materials, A. Gonis, P.E.A. Turchi, A.J. Ardell, 2000, ISBN: 1-55899-488-2
Volume 581— Nanophase and Nanocomposite Materials III, S. Komarneni, J.C. Parker, H. Hahn, 2000, ISBN: 1-55899-489-0
Volume 582— Molecular Electronics, S.T. Pantelides, M.A. Reed, J. Murday, A. Aviram, 2000, ISBN: 1-55899-490-4
Volume 583— Self-Organized Processes in Semiconductor Alloys, A. Mascarenhas, D. Follstaedt, T. Suzuki, B. Joyce, 2000, ISBN: 1-55899-491-2
Volume 584— Materials Issues and Modeling for Device Nanofabrication, L. Merhari, L.T. Wille, K.E. Gonsalves, M.F. Gyure, S. Matsui, L.J. Whitman, 2000, ISBN: 1-55899-492-0
Volume 585— Fundamental Mechanisms of Low-Energy-Beam-Modified Surface Growth and Processing, S. Moss, E.H. Chason, B.H. Cooper, T. Diaz de la Rubia, J.M.E. Harper, R. Murti, 2000, ISBN: 1-55899-493-9
Volume 586— Interfacial Engineering for Optimized Properties II, C.B. Carter, E.L. Hall, S.R. Nutt, C.L. Briant, 2000, ISBN: 1-55899-494-7
Volume 587— Substrate Engineering—Paving the Way to Epitaxy, D. Norton, D. Schlom, N. Newman, D. Matthiesen, 2000, ISBN: 1-55899-495-5
Volume 588— Optical Microstructural Characterization of Semiconductors, M.S. Unlu, J. Piqueras, N.M. Kalkhoran, T. Sekiguchi, 2000, ISBN: 1-55899-496-3
Volume 589— Advances in Materials Problem Solving with the Electron Microscope, J. Bentley, U. Dahmen, C. Allen, I. Petrov, 2000, ISBN: 1-55899-497-1
Volume 590— Applications of Synchrotron Radiation Techniques to Materials Science V, S.R. Stock, S.M. Mini, D.L. Perry, 2000, ISBN: 1-55899-498-X
Volume 591— Nondestructive Methods for Materials Characterization, G.Y. Baaklini, N. Meyendorf, T.E. Matikas, R.S. Gilmore, 2000, ISBN: 1-55899-499-8
Volume 592— Structure and Electronic Properties of Ultrathin Dielectric Films on Silicon and Related Structures, D.A. Buchanan, A.H. Edwards, H.J. von Bardeleben, T. Hattori, 2000, ISBN: 1-55899-500-5
Volume 593— Amorphous and Nanostructured Carbon, J.P. Sullivan, J. Robertson, O. Zhou, T.B. Allen, B.F. Coll, 2000, ISBN: 1-55899-501-3
Volume 594— Thin Films—Stresses and Mechanical Properties VIII, R. Vinci, O. Kraft, N. Moody, P. Besser, E. Shaffer II, 2000, ISBN: 1-55899-502-1
Volume 595— GaN and Related Alloys—1999, T.H. Myers, R.M. Feenstra, M.S. Shur, H. Amano, 2000, ISBN: 1-55899-503-X
Volume 596— Ferroelectric Thin Films VIII, R.W. Schwartz, P.C. McIntyre, Y. Miyasaka, S.R. Summerfelt, D. Wouters, 2000, ISBN: 1-55899-504-8
Volume 597— Thin Films for Optical Waveguide Devices and Materials for Optical Limiting, K. Nashimoto, R. Pachter, B.W. Wessels, J. Shmulovich, A.K-Y. Jen, K. Lewis, R. Sutherland, J.W. Perry, 2000, ISBN: 1-55899-505-6
Volume 598— Electrical, Optical, and Magnetic Properties of Organic Solid-State Materials V, S. Ermer, J.R. Reynolds, J.W. Perry, A.K-Y. Jen, Z. Bao, 2000, ISBN: 1-55899-506-4
Volume 599— Mineralization in Natural and Synthetic Biomaterials, P. Li, P. Calvert, T. Kokubo, R.J. Levy, C. Scheid, 2000, ISBN: 1-55899-507-2
Volume 600— Electroactive Polymers (EAP), Q.M. Zhang, T. Furukawa, Y. Bar-Cohen, J. Scheinbeim, 2000, ISBN: 1-55899-508-0
Volume 601— Superplasticity—Current Status and Future Potential, P.B. Berbon, M.Z. Berbon, T. Sakuma, T.G. Langdon, 2000, ISBN: 1-55899-509-9
Volume 602— Magnetoresistive Oxides and Related Materials, M. Rzchowski, M. Kawasaki, A.J. Millis, M. Rajeswari, S. von Molnár, 2000, ISBN: 1-55899-510-2
Volume 603— Materials Issues for Tunable RF and Microwave Devices, Q. Jia, F.A. Miranda, D.E. Oates, X. Xi, 2000, ISBN: 1-55899-511-0
Volume 604— Materials for Smart Systems III, M. Wun-Fogle, K. Uchino, Y. Ito, R. Gotthardt, 2000, ISBN: 1-55899-512-9

MATERIALS RESEARCH SOCIETY SYMPOSIUM PROCEEDINGS

Volume 605— Materials Science of Microelectromechanical Systems (MEMS) Devices II, M.P. deBoer, A.H. Heuer, S.J. Jacobs, E. Peeters, 2000, ISBN: 1-55899-513-7
Volume 606— Chemical Processing of Dielectrics, Insulators and Electronic Ceramics, A.C. Jones, J. Veteran, D. Mullin, R. Cooper, S. Kaushal, 2000, ISBN: 1-55899-514-5
Volume 607— Infrared Applications of Semiconductors III, M.O. Manasreh, B.J.H. Stadler, I. Ferguson, Y-H. Zhang, 2000, ISBN: 1-55899-515-3
Volume 608— Scientific Basis for Nuclear Waste Management XXIII, R.W. Smith, D.W. Shoesmith, 2000, ISBN: 1-55899-516-1
Volume 609— Amorphous and Heterogeneous Silicon Thin Films—2000, R.W. Collins, H.M. Branz, S. Guha, H. Okamoto, M. Stutzmann, 2000, ISBN: 1-55899-517-X
Volume 610— Si Front-End Processing—Physics and Technology of Dopant-Defect Interactions II, A. Agarwal, L. Pelaz, H-H. Vuong, P. Packan, M. Kase, 2000, ISBN: 1-55899-518-8
Volume 611— Gate Stack and Silicide Issues in Silicon Processing, L. Clevenger, S.A. Campbell, B. Herner, J. Kittl, P.R. Besser, 2000, ISBN: 1-55899-519-6
Volume 612— Materials, Technology and Reliability for Advanced Interconnects and Low-k Dielectrics, K. Maex, Y-C. Joo, G.S. Oehrlein, S. Ogawa, J.T. Wetzel, 2000, ISBN: 1-55899-520-X
Volume 613— Chemical-Mechanical Polishing 2000—Fundamentals and Materials Issues, R.K. Singh, R. Bajaj, M. Meuris, M. Moinpour, 2000, ISBN: 1-55899-521-8
Volume 614— Magnetic Materials, Structures and Processing for Information Storage, B.J. Daniels, M.A. Seigler, T.P. Nolan, S.X. Wang, C.B. Murray, 2000, ISBN: 1-55899-522-6
Volume 615— Polycrystalline Metal and Magnetic Thin Films—2000, L. Gignac, O. Thomas, J. MacLaren, B. Clemens, 2000, ISBN: 1-55899-523-4
Volume 616— New Methods, Mechanisms and Models of Vapor Deposition, H.N.G. Wadley, G.H. Gilmer, W.G. Barker, 2000, ISBN: 1-55899-524-2
Volume 617— Laser-Solid Interactions for Materials Processing, D. Kumar, D.P. Norton, C.B. Lee, K. Ebihara, X. Xi, 2000, ISBN: 1-55899-525-0
Volume 618— Morphological and Compositional Evolution of Heteroepitaxial Semiconductor Thin Films, J.M. Millunchick, A-L. Barabasi, E.D. Jones, N. Modine, 2000, ISBN: 1-55899-526-9
Volume 619— Recent Developments in Oxide and Metal Epitaxy—Theory and Experiment, M. Yeadon, S. Chiang, R.F.C. Farrow, J.W. Evans, O. Auciello, 2000, ISBN: 1-55899-527-7
Volume 620— Morphology and Dynamics of Crystal Surfaces in Complex Molecular Systems, J. DeYoreo, W. Casey, A. Malkin, E. Vlieg, M. Ward, 2000, ISBN: 1-55899-528-5
Volume 621— Electron-Emissive Materials, Vacuum Microelectronics and Flat-Panel Displays, K.L. Jensen, W. Mackie, D. Temple, J. Itoh, R. Nemanich, T. Trottier, P. Holloway, 2000, ISBN: 1-55899-529-3
Volume 622— Wide-Bandgap Electronic Devices, R.J. Shul, F. Ren, M. Murakami, W. Pletschen, 2000, ISBN: 1-55899-530-7
Volume 623— Materials Science of Novel Oxide-Based Electronics, D.S. Ginley, D.M. Newns, H. Kawazoe, A.B. Kozyrev, J.D. Perkins, 2000, ISBN: 1-55899-531-5
Volume 624— Materials Development for Direct Write Technologies, D.B. Chrisey, D.R. Gamota, H. Helvajian, D.P. Taylor, 2000, ISBN: 1-55899-532-3
Volume 625— Solid Freeform and Additive Fabrication—2000, S.C. Danforth, D. Dimos, F.B. Prinz, 2000, ISBN: 1-55899-533-1
Volume 626— Thermoelectric Materials 2000—The Next Generation Materials for Small-Scale Refrigeration and Power Generation Applications, T.M. Tritt, G.S. Nolas, G. Mahan, M.G. Kanatzidis, D. Mandrus, 2000, ISBN: 1-55899-534-X
Volume 627— The Granular State, S. Sen, M. Hunt, 2000, ISBN: 1-55899-535-8
Volume 628— Organic/Inorganic Hybrid Materials—2000, R.M. Laine, C. Sanchez, E. Giannelis, C.J. Brinker, 2000, ISBN: 1-55899-536-6
Volume 629— Interfaces, Adhesion and Processing in Polymer Systems, S.H. Anastasiadis, A. Karim, G.S. Ferguson, 2000, ISBN: 1-55899-537-4
Volume 630— When Materials Matter—Analyzing, Predicting and Preventing Disasters, M. Ausloos, A.J. Hurd, M.P. Marder, 2000, ISBN: 1-55899-538-2

Prior Materials Research Society Symposium Proceedings available by contacting Materials Research Society

Skutterudites

The Synthesis of Metastable Skutterudites and Crystalline Superlattices

Heike Sellinschegg, Joshua R. Williams, Gene Yoon and David C. Johnson[1]
[1]Materials Science Institute and Department of Chemistry, University of Oregon, Eugene, OR 97403-1253
Mike Kaeser and Terry Tritt
Department of Physics and Astronomy, Clemson University, Clemson, SC 29634-1905
George Nolas
Research and Development Division, Marlow Industries, Dallas, TX 75238-1645
E. Nelson
U. S. Army Research Laboratory, Adelphi, MD 20783

ABSTRACT

We have used controlled crystallization of elementally modulated reactants to prepare a series of kinetically stable, crystalline skutterudites ($M'_xM_4Sb_{12}$ where M' = vacancy, RE, Hf,...; M = Ni, Fe, Co) and crystalline superlattices composed of promising thermoelectric materials. For the bulk synthesis of skutterudites, low angle diffraction data demonstrates that the elemental layers interdiffuse at temperatures below 150°C. Nucleation of the skutterudite structure occurs with at large exotherm on annealing at temperatures below 200°C regardless of the ternary metal. All of the metastable ternary compounds and the new metastable binary compounds were found to decompose exothermically on higher temperature annealing. The decomposition temperature ranged from 250°C for the binary compound $NiSb_3$ to above 550°C for the rare earth containing cobalt compounds. The occupation of the ternary site was found to depend on the composition of the initial reactant and was varied from 0 to 1. Full occupancy typically required an excess of the filling cation. The lattice parameters of the compounds prepared at low temperatures are distinctly smaller than those prepared using traditional synthetic approaches. High temperature annealing converts the lattice parameters of the low temperature compounds to those prepared at higher temperatures using traditional synthetic approaches. Diffraction patterns of crystalline superlattices containing skutterudites prepared using elementally modulated reactants show splitting of high angle diffraction maxima as well as the presence of the expected low angle diffraction pattern from a supperlattice. The skutterudite superlattices are stable with respect to low temperature annealing.

INTRODUCTION

The resurgence of interest in new thermoelectric materials results in part from recent new ideas on how to potentially improve performance. Hicks and Dresselhaus [1,2] suggested that quantum confinement effects resulting from superlattices and quantum wires can result in dramatic increases in the figure of merit. Slack has suggested that materials that are "electron crystals but phonon glasses" would allow optimizing electrical properties while minimizing thermal conductivity [3]. The synthesis of compounds meeting the demands of these proposals provides distinct challenges. Dresselhaus' idea requires the preparation of nanostructures where the dimensions of the conducting layer or wire are on the order of Å or tens of Å. Such nanostructured materials are typically made via molecular beam epitaxy approaches at relatively low temperatures to prevent mixing of the layers expected from entropy considerations. Slack's "electron crystal – phonon glass" materials require one atom in the structure to be very weakly bound in a large cavity which permits this atom to effectively scatter phonons. As a consequence of the weak bonding of these "rattlers", many potential compounds with these characteristics are metastable with respect to disproportionation into a mixture of binary compounds. There are undoubtedly a great number of very interesting compounds which could be used to test these ideas that would be kinetically stable. The challenge is discovering techniques to make them in bulk amounts.

Our approach to this challenge has been the development of modulated elemental reactants [4]. In this approach, we use high vacuum deposition techniques to deposit alternating layers of the desired elements. This approach gives Å level control of the thickness of each layer, allowing us to control the diffusion distances and composition within each reactant. We have shown that this approach has the ability to prepare metastable compounds if these compounds nucleate before the competing thermodynamically more stable phases in the relevant phase diagrams. The controlled and designed nature of the starting reactant also permits us to explore the reaction mechanism as a function of the structure of the initial reactant

In this paper we discuss the application of modulated elemental reactants to the bulk synthesis of new compounds with the skutterudite structure type and crystalline superlattices containing skutterudites.

EXPERIMENT

The elementally modulated reactants were prepared in a high-vacuum evaporation system that has been described in detail elsewhere [5]. Briefly, the elements were sequentially deposited in high vacuum (lower than. 5×10^{-7} torr) under the control of a personal computer. The pumping of the deposition system is by a CTI cryopump. The gas pressure during deposition was limited by the outgassing of the deposition sources but was typically below 1×10^{-7} torr. Iron was deposited from electron beam evaporation sources at a rate of 0.5 Å/sec. Antimony was deposited by an effusion cell at approximately the same rate. Sources were independently monitored and controlled by quartz crystal thickness monitors. The thickness of each elemental layer was controlled to better than the nearest Ångstrom. The repeat thickness (sum of the individual elemental thicknesses) for all of the samples was varied to determine the critical thickness where the system avoided interfacial nucleation. All subsequent films were prepared with a repeat layer thickness less than this critical thickness (25Å) to increase the probability that the films would interdiffuse at low temperatures without nucleating binary compounds at the reacting interfaces. The elemental composition of the samples was found to be repeatable to within about 3%. The films were simultaneously deposited on silicon, off-cut quartz and photoresist-coated silicon wafers. The silicon substrates were used for low angle diffraction studies. The films on the off-cut quartz were annealed and used to collect diffraction data for structural refinement. Free-standing films were prepared from the coated substrates by dissolving the underlying photoresist in acetone. The elementally modulated film particles were collected from the acetone by filtration with Teflon filters. A potion of these free-standing films were used in calorimetry investigations. Measured exotherms were correlated with X-ray results to identify and track the interdiffusion of the elements and the crystallization of any compounds. Samples were annealed in a nitrogen atmosphere or in a dynamic vacuum better than 10^{-5} torr.

Low angle X-ray diffraction was used to characterize the multilayer periodicity and to study the interdiffusion of the elements. The repeat thickness was determined from the position of the Bragg diffraction maxima. The total film thickness was determined from the higher frequency oscillations resulting from interference of the x-rays between the front and back of the films. High angle X-ray diffraction was used to identify crystalline compounds. Copper Kα radiation was used in both the low and high angle diffraction studies that were done on a Scintag XDS 2000 or a Philips X'pert MRD diffractometer. The average composition of the multilayer films was determined by electron microprobe analysis using an energy dispersive X-ray detector.

RESULTS

Most binary phase diagrams have been thoroughly investigated making the preparation of new binary compounds a significant achievement. Indeed, one of the triumphs of the low temperature "chemie-duce" synthesis approach to oxides is the preparation of new binary transition metal oxides [6]. With respect to thermoelectric materials, the parent skutterudite

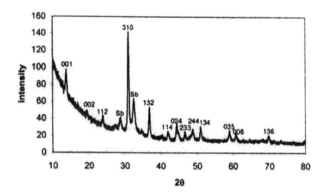

Figure 1. High angle diffraction data collected on a NiSb$_3$ binary sample.

binary compound CoSb$_3$ is a thermodynamically stable compound but the corresponding nickel and iron compounds are not present in the equilibrium phase diagrams. We have previously reported the synthesis of the new binary compound FeSb$_3$ [7] and have recently been successful in preparing the new binary compound NiSb$_3$. This compound exothermically forms from modulated reactants with compositions near 1:3 nickel to antimony below 200°C. The diffraction pattern of this new compound is shown in Figure 1. NiSb$_3$ decomposes on heating past 250°C. Both the iron and nickel antimony skutterudites decompose exothermically to a mixture of the respective diantimonide and antimony, demonstrating that they were prepared under conditions where they were unstable with respect to disproportionation.

While the binary compounds are of academic interest, optimizing the electrical and thermal properties of the ternary filled skutterudites offers the potential to increase the performance of thermoelectric materials through the "electronic crystal-phonon glass" concept originally proposed by Slack. This has resulted in several studies investigating properties as a function of occupancy of the ternary cation. Previous attempts to increase the occupancy of lanthanum in La$_x$Co$_4$Sb$_{12}$ above 0.1 results in samples with impurity phases containing lanthanum. Samples with higher lanthanum content can only be prepared by charge compensating for the addition of lanthanum by replacing some of the cobalt with iron or some of the antimony with tin [8]. Figure 2 contains a graph of the lattice parameter as a function of nominal composition for a series of La$_x$Co$_4$Sb$_{12}$ samples prepared using modulated reactants. Also shown in Figure 2 are samples with similar lanthanum content prepared using traditional high temperature synthesis and compensating for the lanthanum. The lattice parameters of the samples prepared at low temperature are consistently lower than those prepared at higher temperatures. On further annealing to higher temperatures (600°C) the samples prepared using modulated elemental reactants at low temperatures, the lattice parameters approach those prepared by conventional synthesis techniques. Significantly, the samples annealed at higher temperatures still appear to retain the lanthanum within the structure. Preliminary analysis of the diffraction data suggests that the observed difference in lattice parameter is due to a rotation of the $CoSb_6$ octahedra in the compounds prepared at low temperature, resulting in an expanded unit cell parameter.

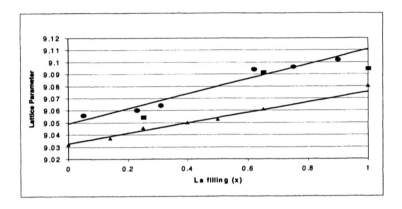

Figure 2. Plot of lattice parameters of $La_xCo_4Sb_{12}$ samples as a function of filling amount x. The triangles represent samples synthesized using the modulated elemental layer method, the circles represent samples made using traditional solid state synthesis and the squares denote samples which have been synthesized using the multilayer approach but have been annealed to high temperatures (600C).

Modulated elemental reactants also permit a host of new ternary skutterudites to be prepared directly by simply including the ternary element as a third layer within the repeating unit of the initial reactant. In the iron and cobalt antimonide systems, for example, all of the rare earth containing skutterudites can be prepared using modulated elemental reactants while conventional synthesis approaches can prepare only skutterudites containing the larger, lighter rare earth elements. Skutterudites containing post-transition elements such as indium, tin and lead can also be prepared from modulated elemental reactants [9]. The later rare earth cations have more room for motion in the structure and, due to their higher mass, vibrate at lower frequencies. Measurements of bulk samples discussed below suggest that the smaller, heavier rare earth cations may be more effective scatters of heat conducting phonons.

Our recent focus has been on the preparation of bulk samples required to measure the properties of these new compounds. To increase the amount of material deposited we have increased the surface area

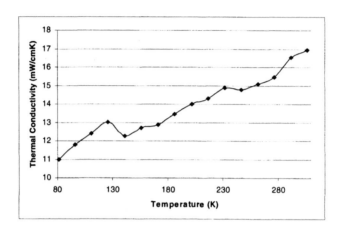

Figure 3. Thermal Conductivity as a function of temperature for a $Gd_{0.45}Co_4Sb_{12}$ sample.

of the deposition, increased the number of repeating layers deposited and reduced the time required to move samples in our deposition system. The net effect of these changes is that we are able to deposit approximately half a gram of skutterudite precursor in twelve hours. To limit the formation of oxides at surfaces, we have found that it is best to anneal the sample, forming the crystalline skutterudite, before removing the sample from the substrate. The resulting powder is then hot pressed to form a dense pellet. Figures 3 and 4 contain the measured properties of a pressed pellet of $Gd_{0.45}Co_4Sb_{12}$ prepared using this approach.

While Slack's "electron crystal – phonon glass" model for new thermoelectrics presents synthetic challenges, Dresselhaus's proposals are even more synthetically daunting. Very thin layer thicknesses, on the order of 10 Å, are required to achieve the dramatic increases in performance predicted. Crystalline superlattices on these dimensions are typically prepared using molecular beam epitaxy techniques that are very difficult and expensive to scale up. Previous work in our laboratories has shown that crystalline superlattices on these lengthscales can be prepared using modulated elemental reactants by preparing an initial reactant containing several different diffusion lengths. Figure 5 shows the high angle diffraction data collected around the 310 Bragg diffraction maxima of a crystalline superlattice prepared by annealing a modulated reactant containing 7 repeating iridium/antimony bilayers followed by 7 cobalt/antimony bilayers in a ~ 245Å repeating unit. The splitting of the parent Bragg diffraction peak of the skutterudite structure by the superlattice structure is clearly evident and agrees with that expected from the low angle diffraction pattern.

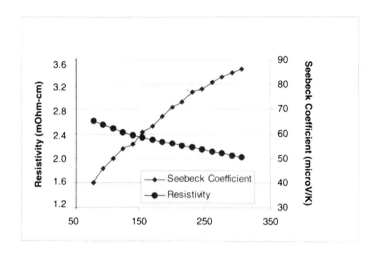

Figure 4: Resistivity and Seebeck coefficient as a function of temperature for a $Gd_{0.45}Co_4Sb_{12}$

To follow the evolution of the initially amorphous modulated reactant into the observed crystalline superlattice, the low angle diffraction data was collected as a function of annealing temperature and time. This data is shown in Figure 6. The superlattice structure contracts slowly on low temperature annealing as mobile vacancies are diffused out of the structure. The diffraction patterns after low temperature annealing still contain many low angle diffraction maxima because the slow interdiffusion of the iridium-antimony and the cobalt-antimony layers in the repeating unit. Extensive crystal growth during the anneal at 250°C results in a significant contraction of the superlattice and the appearance of splitting in the high angle diffraction pattern of the forming crystalline superlattice The low angle diffraction maxima remain after the sample crystallizes, providing further evidence that a crystalline superlattice has formed. Preparation of bulk samples of skutterudite superlattices with varying superlattice periods are currently underway.

SUMMARY

Modulated elemental reactants permit the formation of a host of new metastable skutterudite compounds. The ability to scale the synthesis to make bulk samples permits the properties of these new compounds to be evaluated. Measurements on $Gd_{0.2}Co_4Sb_{12}$ prepared using this approach show that increasing the mass and decreasing the size of the ternary cation may further reduce the lattice thermal conductivity of these materials. This synthesis approach also permits the formation of crystalline superlattices consisting of interwoven skutterudites.

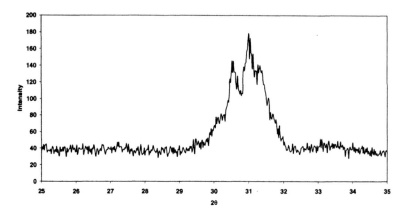

Figure 5. Diffraction data collected around the 310 Bragg diffraction maxima of a crystalline skutterudite structure.

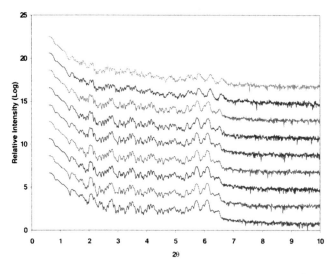

Figure 6. Low angle X-ray spectra of $IrSb_2/CoSb_3$ superlattice with 250 Å period taken at different annealing conditions.

ACKNOWLEDGMENTS

Support of this research by the Office of Naval Research and the Defense Advanced Research Projects Agency through grant N00014-98-1-0447 and the support from the U.S. Army Research Laboratory under contract number DAAD17-99-C-0006 is greatly appreciated.

REFERENCES

[1] L. D. Hicks, M. S. Dresselhaus, *Physical Review B*, **47**, 12727-12731 (1993).
[2] L. D. Hicks, T. C. Harmon, M. S. Dresselhaus, *Appl. Phys. Lett.*, **63**, 3230-3232 (1993).
[3] G. A. Slack, CRC Handbook of Thermoelectrics, ed. D. M. Rowe, CRC Press: Boca Raton, FL, (1995) pp 407-440.
[4] D. C. Johnson, *Current Opinion in Solid State and Materials Science*, **3**, 159-167 (1998).
[5] L. Fister, X. M. Li, T. Novet, J. McConnell, D. C. Johnson, *J. Vac. Sci. & Technol. A*, **11**, 3014–3019 (1993).
[6] R. Marchand, L. Brohan, M. Tournoux, *Mat. Res. Bull.*, **15**, 1129-1133 (1980).
[7] M. D. Hornbostel, E. J. Hyer, J. P. Thiel, D. C. Johnson, *J. Amer. Chem. Soc.* **119**, 2665-2668 (1997).
[8] B. C. Sales, D. Mandrus, R. K. Williams, *Science*, **272**, 1325-1328 (1996).
[9] S. L. Stuckmeyer, H. Sellinschegg, M. D. Hornbostel, D. C. Johnson, *Chemistry of Materials*, **10**, 1096-1101 (1998).

How cerium filling fraction influences thermal factors and magnetism in $Ce_yFe_{4-x}Ni_xSb_{12}$.

L. CHAPON*, D. RAVOT*, J.C. TEDENAC*, F. BOUREE-VIGNERON**
LPMC, Université Montpellier II, 34090 Montpellier, France
LLB-CEA-Saclay, 91000 Gif sur Yvette, France

INTRODUCTION :

Since few years, cerium filled and partially filled skutterudites are intensively studied because they show a wide variety of fundamental and applied properties. One of them consists in high values of thermal factors for rare earth atom in antimony skutterudites [1,2]. Slack suggests [3,4] a incoherent rattling of this ion in the oversized cage "Sb_{12}" surrounding the cerium which affects highly the phonon motion and thus lowers the lattice thermal conductivity (k_l). As a rule, the lattice thermal conductivity is decreased by a factor of 5 or greater by filling entirely the voids of the binary filled skutterudites with rare earth atoms [5]. Besides, k_l decreases for partially filled compounds in respect with totally filled ones [6,7]. Mass fluctuation mechanism between cerium atom and vacancy is obviously involved as the origin of this last reduction. On that purpose, theoretical calculations [7] demonstrate that the reduction belonging to mass fluctuation mechanism is an order of magnitude lower than the measured decrease. As the mass fluctuation added to the "rattling" on the cerium site is not sufficient to explain such low values of thermal conductivity, another phonon scattering mechanism must exist. In order to find another mechanism we present the influence of the filling fraction of cerium on thermal factors and the temperature dependence of this factor for a partially filled compound.

On the other hand, $CeFe_4Sb_{12}$ compound and related alloys have been previously described as heavy fermion or moderated heavy fermion system [8]. Magnetic properties have been investigated and lead to very controversal results. Even if iron contributes obviously to magnetism in $CeFe_4Sb_{12}$ because of a total effective moment above the trivalent cerium value, there are no yet direct evidences of the respective contributions of cerium and iron in that compounds. In order to elucidate magnetic behavior of iron, nickel and cerium, X.A.N.E.S experiments have been performed at different edges of the elements.

EXPERIMENTAL DETAILS :

Polycrystalline compounds were prepared by the method presented elsewhere [9]. Powder neutron diffraction were recorded using the high-resolution diffractometer (3T2) in the LLB laboratory (Saclay-CEA, France). Diffraction patterns of the $Ce_{0.92}Fe_4Sb_{12}$ compound were collected at 10, 100, 200 and 300 K over the range 7°-125° (2θ) with a step size of 0.05°. The diffraction patterns for other samples have been recorded at the room temperature. The used wavelength is 1.2251 Å for all the experiments.

X.A.N.E.S. spectrums were recorded at the LURE laboratory (France) in DCI line D21 for high energy experiments (cerium L_3 edge, Fe and Ni K edge) and in Super ACO for lower energy (cerium M edge). The step size used is 0.2 eV for L_3 and K edges and 0.1 eV for M edges.

Magnetic susceptibility have been measured in the range 2-300K under 0.2 T. for each sample with a Vibrating Sample Magnetometer from Oxford Instruments.

RESULTS:

Powder neutron diffraction on $Ce_{0.92}Fe_4Sb_{12.1}$ compound (final composition deduced from EDX analysis) doesn't revealed any extra phases. All the diffraction lines belong to the skutterudite phase according to Im3 space group with atomic positions in agreement with previously published paper. Rietveld analysis allows the extraction of thermal factors. We choose isotropic thermal factors for all the atoms. Even if the symmetry prescribes isotropic thermal factors for cerium, that is not the case for other atoms. Nevertheless, refine the structure with anisotropic displacement parameters add error proceeding to the increase of adjustable parameters.

Thermal factors for Ce, Fe and Sb are proportional to T as expected for a dynamic contribution. The figure 1 displays the temperature dependence of thermal factors. At room temperature the cerium thermal factor is 4 times greater than the antimony and iron ones. Obviously, cerium atom is poorly bounded to the framework Fe-Sb and free to "rattle" inside the oversized cage formed by the twelve surrounding antimony atoms. As for iron and antimony the assumption of covalent bounding between atoms seems to be confirmed by low values of thermal factors even at room temperature. The low value of the coefficient of linear expansion 1.06×10^{-5} K^{-1} strengthened this idea of a rigid framework. Low temperature thermal factors are nearly zero for antimony and iron unlike cerium one which shows a residual non zero term even at the lowest temperature experiment (10K). As a rule this contribution knew as a static component of the thermal factor is temperature independent adverse to the dynamic contribution. Before discussing the origin of this static component, the exact static contribution is deduced from Housley and Hess theory [10] allowing to remove the dynamic contribution du to the first normal mode at 0K.

For the dynamic contribution of a k type atom to thermal factor (B_k^{dyn}), the following inequality reaches an equality for $T \rightarrow \infty$ and allows to estimated the maximal dynamic

$$\left(B_k^{dyn}(0)\right)^2 \leq \frac{h^2}{2m_k k_B T} B_k^{dyn}(T) \qquad (1)$$

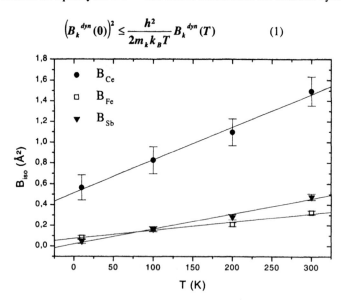

Figure 1: Isotropic thermal factors for Ce, Fe and Sb vs. Temperature for $Ce_{0.92}Fe_4Sb_{12.1}$.

contribution at 0K. The minimum static contribution is obtained by subtracting the maximal dynamic contribution from experimental measurements close to 0K (10K in that case).

As the maximal dynamic contribution for Fe and Sb is greater than the experimental values at 10K, only dynamic contribution can be expected for these atoms. For cerium atom the minimal static contribution is evaluated to 0.42 Å^2. This value does not represent a negligible part of the total thermal factor even at room temperature (28%). To explained this important static contribution, a relaxation of the framework around the empty sites and a displacement of rare earth atom away from their crystallographic positions have been involved by Chakoumakos & al [1]. Even if the refinement is not stable with cerium away from their special positions because of the random behavior of that displacement, the existence of a static displacement could be, nevertheless the right hypothesis. The disorder introduced by this displacement added to the dynamic one could be a reason of the decreasing lattice thermal conductivity of partially filled compounds in respect with totally filled skutterudites [6,7]. This way, the mass fluctuation mechanism between cerium and vacancy would not be the only scattering event. Besides, at room temperature the displacement of the cerium atom is principally dynamic (72% in that case). It may explained the greater reduction of k_l for totally filled compounds in respect with binary skutterudite than the one observed between totally filled and partially filled skutterudite. Thus, rattling stays as the main cause to reduce the lattice thermal conductivity at room temperature.

$CeFe_4Sb_{12}$ has been described as a moderated heavy fermion system [8]. The dependence of the magnetic susceptibility with temperature published elsewhere [11] showed a important discrepancy to the Curie-Weiss law at low temperature (below 200K) comparable to what happens in $CeAl_3$ compound [12]. On the other hand specific heat measurement reveals a moderated heavy fermion ground state with relatively high value of γ and an integrated entropy about 10% of Rln2 expected for heavy fermion system [11]. Nevertheless, very few information about cerium valence are available and low temperature susceptibility is often interpreted by a fluctuating valence of cerium leading to intermediate value such as 3.74 at 4K [13].

Insofar as X.A.N.E.S experiments is a good probe of cerium valence, we have performed XAS measurements at the L_3 edge and M edges of cerium for $Ce_yFe_{4-x}Ni_xSb_{12}$ and 0.16<y<1. For $CeFe_4Sb_{12}$, the XANES spectrums show without any doubt a entire 3+ valence for cerium [11]. The edge is characterized by a white line appearing at 5722.2 eV whatever the temperature down to 10K. This white line results from the transition from the ground state ($2p_{3/2}$) to the empty 5d level of cerium. The second derivative of μ with energy demonstrates the existence of two transitions within the white line separated by 1.6 eV. Fitting the edge[1] spectrum with two Lorentzian functions (convoluted by a Gaussian function modeling the experimental resolution) and an arctan component to model the transition to the continuum show very good results. The core hole lifetime is coherent to our knowledge of rare earth (3.1 eV). On the other hand the double transition could represent the transitions to two different final empty states separated by the crystal field as in $CeRu_4P_{12}$ [14]. The larger separation of the two transitions in the case of $CeRu_4P_{12}$ (3.3 eV) seems to be a direct effect of the shorter distance between cerium and phosphorus. No other transition appears still few eV after the white line as in CeO_2. Thus, cerium is trivalent in $CeFe_4Sb_{12}$ to our point of view. The partial filling of cerium and the substitution of Ni for Fe in the structure has no effect on cerium valence. The white line still appears at the same energy without changes in the shape or in the magnitude of the transition. The second derivatives show the same splitting of the transition into two components with a gap close to the value of 1.6 eV. If we assume that crystal field effect is the origin of this behavior and that cerium environment does not change with filling

---

[1] To be published elsewhere

fraction according to EXAFS and neutron diffraction experiments, crystal field should be quite equal in the range of composition.

Experiments at the M_4 and M_5 edge of cerium confirms the hypothesis of trivalent cerium. In the case of trivalent cerium, the M edges correspond both to the transition from the initial state ($3d^{10}$) to the $3d^9 4f^2$ excited state. As the spin orbit splitting is large, M_4 and M_5 edges spectra are well separated into two absorption lines. The figure 2 displays this spectrum for $CeFe_4Sb_{12}$ and partially filled compounds. Two lines appear respectively at 887 and 904 eV. Energy of the transitions are similar to what is founded in Ce^{3+} compounds such as $CePO_4$. No extra line exists few eV after the first ones which should be characteristic of intermediate valence. The fine structure within each absorption transition is a result of the $3d^9 4f^2$ multiplet splitting allowing several transitions to the final states.

Because the M edges are a direct probe of the f states and confirm the experiments done at the L_3 edge, we conclude the entire trivalent state for cerium. By no means from our results, intermediate valence can be putted forward in that type of compounds even in $CeFe_4Sb_{12}$. Thus the inverse susceptibility variation versus temperature for $CeFe_4Sb_{12}$ [11] can be explained following two different ways: at high temperature the trivalent cerium behaves like a localized ion and follows the Curie-Weiss law above 200K. The resulting effective moment (3.36 μ_B), greater than the moment of Ce^{3+} imply a contribution of iron to the magnetism. As the cerium is trivalent, there is a hole in the top of the valence band formed by Fe-d and Sb-p states. Statically, one iron has the d^5 electronic configuration and contributes to magnetism while the three others keep the d^6 configuration like Co in $CoSb_3$ and lye in a non magnetic state.

The resulting theoretical effective moment is thus 3.07μ_B by taking into account an only spin contribution for iron (1.73 μ_B). Two phenomena can explain the greater experimental moment : at first, the assumption of a localized magnetism on iron falls down and itinerant magnetism takes place in such a compound. Secondly, the orbital locking by the crystal field is not entire.. Moreover in $CeFe_4Sb_{12}$ the contribution extracted for iron is 2.19μ_B assuming an effective moment of 2.54μ_B for cerium atom is very close to the contribution of iron in $LaFe_4Sb_{12}$ (2.23μ_B). This fact strengthens the ideas of a delocalized entity and/or the fact that S is not a good quantum number.

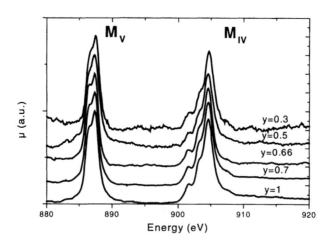

Figure 2 : M edges of cerium for $Ce_yFe_{4-x}Ni_xSb_{12}$ compounds at room temperature.

At lower temperature, down to 2K the inverse susceptibility of $CeFe_4Sb_{12}$ shows a important discrepancy to the Curie-Weiss law. Seeing that the curvature of the discrepancy is negative, it could be attributed firstly to a splitting of the magnetic levels of the cerium atom into a doublet as a ground state and a quadruplet at higher energy. Fitting the data with the Jones formula leads to a value of the crystal field larger than 700K and to an relative disagreement between experimental and fit curve. Though the hypothesis of a high crystal field can not be turn down, it doesn't seem to be the only cause of the discrepancy. As the theoretical calculations found f states of cerium just below the Fermi level[2], the f electrons can be correlated to conduction electrons. To our point of view, below a sufficient low temperature defined as the spin fluctuation temperature (T_K), f electrons are in a singlet state and hybridized with conduction electrons. The interaction between f electrons and conduction electrons decreases the magnetic moment if we suppose negative interaction. At lowest temperatures, other interactions restore magnetism and the inverse susceptibility fall down. Theoretical calculations are actually in progress to propose a model open to fit the data.

For partially filled compounds the discrepancy at low temperature decreases with decreasing cerium filling fraction. At high temperature, the total effective moment is close to theoretical moment assuming trivalent cerium and an unchanged contribution of iron. Nickel seems to carry no magnetic moment in these compounds. Intuitively, because the 3d states of nickel lie below the ones of iron, t_{2g} states would be full unlike the partially filled 3d states for iron. On that purpose, X.A.N.E.S experiments at the K edge of these elements can provide an indirect proof of this assumption. Figure 3 displays K edges of these elements. Both edges are characterized by an extensive structure belonging to the multiple scattering with the firsts neighboring atoms. This part of the edge is identical for iron and nickel which is coherent with the crystallographic structure and the isotypic environment for both atoms. The basement of the edge is characterized by a small intensity peak. This pre-edge peak is assigned to the 1s→ 3d transition. This transition is forbidden by the selection rules but the coupling between the 3d and 4p bands of transition metal allows this transition in agreement with our first band structure calculations. As the intensity of the peak doesn't change with cerium filling fraction, the density of the empty states seems to be constant and an unchanged contribution of nickel and iron to magnetism can be expected in concordance with previous results. Moreover, the

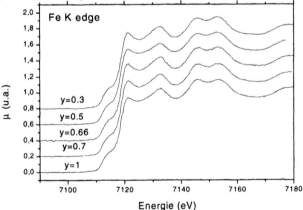

---

[2] to be published elsewhere

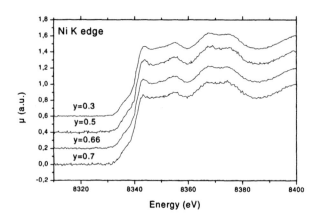

Figure 3 : K edges for iron and nickel at room temperature for different values of cerium filling fraction.

intensity of this peak for Ni K edge is lower than the one appearing in the iron edge. This is coherent with the model of transition metal d bands partially filled in the case of iron and totally filled in the case of nickel. In the first case, transition $1s \rightarrow t_{2g}$ and $1s \rightarrow e_g$ are allowed while in the second one the only transition $1s \rightarrow e_g$ exists. Then, examination of the K edge emphasizes the intuitive hypothesis of a non magnetic state for nickel in $Ce_yFe_{4-x}Ni_xSb_{12}$ compounds.

ACKNOWLEDGEMENTS :

The authors thanks R. Cortes and A. M. Flanck from the LURE laboratory for assistance during experiments and helpful discussion. We will thank A. Mauger, M. L. Doublet and F. Lemoigno for help with theoretical considerations and their involvement in our work.

REFERENCES :

[1] B. C. Chakoumakos, B.C. Sales, D. Mandrus, V. Keppens, Act. Cryst. (1999). B55, 341
[2] B. C. Sales, B. C. Chakoumakos, D. Mandrus, Phys. Rev. B (2000). 61(4), 2475
[3] G. A. Slack, V. G. Tsoukala, J. Appl. Phys. (1994). 76, 1665
[4] G. A. Slack, in Thermoelectric Handbook, edited by D. M. Rowe (CRC, Boca Raton, FL, 1995), 407.
[5] G. S. Nolas, G. A. Slack , D.T. Morelli, T.M. Tritt, A.C. Erlich, J. Appl. Phys. (1996). 79(8), 4002.
[6] G. P. Meisner, D. T. Morelli, S. Hu, J. Yang, C. Uher, Phys. Rev. Letters (1998). 80(16), 3551.
[7] G. S. Nolas, J. L. Cohn, G. A. Slack, Phys. Rev. B (1998). 58(1), 164.
[8] D. A. Gajewski & al., J. Phys. Condens. Matter (1998). 10, 6973.
[9] L. Chapon, D. Ravot, J. C. Tedenac, J. Alloys and Compounds (1999). 282, 58.
[10] R. M. Housley, F. Hess, Phys. Rev. (1966). 146, 517.
[11] L. Chapon, D. Ravot, J.C. Tedenac, J. Alloys and Compounds (2000). 299, 68.
[12] G. R. Stewart, Rev. Mod. Phys. (1984).56, 755.
[13] B. Chen & al., Phys. Rev. B,(1997). 55(3), 1476.
[14] C. H. Lee & al., Phys. Rev. B (1999). 60(19), 13253.

Thermoelectric Properties of Some Cobalt Phosphide-Arsenide Compounds

Anucha Watcharapasorn[*], Robert C. DeMattei[†], and Robert S. Feigelson[*†]
[*]Department of Materials Science and Engineering, Stanford University, Stanford, CA 94305, U.S.A.
[†]Laboratory for Advanced Materials, McCullough Bldg., Rm. 119, Stanford University, Stanford, CA 94305-4045, U.S.A.

Thierry Caillat, Alexander Borshchevsky, G. Jeffrey Snyder, and Jean-Pierre Fleurial
Jet Propulsion Laboratory, California Institute of Technology, MS 277/207, 4800 Oak Grove Drive, Pasadena, CA 91109, U.S.A.

ABSTRACT

Samples of CoP_3, $CoAs_3$ and $CoP_{1.5}As_{1.5}$ have been synthesized and their thermoelectric properties measured. All three samples show semiconducting behavior. The Seebeck coefficients of CoP_3 and $CoAs_3$ are weakly dependent on temperature and are relatively small with maximum values of about 40 and 50 μV/K, respectively. The Seebeck coefficient of the solid solution gradually decreases with increasing temperature and the values are larger than those of CoP_3 and $CoAs_3$ in the temperature range investigated, with a maximum value of about 89 μV/K near room temperature. The thermal conductivity of CoP_3 and $CoAs_3$ are higher than that of $CoSb_3$, as can be expected from the effect of anionic size on lattice vibration. A substantial reduction in thermal conductivity was observed for the solid solution compared to the constituent binary compounds due to additional phonon scattering from lattice disorder and other possible point defects such as vacancies. Other compositions in the $CoP_{3-x}As_x$ system have also been synthesized and their thermoelectric properties are currently being investigated to provide essential information about lattice thermal conductivity reduction by point defect scattering and to further develop strategies for optimizing the thermoelectric properties of skutterudite materials.

INTRODUCTION

Over the last decade, a large number of skutterudite materials have been synthesized and their thermoelectric properties studied. Large ZT values have been obtained for $CeFe_{4-x}Co_xSb_{12}$ (0<x<1) samples near 600 °C [1,2]. Other antimonide skutterudites have also been prepared by substituting Fe with other types of cations such as Ru and, in some cases, with Ge or As for anion substitution [3]. However, these materials show semimetallic behavior and their electrical properties are difficult to optimize. The preparation of semiconducting skutterudite compositions are of interest because they can potentially be doped n- and p-type as are most state-of-the-art thermoelectric materials [4]. Some semiconducting phosphide compounds such as CoP_3, $CeFe_4P_{12}$, and $CeRu_4P_{12}$ have been synthesized and their thermoelectric properties measured [5, 6]. A systematic study of phosphide and arsenide skutterudite compounds would help establish trends leading to the optimization of the transport properties of this class of materials. In particular, information about lowering the lattice thermal conductivity by incorporating filling ions and through anion substitutions would provide the critical information necessary to improve further the ZT values in these materials.

Lutz and Kliche [7] have found that CoP_3 and $CoAs_3$ can form complete solid solutions which renders this system very attractive. They also reported that an extra vibrational mode was observed when some amount of P was substituted into As sites and this would indicate the possibility of a reduction in thermal conductivity. In this paper, the synthesis of CoP_3, $CoAs_3$ and $CoP_{1.5}As_{1.5}$ samples is discussed along with their thermoelectric properties.

EXPERIMENTAL PROCEDURES

Phase pure samples of CoP_3, $CoAs_3$, and $CoP_{1.5}As_{1.5}$ have been made by direct synthesis of the pure elements. Cobalt powder (99.999%), crushed red phosphorus (99.995%), arsenic powder (99.999+%) and arsenic pieces (99.9998%) were used. These materials were mixed together in stoichiometric amounts, placed in evacuated and sealed fused silica tubes, and then slowly heated to a temperature between 700 - 950 °C for about one week. In all experiments, small amount of excess phosphorus and arsenic were used to compensate for decomposition of the samples. The lattice constants were calculated using least square fit program from powder x-ray diffraction experiments.

The synthesized powders were hot-pressed in graphite dies at temperatures between 800 - 1000 °C under an Ar atmosphere. Dense samples having dimensions of approximately 2.0 mm long and 12.0 mm in diameter were prepared. Chemical composition was determined by microprobe analysis. The density of the samples was calculated from the measured weight and dimensions. One mm thick disks were then cut from these samples for Seebeck coefficient, Hall effect, electrical resistivity and thermal conductivity measurements in the 300 - 900 K temperature range. Detailed measurement techniques have been described elsewhere [5].

RESULTS AND DISCUSSION

Some of the room temperature physical properties of the samples synthesized are listed in table I. The properties are compared to those of an undoped $CoSb_3$ sample [8]. Except for the solid solution, all other samples had a measured density close to theoretical. Electron probe microanalysis showed that the CoP_3 sample contained about 99.5% skutterudite phase having composition of $Co_{1.00}P_{2.91}$. $CoAs_3$ was also single phase, having the composition, $Co_{1.00}As_{2.98}$. X-ray diffraction analysis of the hot-pressed $CoP_{1.5}As_{1.5}$ sample showed that it contained about 90-95% skutterudite phase and had a lattice constant of 7.9624 Å, which was slightly smaller than that of the as-synthesized sample.

All samples exhibited p-type conduction. The relatively high electrical resistivity of $CoAs_3$ is due mainly to its relatively low carrier concentration, which was about one order of magnitude less than that of the other samples. The room temperature Seebeck coefficients of CoP_3 and $CoAs_3$ are relatively low when compared to the solid solution and $CoSb_3$. Carrier

Table I. Some room temperature properties of CoP_3, $CoAs_3$, $CoP_{1.5}As_{1.5}$ and $CoSb_3$ [8].

	Units	CoP_3	$CoAs_3$	$CoP_{1.5}As_{1.5}$	$CoSb_3$
Lattice constant	(Å)	7.7073	8.2045	7.9645	9.0345
Percentage of theoretical density	%	97.5	99.6	86.5	99.9
Type of conductivity		p	p	p	p
Electrical resistivity	mΩ-cm	0.47	8.40	1.57	0.44
Seebeck coefficient	μV/K	30	26	89	108
Hall carrier concentration	10^{19} cm^{-3}	2.69	0.057	0.51	1.00
Hall mobility	cm^2/Vs	493	1316	783	1432
Thermal conductivity	mW/cmK	258	125	44	100

effective masses were estimated to be $0.135m_0$, $0.009m_0$ and $0.132m_0$ (where m_0 is the free electron mass) for CoP_3, $CoAs_3$ and $CoP_{1.5}As_{1.5}$ samples, respectively, using a single band model with the acoustic phonon scattering approximation [8]. The calculated effective mass of CoP_3 and $CoP_{1.5}As_{1.5}$ was about two times less than that of $CoSb_3$. The room temperature thermal conductivity decreased from CoP_3 to $CoAs_3$ to $CoSb_3$. The thermal conductivity of the solid solution sample showed an even lower value than $CoSb_3$ and is in agreement with the far-infrared spectra study of this solid solution [7], where an extra lattice vibration mode was identified.

The variation of the electrical resistivity as a function of temperature is shown in figure 1. The values for undoped $CoSb_3$ are also shown for comparison. CoP_3, $CoAs_3$ and $CoP_{1.5}As_{1.5}$ show semiconducting behavior. Above 580K, the electrical resistivity of $CoP_{1.5}As_{1.5}$ becomes higher than that of CoP_3 and $CoAs_3$. The weak temperature dependence of the electrical

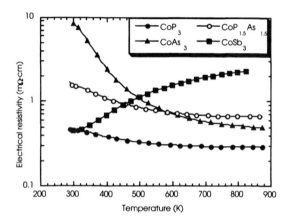

Figure 1. Electrical resistivity of CoP_3, $CoAs_3$, $CoP_{1.5}As_{1.5}$ and $CoSb_3$ [8] as a function of temperature.

resistivity for the CoP_3 and $CoP_{1.5}As_{1.5}$ sample suggest that these materials may be narrow indirect band gap semiconductors in agreement with theoretical predictions by Llunell et al. who calculated a band gap of 0.07 eV for CoP_3 [9].

The Seebeck coefficient as a function of temperature for these three compounds are plotted in figure 2. $CoP_{1.5}As_{1.5}$ had larger Seebeck coefficient in the entire temperature range of measurement compared to CoP_3 and $CoAs_3$.

Figure 3 shows the temperature dependence of thermal conductivity. For binary compounds, the thermal conductivity decreases from CoP_3 to $CoAs_3$ to $CoSb_3$, which is in

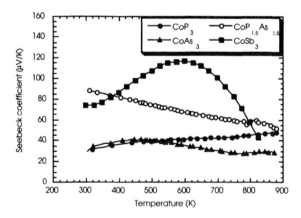

Figure 2. Seebeck coefficient of CoP_3, $CoAs_3$, $CoP_{1.5}As_{1.5}$ and $CoSb_3$ [8] as a function of temperature.

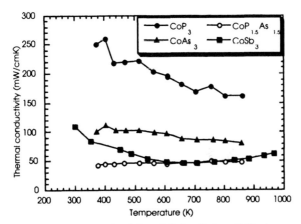

Figure 3. Thermal conductivity of CoP_3, $CoAs_3$, $CoP_{1.5}As_{1.5}$ and $CoSb_3$ [8] as a function of temperature.

agreement with the observed trends in other binary compounds having the same crystal structure for the dependence of lattice vibration on the mass of atoms in the unit cell, i.e. a decrease in lattice thermal conductivity with increasing anionic atomic mass. A $1/T$ temperature dependence of the thermal conductivity was also observed for the binary compounds.

For the $CoP_{1.5}As_{1.5}$ sample, the thermal conductivity was substantially lower than that for CoP_3 and $CoAs_3$ in the whole temperature range of measurement and it is comparable to the undoped $CoSb_3$ at high temperatures. The substantial reduction in lattice thermal conductivity for the solid solution is primarily due to phonon scattering by point defects. It is to be noted that the experimental density of this sample was only about 86.5% of the theoretical density.

CONCLUSIONS

CoP_3, $CoAs_3$ and $CoP_{1.5}As_{1.5}$ have been synthesized and their thermoelectric properties measured. All three compounds show semiconducting behavior. Larger values of the Seebeck coefficient were observed for the solid solution compared to the two end compounds. The substantial reduction in thermal conductivity for the solid solution was in agreement with earlier lattice vibration studies in the $CoP_{3-x}As_x$ system. More $CoP_{3-x}As_x$ samples are being synthesized and their thermoelectric properties measured in order to fully investigate the variations of the thermoelectric properties as a function of composition. This will provide critical information for the optimization of the thermoelectric efficiency of this class of materials.

ACKNOWLEDGMENTS

This work was supported by the Defense Advanced Research Projects Agency, Grants No. N00014-97-1-0524 and No. E754. Work performed at the Jet propulsion Laboratory/California Institute of Technology was sponsored by DARPA, through an agreement with the National Aeronautics and Space Administration.

REFERENCES

1. C. Sales, D. Mandrus, and R. K. Williams, *Science* **272**, 1325 (1996).
2. J-P. Fleurial, T. Caillat, A. Borshchevsky, D. T. Morelli, and G. P. Meisner, in *Proceedings of the 15th International Conference on Thermoelectrics*, edited by T. Caillat (Institute of Electrical Engineers, Piscataway, NJ, 1996), p.91
3. J-P. Fleurial, T. Caillat, G.J. Snyder and A. Borshchevsky, Thermoelectric Workshop, Herndorn, VA, 1997
4. A.F. Ioffe, *Semiconductor Thermoelements and Thermoelectric Cooling*, Infosearch, London, 1957
5. A. Watcharapasorn, R.C. DeMattei, R.S. Feigelson, J-P. Fleurial, T. Caillat, G.J. Snyder and A. Borshchevsky, *J. Appl. Phys.*, **86**, 11, 1 (1999)
6. A. Watcharapasorn, R.C. DeMattei, R.S. Feigelson, J-P. Fleurial, T. Caillat, G.J. Snyder and A. Borshchevsky, to be published in the International Conference on Thermoelectrics proceedings, 1999
7. H.D. Lutz and G. Kliche, *J. Solid State Chem.*, **40**, 64-68 (1981)
8. T. Caillat, A. Borshchevsky, and J. -P. Fleurial, *J. Appl. Phys.*, **80**, 4442 (1996)
9. M. Llunell, P. Alemany, S. Alvarez, and V. Zhukov, *Phys Rev. B* **53**, 10605 (1996)

Superlattice

Epitaxial Growth and Thermoelectric Properties of Bi_2Te_3 Based Low Dimensional Structures

Joachim Nurnus, Harald Beyer, Armin Lambrecht, Harald Böttner

Fraunhofer Institut Physikalische Messtechnik, Heidenhofstr. 8
D-79110 Freiburg i. Br., Germany

ABSTRACT

Bi_2Te_3 based low dimensional structures are interesting material systems to increase the thermoelectric figure of merit ZT by either the expected reduction of the thermal conductivity or by a possible power factor enhancement due to quantum confinement. Due to low lattice mismatch $Bi_2(Te_{1-x}Se_x)_3$, $PbSe_{1-x}Te_x$, as well as $Pb_{1-x}Sr_xTe$, and BaF_2 are suitable for Bi_2Te_3 based low dimensional structures. Especially due to their significantly enhanced band gap lead chalcogenide compounds like $Pb_{1-x}Sr_xTe$ ($Pb_{0.87}Sr_{0.13}Te$: 0.6 eV) are well-suited barrier materials in MQW structures. Alternatively the insulator BaF_2 can be used for that purpose.

Here we report mainly on results of different superlattice structures mentioned above grown by molecular beam epitaxy (MBE) on $BaF_2(111)$. The structural properties of these layers were investigated by X-ray diffractometry (XRD), scanning electron microscopy (SEM) and secondary ion mass spectroscopy (SIMS). Structural performance and thermoelectric properties of different Bi_2Te_3 based superlattices were reported and compared with regard to their superlattice parameters.

INTRODUCTION

Calculations predict an enhancement of the thermoelectric figure of merit in low dimensional-structures -multi-quantum wells (MQW) and super-lattices (SL)- in comparison to bulk materials [1, 2]. Further calculations deal with those structures based on Bi_2Te_3 and PbTe [3]. Because of their high bulk figures of merit both materials are well-suited for thermoelectric applications. The predicted huge increase in ZT for 2D structures results on one hand from the large density of electron states due to quantum confinement. On the other hand the thermal conductivity is reduced by interface scattering in the boundary regions between the low dimensional structure forming layers [4].

A prerequisite for proving both approaches experimentally is the availability of epitaxial systems with compatible properties. These can be derived in a first approach from only a few physical material properties of the compounds. Figure 1– data taken from [5, 6] - shows the basic epitaxial map of the interesting compounds. In analogy to the experimental results concerning epitaxial deposition in the lead chalkogenide based pseudomorphic systems [6] one can expect similar behaviour for Bi_2Te_3-based epitaxy for the continuous series of Bi_2Te_3/Bi_2Se_3 solid solutions. In spite of extended miscibility gaps, which are partially known for the Sr-free systems of $Bi_2Te_3/PbTe$ and $Bi_2Te_3/PbSe$ [5] and which should be expected for BaF_2/Bi_2Te_3 system, in all likelihood epitaxial growth will be possible as lattice parameters fit well in the hexagonal growth planes accordingly, see figure 1. Our concept is to structure these multifaceted possibilities into systems either suited for MQW's or SL's as outlined in figure 2.

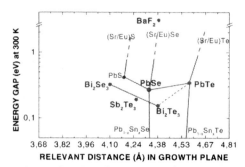

Figure 1. Epitaxial map of semiconductor materials (room temperature) which could be suited for a combination with Bi_2Te_3 due to their relevant atomic distances with respect to the a-plane of Bi_2Te_3.

As far as the bandgaps are concerned (figure 1), it is evident, that all these materials could be used as MQW-barrier materials for Bi_2Te_3 wells if they have bandgaps like PbTe or larger as in the case of $Pb_{1-x}Sr_xTe$ or BaF_2. To our knowledge up to now no suitable barrier materials for realising MQW structures with Bi_2Te_3 wells were reported or experimentally shown. Here we report on basic structural data for possible multilayers in the BaF_2/Bi_2Te_3 and $Bi_2Te_3/Pb_{1-x}Sr_xTe$ systems for future use in MQW systems.

The most common way to decrease the thermal conductivity of thermoelectric materials is to decrease the lattice thermal conductivity by alloy scattering, well known from the material system Bi_2Te_3-Bi_2Se_3-Sb_2Te_3. A possibility to further reduce the thermal conductivity is the use of superlattice structures. It has been predicted and shown by Venkatasubramanian that using Bi_2Te_3-Sb_2Te_3 SL structures the thermal conductivity along the SL axis can be reduced by about a factor of 7 with respect to the binary compounds [7, 8]. The reduction of the thermal conductivity has a minimum for SL dimensions of about 2.5 nm. For larger SL dimensions the thermal conductivity is still lower, e.g. for SL dimensions of 25 nm the thermal conductivity is still four times smaller than that of the corresponding alloy. Similar results have been obtained using different material systems or different preparation- and measurement methods [9].

For the investigations concerning the SL concept we have chosen combinations of the well know materials Bi_2Te_3 and $Bi_2(Te_{1-x}Se_x)_3$ and the heteroepitaxial system $Bi_2Te_3/PbSe$. Here we report mainly on structural and thermoelectric properties of these combinations and the MBE grown bulk material $Bi_2(Te_{1-x}Se_x)_3$.

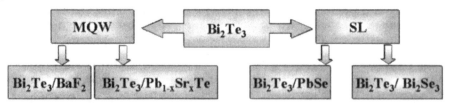

Figure 2. Possible material combinations for realising low dimensional MQW and SL structures based on Bi_2Te_3.

EXPERIMENTAL DETAILS

The layers presented here were grown in two different MBE systems, a custom built IPM-MBE (1" substrate holder) and a EPI 930 MBE system (3" substrate holder). Both systems are equipped with bismuth and tellurium element sources, a Se-valved cracker cell (cracking temperature 600°C) to transform the Se-beam mainly into Se_2-molecules for a better incorporation during layer growth, IV-VI (PbTe, PbSe) compound sources and a strontium element source for the growth of ternary IV-VI compounds. The EPI MBE further is equipped with a high temperature cell used for the BaF_2 deposition.

Usually freshly cleaved BaF_2 substrates with a (111) surface with a area of ~1 cm^2 and a thickness of about 1 mm were used. The BaF_2 substrates were glued onto molybdenum substrate holders using a liquid InGa-eutectic and then transferred into the respective MBE system. Prior to the growth the substrates were baked at temperatures of ~400°C to remove adsorbed water from the substrate surface.

The growth rates were about 1.5 µm h^{-1} for the IV-VI-materials, ~0.4 µm h^{-1} for BaF_2 and typically 0.7 µm h^{-1} for the $Bi_2(Te_{1-x}Se_x)_3$ layers. In the case of the V-VI layers the given rate strongly depends on the substrate temperature as well as on the flux ratios of bismuth and tellurium. The value mentioned before is valid using a flux ratio from Bi/(Te,Se)~5/12 up to 5/15 for substrate temperatures between 290°C and 325°C. The dependence on the substrate temperature can be visualised by the following example. For the EPI MBE after baking the BaF_2 substrate at ~400°C a recovery time of about 60 minutes is required until equilibrium condition for the preset growth temperature of the substrate holder is reached. In figure 3 a SEM-cross-sections of Bi_2Te_3 grown on BaF_2 (outside the optimum temperature range) using a standby time of ~10 minutes is shown. If the growth starts soon after the prebaking process the low limit epitaxial growth temperature will be reached later than after a standby time of 60 minutes. The thickness ratio of polycrystalline/epitaxial thickness of 1.5 as derived from figure 3 raises to 3.5 using a preset time of 60 minutes.

Again the enormous influence of the substrate temperature known from previous experiments becomes obvious looking at the tellurium content determined by EDX analysis. The tellurium content changes from about 60 at% Te within the crystalline region to almost 80 at% in the polycrystalline top layers (figure 3). Analogous results already have been reported for MBE

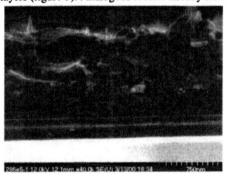

Figure 3. Bi_2Te_3 *layer grown on a freshly cleaved* BaF_2 *substrate outside the optimum temperature range (substrate temperature preset time: ~10 min after reaching the nominal temperature of 260°C).*

grown Bi_2Te_3 layers [10].

As the optimum substrate temperatures for the growth of IV-VI (~370°C) and Bi_2Te_3 (~300°C) differ by about 100°C, the layered IV-VI/V-VI samples have to be prepared in a special way. Mainly one has to alter the growth parameters of $PbSe_{1-x}Te_x$. Reducing the substrate temperature leads to a higher sticking coefficient of both chalcogens. Therefore the additional chalcogen flux, used under standard growth conditions, has to be reduced to preserve the material quality. Because chalcogen rich $PbSe_{1-x}Te_x$ is p-type and lead rich $PbSe_{1-x}Te_x$ is n-type one has to adjust the additional chalcogen flux carefully if a certain carrier concentration is needed. To keep optimum structural properties the additional chalcogen flux is not that important. As the doping behaviour of IV-VI compounds and Bi_2Te_3 show opposite characteristics as far as Bi and chalcogen (Se,Te) are addressed this has to be taken into account for the preparation of throughout p or n-type IV-VI/V-VI layers. Up to now all (IV-VI)-(V-VI) material combinations were grown without additional bismuth doping of the IV-VI-layers.

The structural performance of the films was investigated by X-ray diffractometry (XRD), scanning electron microscopy (SEM) and secondary ion mass spectroscopy (SIMS). The thermoelectric properties were examined by Hall and Seebeck measurements. Bandgap determination was performed by infrared transmission measurements (FTIR).

DISCUSSION

BaF_2/Bi_2Te_3: As outlined in figure 1 BaF_2 may act as a barrier for MQW's with Bi_2Te_3 as well material. To date to our knowledge nothing is known about the epitaxial growth of group-II-fluorides on Bi_2Te_3. Taking into account on one hand that typical growth temperatures for BaF_2 are much higher than those for V-VI-layers and on the other hand the relative delicate adjustment of the growth temperature for Bi_2Te_3 growth on BaF_2 [11] one has to determine growth rates and growth temperatures compatible with both materials. For a first attempt we chose a substrate temperature of ~305°C for a 20 fold stack of nominal 20 nm Bi_2Te_3 and 20 nm BaF_2. SEM pictures of the a cross section and the surface of this stack are shown in figure 4.

The stack can be divided mainly into two parts. About the first half shows layer by layer growth, which more or less suddenly changes into polycrystalline behaviour. The first result is,

 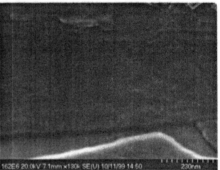

Figure 4. *Left: Cross section of a Bi_2Te_3-BaF_2 structure (Bottom: BaF_2 substrate, middle: SL-region, Top: polycrystalline region). Right: Other region of the same sample showing the surface of a Bi_2Te_3 layer near the substrate.*

that obviously a layer by layer growth at rather low temperatures for II-fluorides is possible. In the lower part of the cross section, shown in figure 4, the desired structure of alternating layers of Bi_2Te_3 and BaF_2 can easily be observed. Also sharp interfaces between the layers can be seen in this region of the sample. The right part of figure 4 shows a other region of the same sample, this time the sample is tilt about 20°. This figure show the surface of a Bi_2Te_3 layer located in the lower part of the sample. Triangular shaped islands consisting of several stacked layers are visible on the surface of this Bi_2Te_3 layer near the BaF_2 substrate. Similar structures were already reported for Bi_2Te_3 bulk layers grown on (111) BaF_2 substrates [12]. The height of the single layers forming the islands was found to be 1 nm, this is in good agreement with the Bi_2Te_3 crystal structure [5]. Because of this the Bi_2Te_3 layers in the lower region of this sample are expected to have a good structural quality.

On the other hand these islands could be the reason for the structural change during the growth of this sample. As the roughness increases with increasing number of periods grown either BaF_2 or Bi_2Te_3 start growing in a three dimensional way resulting in the polycrystalline region of the sample shown in figure 4. So reducing the thickness of the Bi_2Te_3 layers should decrease the roughness and result better SL structures. Beside the gain in structural perfection the reduction of the Bi_2Te_3 layer thickness is necessary since ZT enhancement in the quantum wells are predicted for wells thinner than 3 nm [1].

Since the isotropic linear thermal expansion coefficient of BaF_2 is almost the same as the one of Bi_2Te_3 perpendicular to the growth axis [13] and the relevant atomic distances of both materials fit almost perfect interfacial stress in not thought to be the reason for the epitaxial to polycrystalline changes observed here.

$Pb_{1-x}Sr_xTe/Bi_2Te_3$: As already outlined in the introduction also $Pb_{1-x}Sr_xTe$ may act as a barrier material for MQW's with Bi_2Te_3 as well material. In contrast to the BaF_2 layers discussed before the electric properties of this barrier material can be tuned by varying the strontium content. The band gap can be increased from about 0.3 eV (x=0) up to 0.6 eV (x=0.13) without reducing the structural quality of these compounds as know from investigations on $PbTe/Pb_{1-x}Sr_xTe$ MQW structures [6]. An additional advantage compared to the former system is the less sensitive behavior of this system resulting in a stable epitaxial growth of IV-VI-compounds when changing the growth temperature. Thus the growth temperature was adjusted for Bi_2Te_3 deposition. In figure 5 a SEM cross sections of a exemplary structure having 50 periods of $Bi_2Te_3/Pb_{1-x}Sr_xTe$ is shown. The periods can be observed suggesting a regular growth. The thickness of the Bi_2Te_3 wells estimated from the growth rates is about 10 nm, the $Pb_{1-x}Sr_xTe$ barriers have a thickness of 54 nm. The power factors determined for samples like the one shown above with well thickness down to ~10 nm range from 0.1 to 1.0 $\mu W\ cm^{-1}\ K^{-2}$ at room temperature. Since the dependence of the carrier concentration on the tellurium stabilisation flux is somewhat more sensitive than in the case of binary (IV-VI)-compounds the thermoelectric results will not be discussed in further detail.

The recording and interpretation of the XRD measurements of these samples is more complicated than for homoepitaxial systems like $PbTe/Pb_{1-x}Sr_xTe$ or $Bi_2Te_3/Bi_2(Te_{1-x}Se_x)_3$ due to the fact that the single layers making up the SL have different crystal structures. This work is still in progress.

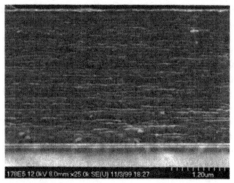

Figure 5. *Cross section of a $Bi_2Te_3/Pb_{1-x}Sr_xTe$ MQW structure showing 50 periods of 10 nm Bi_2Te_3 and 54 nm $Pb_{1-x}Sr_xTe$.*

PbSe/Bi$_2$Te$_3$: First results of the structural and thermoelectric properties of thick stacks consisting of thick bulk layers of Bi_2Te_3 and bismuth doped PbSe already have been reported [12]. These stacks were obtained by overgrowing the starting layer PbSe with Bi_2Te_3. The measured electrical conductivities and thermopowers of the stacks are in good agreement with those calculated using the respective data of the reference layers over a wide temperature range. These results indicate that there is no significant interdiffusion at the interface. This is very important since the thermoelectric properties of both materials used could be changed drastically. Thus several stacked layers and SL structures were investigated using SIMS analysis (figures 6, 7). N-type PbSe samples with different carrier concentrations of $1.2*10^{19}$ cm^{-3} and $7.8*10^{18}$ cm^{-3} were overgrown with the same Bi_2Te_3 layer at a substrate temperature of ~290°C in the IPM MBE system. Analysing the SIMS depth profiles (figure 6) shows that there is no dependence of the steepness of the bismuth signal at the interface with respect to the bismuth doping level of the overgrown bismuth doped PbSe layers.

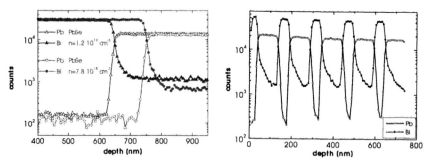

Figure 6. *Left: SIMS depth profile of differently bismuth doped PbSe bulk layers overgrown with Bi_2Te_3. Right: stack of 5 periods of undoped PbSe (nominal thickness 100 nm) and Bi_2Te_3 (nominal thickness: 50 nm).*

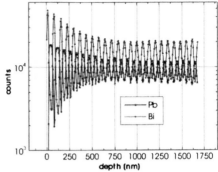

Figure 7. SEM cross section (left) and SIMS depth profile (right) of a SL structure consisting of 25 periods of undoped PbSe(~40 nm) and Bi_2Te_3 (~30 nm).

For structural investigations it is not necessary to have PbSe with a certain carrier concentration. Therefore SL structures in this system were grown using Bi_2Te_3 and undoped PbSe with and without selenium stabilisation yielding PbSe with carrier concentrations of about 10^{17} cm^{-3}. A SEM cross section of an other PbSe (~40 nm)-Bi_2Te_3 (~30 nm) SL sample is shown in figure 7 together with the SIMS depth profile of this sample. The number of periods (25) can be easily observed in the SEM picture and in the SIMS depth profile.

Despite of the problem of having used PbSe with carrier concentrations initially lying far from optimum (~10^{19} cm^{-3}) for thermoelectric applications thermopower and hall effect measurements were made for both samples prepared using undoped PbSe. Both of the samples were n-type in the in the temperature range from 320 down to 50°K. Anyhow power factors of about 10 µW cm^{-1} K^{-2} at room temperature were achieved for both samples (figure 8). Since the EPI MBE recently was equipped with an additional bismuth cell further investigations on n(Bi)-PbSe/Bi_2Te_3 SL structures can be made taking advantage of the availability of both basic materials Bi_2Te_3 and PbSe having optimum starting conditions.

Bi_2Te_3/Bi_2Se_3: The last system outlined in the introduction (figure 2) was investigated mainly because of the SL approach. We decided to grow n-type Bi_2Te_3-$Bi_2(Te_{1-x}Se_x)_3$ SL structures because of the need of two materials both having good thermoelectric properties. As it

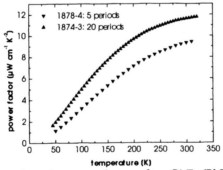

Figure 8: Temperature dependent power factors of two Bi_2Te_3/PbSe stacks

has been already shown that MBE grown Bi_2Te_3-layers should be suitable for SL structures [12] we decided to use Bi_2Te_3 also as a starting point for the V-VI-SL structures. To get a entire n-type stack Bi_2Te_3 was chosen as second material as the growth parameters for preparing these quasiternary materials should be almost the same as for Bi_2Te_3.

First $Bi_2(Te_{1-x}Se_x)_3$ bulk layers were grown to quantify the Se incorporation in the Bi_2Te_3 layers. The $Bi_2(Te_{1-x}Se_x)_3$ layers were grown at a substrate temperature of 290°C using a Bi/Te ratio of 5/12. The amount of Se was chosen by opening or closing the valve of the Se cracker cell. The results of thermoelectric measurements of these samples are listed in table I. Further the absorption edges of these samples were determined by IR-transmission measurements (figure 9). In this first approach the corrections due to the Burstein-Moss-shift and the free carrier absorption, necessary for the optical determination of the band edge, were neglected. By a variation of the growth conditions of the $Bi_2(Te_{1-x}Se_x)_3$ layers the thermoelectric properties could be adjusted, in order to test the feasibility of such V-VI-SL structures it is sufficient to know the properties of the starting materials used.

On this basis first SL layers of 2.5 times (200 nm Bi_2Te_3 - 200 nm $Bi_2(Te_{1-x}Se_x)_3$, entire thickness 1 µm) and 25 times (20 nm Bi_2Te_3 - 20 nm $Bi_2(Te_{1-x}Se_x)_3$, entire thickness 1 µm) were grown. Using the Se signals of SIMS depth profiles measured for these samples the nominal thickness of the layers mentioned above was verified.

In further experiments the thickness of the single layers was reduced while the number of layers was increased to obtain a overall thickness of 1 µm. A comparison between the IR transmission spectra of the different SL structures shows that the average cutoff edges are the same within the experimental error.

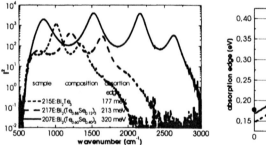 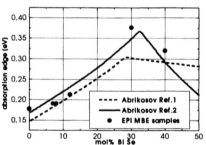

Figure 9. *Selected FTIR transmission spectra (left) and determined absorption edges (right) of $Bi_2(Te_{1-x}Se_x)_3$ samples in comparison with different curves taken from [5]*

Table I. *Room temperature properties of $Bi_2(Te_{1-x}Se_x)_3$ samples having different compositions*

Sample-no.	$Bi_2(Te_{1-x}Se_x)_3$ X	E_{Gap} [meV]	n [10^{19} cm^{-3}]	σ [(Ω cm)$^{-1}$]	S [µV K^{-1}]	σS^2 [µW cm^{-1}K^2]
215E	0	177	2.7	610	-207	26
216E	0.07	191	3.4	444	-208	19
246E	0.08	191	7.3	337	-196	13
217E	0.12	213	3.3	589	-215	27
207E	0.30	376	0.9	164	-276	13
208E	0.40	320	1.1	251	-234	14

High resolution X-ray diffraction (HRXRD) measurements were made at Fraunhofer IAF to determine the period of the SL structures. HRXRD spectra of exemplary samples investigated are shown in figure 10, these samples all have 50 or more periods and show a well defined XRD spectrum of the SL. The periods determined from these measurements are listed in table II together with the sample dimensions and compositions. The measured XRD periods are in good agreement with the nominal ones estimated from the growth rates of the respective V-VI-layers. Further the width of the SL0 peaks at half maximum is similar to those of Bi_2Te_3 layers grown on freshly cleaved BaF_2 substrates (about 0.2 deg). The influence of the substrate material used on the structural quality of the V-VI-SL's is evident looking at figure 10. This fact has already been investigated for Bi_2Te_3 grown on polished BaF_2 substrates using AFM measurements. In this case a decrease of the width of the atomically flat growth terraces were observed [10]. Thermopower measurements in the temperature range between 120 and 320°K were performed for some of the SL samples and the corresponding Bi_2Te_3 and $Bi_2(Te_{1-x}Se_x)_3$ layers. The temperature dependence of the thermopower is comparable to other Bi_2Te_3 n-type samples given in the literature [14]. The dependence of the thermopower on the carrier concentration at room temperature is the same for all samples: increasing the carrier concentration at a given temperature decreases the thermopower (table III). With increasing temperatures also the thermopower increases and tends towards a maximum. The samples investigated all reach their expected maximum values at temperatures higher than 320°K. At temperatures higher than

Table II. *Nominal and measured structures of Bi_2Te_3-$Bi_2(Te_{1-x}Se_x)_3$ SL. The Bi_2Te_3 and $Bi_2(Te_{1-x}Se_x)_3$ layers used always have the same thickness.*

Sample no.	Nominal single layer thickness [nm]	# periods	Average Se content [at%] EDX	Measured period [nm]
226E	200	2.5	-	400 (SIMS)
245E	20	25	-	37±6 (XRD)
227E	20	25	-	39.5 (SIMS)
237E	10	50	6	19.8±0.5 (XRD)
247E	10	100	6	19.3±0.5 (XRD)
238E	5	100	6	9.6±0.5 (XRD)

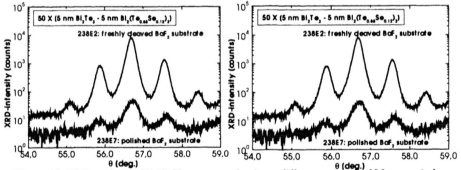

Figure 10. *XRD analysis of V-VI-SL structures having a different number of 20 nm periods (left) and grown on different substrates (right).*

Table III. Thermoelectric properties of Bi_2Te_3-$Bi_2(Te_{1-x}Se_x)_3$ SL's at room temperature

Sample no.	$\sigma\ [(\Omega\ cm)^{-1}]$	$S\ [\mu V\ K^{-1}]$	$n\ [cm^{-3}]$	Power factor $[\mu W\ cm^{-1}\ K^{-2}]$
226E	1118	-175	$5.7*10^{19}$	20.4
227E	747	-192	$4.3*10^{19}$	27.6
237E	812	-199	$4.4*10^{19}$	32.1
238E	639	-204	$4.0*10^{19}$	26.5

that of the thermopower maximum the values should decrease due to intrinsic behavior in this temperature range. The thermopower of all SL samples is similar to the values of the corresponding Bi_2Te_3 and $Bi_2(Te_{1-x}Se_x)_3$ samples.

To determine temperature dependent power factors also Hall measurements using the van der Pauw method were made in the temperature range of 50 to 320 K. The results for the conductivities is shown in figure 11. The temperature dependence of these films in principle show the same behavior as found in reference data taken from publications for bulk Bi_2Te_3 [14]. With increasing temperature the conductivities decrease towards a minimum with different magnitudes depending on the carrier concentration of the sample. At temperatures higher than that of the minimum the curves seem to tend to a common line. Because of the limitations of our Hall measurement equipment concerning temperatures higher than 320 K, measurements in the intrinsic region were not possible, but we assume for our samples the same qualitative behavior.

The corresponding power factors in the temperature range from 120 to 320 K are also shown in figure 11 together with the power factor calculated using a simple stack model where all SL layers are assumed to be connected in parallel. Even when assuming a error of 10% for the conductivity as well as for the thermopower measurements (error bars in calculated line in figure 11), values of the power factors of the SL samples are similar to those of the corresponding bulk samples and well above those of the data of the stack model.

CONCLUSIONS

We investigated thermoelectric and structural properties of several materials and new material combinations based on Bi_2Te_3. Since Bi_2Te_3 is the standard material for applications in the region of room temperature it has been thought of as a possible candidate for testing the theorys basing on different MQW's and SL's as well.

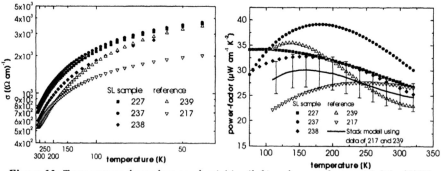

Figure 11: Temperature dependent conductivities (left) and power factors (right) of V-VI-SL structures together with the corresponding Bi_2Te_3 and $Bi_2(Te_{1-x}Se_x)_3$ layers

The predicted enhancement of the thermoelectric figure of merit in 2D MQW system is expected if the Bi_2Te_3 well is 3 nm or thinner. Beside the problem how to realise such thin films for a material with a unit cell of 3 nm size also no suitable barrier materials were known. We investigated BaF_2 and ternary $Pb_{1-x}Sr_xTe$ compounds as barrier materials. In the case of BaF_2 the large temperature difference of optimum growth temperatures for barrier and well makes the optimum multilayer growth difficult. Nevertheless first stacked structures of Bi_2Te_3 and BaF_2 were presented showing that it might be possible to realise such heteroepitaxial structures. In the case of the more adopted material $Pb_{1-x}Sr_xTe$ the wanted structural properties can be expected SEM analysis. Further experiments using n-type doped PbSrTe must be performed to get further information about this material combination.

For testing new superlattice structures thermoelectric and structural properties $PbSe/Bi_2Te_3$ with periods down to 70 nm have been presented. Although also in this case undoped PbSe was used surprisingly high power factor of about 10 $\mu W\ cm^{-1}K^{-2}$ were measured. Therefore this system seems to be the best candidate for further investigations in the heteroepitaxial (IV-VI)-(V-VI) SL systems.

Superlattice structures in the system Bi_2Te_3-Bi_2Se_3 were grown and periods down to 10 nm were confirmed using X-ray measurements. The determined power factors of the SL structures are at least as high as those of the respective bulk materials used. Therefore the material quality perpendicular to the growth direction is preserved for the presented SL structures, thus it is expected that this is also valid along the SL axis. Because of the anisotropy of Bi_2Te_3 the power factors along the SL axis should be about a factor of four smaller than the one measured here. Combining this with the strong reduction of the thermal conductivity along the SL axis these systems are still very promising. A other possibility is a in plane reduction of the thermal conductivity as shown for $PbTe/Pb_{1-x}Sr_xTe$ MQW systems [15]. If this is also true for the Bi_2Te_3-Bi_2Se_3 systems presented here this would result again in a reduced thermal conductivity taking advantage of the preserved higher power factors.

In contrast to the Bi_2Te_3-Sb_2Te_3 system used to show the reduction of the thermal conductivity [8, 9] the Bi_2Te_3-$Bi_2(Te_{1-x}Se_x)_3$ SL consist only of n-type materials making thermoelectric measurements more reliable. Thermal conductivity measurements of these samples have to be done in order to check the predicted reduction also within this system.

ACKNOWLEDGEMENTS

This work was supported by the German Ministry for Education and Research (BMBF), grant No. 03N2014A. We thank N. Herres and H. Güllich of Fraunhofer IAF for performing the XRD measurements and fruitful discussions.

REFERENCES

1. L.D. Hicks, M.S. Dresselhaus, *Phys. Rev. B*, **47**(19), 12727-12731 (1993)
2. M.S. Dresselhaus, T. Koga, X. Sun, S.B. Cronin, K.L. Wang, G. Chen, *Proc. 16th International Conference on Thermoelectrics*, Dresden, Germany, 1997, 12-20
3. L. D. Hicks et al, *Phys. Rev. B*, **53**(16) R10493-R10496 (1996)
4. T. L. Reinecke et al, *16th International Conference on Thermoelectrics*, Dresden, Germany, 1997, 424-428
5. N. Kh. Abrikosov, Semiconducting II-VI, IV-VI and V-VI compounds, *Plenum Press*, New York (1969)

6. A. Lambrecht, to be published in *Encyclopedia of Materials: MBE of IV-VI compounds*
7. R. Venkatasubramanian, T.S. Colpitts, D. Malta, M. Mantini, *Proc. 13th International Conference on Thermoelectrics*, Kansas City, 1994
8. R. Venkatasubramanian, *Mat. Res. Soc. Symp. Proc.*, **478**, 73-84 (1997)
9. I. Yamasaki, R. Yamanaka, M. Mikami, H. Sonobe, Y. Mori, T. Sasaki, *Proc. 17th International Conference on Thermoelectrics*, Nagoya, Japan, 1998, 210-213
10. J. Nurnus, H. Beyer, H. Böttner, A. Lambrecht, *Proc. 5th European Workshop on Thermoelectrics*, Pardubice, Czech Republic, 1999, 25-30
11. X.M. Fang, H.Z. Wu, Z. Shi, P.J. McCann, N. Dai, *J. Vac. Sci. Technol. B*, **17**(3), 1297-1300 (1999)
12. J. Nurnus, H. Böttner, H. Beyer, A. Lambrecht, *Proc. 18th International Conference on Thermoelectrics*, Baltimore, USA, 1999, in press
13. J.O. Barnes, J.A. Rayne, R.W. Ure Jr., *Physics Letters*, **46A**(5), 317-318 (1974)
14. C.H. Champness, A.L. Kippling, *Canadian Journal of Physics*, **44**, 769-788 (1965)
15. H. Beyer, A. Lambrecht, J. Nurnus, H. Böttner, H. Griessmann, A. Heinrich, L. Schmitt, M. Blumers, F. Völklein, *Proc. 18th International Conference on Thermoelectrics*, Baltimore, USA, 1999, in press

Synthesis and Physical Properties of Skutterudite Superlattices

Joshua R. Williams, David C. Johnson[1]
[1]Department of Chemistry and Materials Science Institute, University of Oregon, Eugene, OR 97403-1253
Michael Kaeser, Terry Tritt
Department of Physics and Astronomy, Clemson University, Clemson, SC 29634-1905
George Nolas
Research and Development Division, Marlow Industries, Dallas, TX 75238-1645
E. Nelson
U. S. Army Research Laboratory, Adelphi, MD 20783

ABSTRACT

Predicted and observed reductions in thermal conductivity of materials with superlattice structure have prompted interesting research into the possibility of using these materials as higher efficiency thermoelectrics. Synthesis of superlattice materials is challenging however, as the structure itself is generally not very stable at high temperatures as it is prone to interdiffusion. Presented here is the successful synthesis and characterization of a superlattice composed of two materials with the skutterudite structure, $IrSb_3$ and $CoSb_3$.

INTRODUCTION

Thermoelectric materials have the very valuable potential to be used in two quite different, but technologically important areas: localized cooling applications and power generation. However, the low efficiency of current materials has limited their practical application. There is much ongoing research into different ways to increase the efficiency of current materials, as well as the design of new materials with potentially more favorable properties. Recent theoretical and experimental work [1,2,3,4,5,6,7] has demonstrated that superlattice materials may exhibit lower thermal conductivity than either of the constituent materials alone. This could be used to great advantage for applications such as thermoelectricity where low thermal conductivity materials are of great interest. Synthesis of superlattice materials is a major challenge however, and there have been only a few successful techniques developed. In this study, the modulated elemental reactant synthesis method was used to synthesize a superlattice material composed of the skutterudite compounds $CoSb_3$ and $IrSb_3$. It has been suggested that this particular superlattice may have good transport properties for thermoelectric applications.

EXPERIMENTAL

The sample was deposited onto a silicon wafer, which was previously spin-coated with polymethylmethacrylate (PMMA). The deposition was performed in a high-vacuum deposition chamber described elsewhere [8]. Iridium and cobalt were deposited using electron beam evaporation guns. Antimony was deposited using a custom resistive heating cell. The background pressure in the chamber was lower than 5×10^{-6} during the entire deposition. A piece of zero-background quartz plate was attached to the silicon substrate during deposition, and was used for x-ray analysis of the film. After deposition, the substrate was removed from the chamber and soaked in acetone to dissolve the PMMA and facilitate sample removal. The sample was filtered from the acetone and collected on Teflon filter paper. Differential Scanning Calorimetry experiments were performed using a TA Instruments DSC 2920 Modulated DSC. A Scintag theta-theta diffractometer and a Phillips X'Pert diffractometer were used for x-ray analysis.

RESULTS AND DISCUSSION

The sample was produced by alternately depositing Ir and Sb to form 8 bilayers and then depositing 8 Co and Sb bilayers. This sequence was repeated 5 times to form a superlattice of elemental layers with an intended period of 250Å and an intended total film thickness of 1250Å. Differential Scanning Calorimetry performed on this sample, as shown in Figure 1, shows that there are several exothermic events that occurred upon heating of this sample. A high angle x-ray diffraction spectrum of the sample powder after DSC to 550°C is shown in Figure 2. This clearly shows the formation of the skutterudite pattern. Closer examination of these high angle peaks (Figure 3) reveals splitting caused by the underlying superlattice reflections interfering with the reflections from the lattice planes. The distance between these splits is as expected for a superlattice with a 250Å period. To try to understand the diffusion and nucleation processes occurring in this sample, we performed a step-annealing study using a piece of zero-background quartz that was attached to the substrate during the sample deposition.

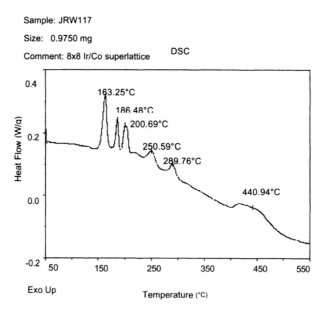

Figure 1. DSC trace of IrSb/CoSb layered sample.

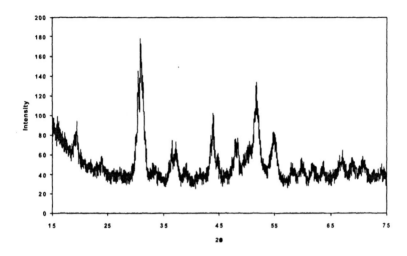

Figure 2. High angle x-ray diffractions spectrum of sample powder after DSC to 550°C.

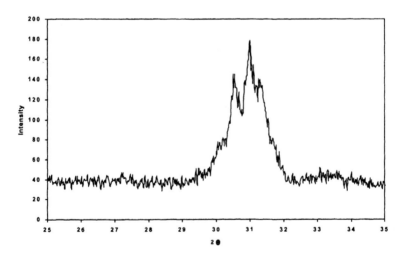

Figure 3. High angle x-ray diffraction spectrum showing splitting due to superlattice structure.

Shown in Figure 4 is a stacked plot showing the evolution of the low angle x-ray diffraction pattern of this sample with annealing. The positions of the Bragg peaks confirm that the actual as-deposited superlattice period is 245Å, very close to the intended period of 250Å. The

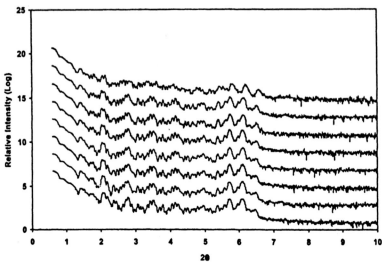

Figure 4. Stacked plot of low angle x-ray diffraction spectra at various annealing conditions.

evolution of this pattern with annealing results from the interdiffusion of the elemental layers as well as other diffusion processes. It is interesting to note that the Bragg peak positions shift to lower angle with increasing annealing temperature showing that the superlattice period is shrinking slightly. This is probably due to the rearrangement of mobile defects and the filling of voids - effects often observed in samples synthesized using the multilayer reactant technique. This hypothesis is supported by the observation that the peak shift tends to approach a limit at a given annealing temperature, and then shift dramatically when a higher annealing temperature is used. At a given annealing temperature and time, only a portion of the defects will have enough energy to rearrange and achieve a lower energy configuration. Shown in Figure 2 is a high angle x-ray diffraction spectra of sample powder different points in the annealing process. The lower temperature scan shows a very broad, low intensity hump characteristic of amorphous materials. The higher temperature spectrum clearly shows the diffraction pattern characteristic of skutterudite compounds, indicating that nucleation has occurred. The Bragg peaks in the low angle spectrum are still visible even after annealing to 300°C, well after nucleation has occurred. This demonstrates that the sample still has a layered structure with a periodicity equivalent to the intended superlattice period even after crystallization of the skutterudite phase.

The splitting of the high angle peaks observed in the powder x-ray is much more pronounced than that observed in the film spectrum. This is likely due to the relatively rapid annealing of the powder in the DSC compared to the long annealings at low temperature that the film underwent temperature that the film underwent as shown in Figure 5. These longer, lower temperature annealing steps may have allowed the different composition regions of the sample to interdiffuse, thereby undermining the integrity of the superlattice interfaces. The impact of annealing conditions on the superlattice formation is still being studied.

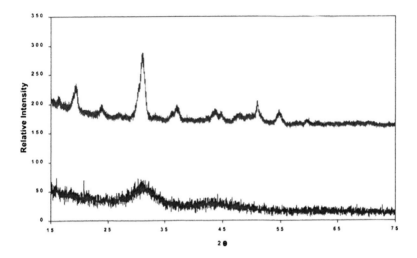

Figure 5. High angle x-ray diffraction spectrum after different annealing conditions.

CONCLUSION

The modulated elemental reactant synthesis method has been successfully used to create an $IrSb_3/CoSb_3$ superlattice. The intended period of the superlattice is confirmed by low angle x-ray diffraction. More rapid annealing processes may be necessary for optimal superlattice formation.

ACKNOWLEDGMENT

Support of this research by the U.S. Army Research Laboratory under contract number DAAD17-99-C-0006 is greatly appreciated.

REFERENCES

[1] L. D. Hicks, T. C. Harmon, M. S. Dresselhaus, *Appl. Phys. Lett.*, **63**, 3230-3232 (1993).
[2] J. O. Sofo, G. D. Mahan, *Applied Physics Letters*, **65**, 2690-2692 (1994).
[3] D. A. Broido, T. L. Reinecke, *Physical Review B*, **51**, 13797-13800 (1995).
[4] P. J. Lin-Chung, T. L. Reinecke, *Physical Review B*, **51**, 13244-13248 (1995).

[5] G. D. Mahan, H. B. Lyon, *Journal of Applied Physics*, **76,** 1899-1901 (1994).
[6] W. S. Capinski, H. J. Maris, *Physica B*, **219/220**, 699-701 (1996).
[7] X. Y. Yu, G. Chen, A. Verma, J. S. Smith, *Applied Physics Letters*, **67**, 3554-3556 1995.
[8] L. Fister, X. M. Li, T. Novet, J. McConnell, D. C. Johnson, *J. Vac. Sci. & Technol. A,* **11,** 3014–3019 (1993).

Artificially Atomic-scale Ordered Superlattice Alloys for Thermoelectric Applications

S. Cho[*], Y. Kim[*], A. DiVenere[*], G. K. L. Wong[*], A. J. Freeman[*], J. B. Ketterson[*],
L. J. Olafsen[**], I. Vurgaftman[**], J. R. Meyer[**], C. A. Hoffman[**], and G. Chen[***]
[*] Dept. of Physics & Astronomy, Northwestern University, Evanston, IL 60208
[**] Naval Research Laboratory, Washington, D.C. 20375-5338
[***] Mechanical & Aerospace Engineering Dept., Univ. of California, Los Angeles, CA 90095.

ABSTRACT

We report artificially atomic-scale ordered superlattice alloy systems, new scheme to pursue high-ZT materials. We have fabricated Bi/Sb superlattice alloys that are artificially ordered on the atomic scale using MBE, confirmed by the presence of XRD superlattice satellites. We have observed that the electronic structure can be modified from semimetal, through zero-gap, to semiconductor by changing the superlattice period and sublayer thicknesses using electrical resistivity, thermopower, and magneto-transport measurements. InSb/Bi superlattice alloys have also been prepared and studied using XRD and thermopower measurements, which shows that their thermoelectric transport properties can be modified in accordance with structural modification. This superlattice alloy scheme gives us one more tool to control and tune the electronic structure and consequently the thermoelectric properties.

INTRODUCTION

Recently, a major effort has been directed to finding high ZT materials (here $ZT=(S^2\sigma/\kappa)T$, where S is the thermopower, σ is the electrical conductivity, and κ is the thermal conductivity). On one side, new bulk materials have been synthesized and tested, and on the other side, (quantum well) superlattice or nanowire structures have been studied. In a conventional quantum well superlattice geometry, enhanced ZT values due to quantum confinement have been reported. [1,2] However, the effective overall ZT is then reduced due to thermal back-flow through the barrier layers in quantum well structures. [3]

An alternative approach for achieving high ZT materials, atomic-scale ordered superlattice alloys, is presented in this article. The advantage of this approach is that it allows engineering the electronic band structure by forming an ordered alloy with the goal of achieving better thermoelectric properties and/or reduced lattice thermal conductivity due to the boundary scattering from the layers.

In an effort to explore this approach, Bi/Sb superlattice alloys have been fabricated using MBE. The atomic arrangement of an ordered superlattice alloy of Bi and Sb, in comparison with a random alloy, is shown in Fig. 1. Bi and Sb are semimetals with a rhombohedral structure. Bi has a small energy overlap between the conduction and valence bands, high carrier mobilities, and small effective masses. These properties have encouraged many researchers to use Bi for quantum size effect studies. The $Bi_{1-x}Sb_x$ random (conventional) alloy is a semiconductor for $0.07 \leq x \leq 0.22$ ($E_{g,max} \sim 18$ meV), but otherwise is a semimetal. [4-15] With properties such as a small bandgap, high mobility and a reduced lattice thermal conductivity, semiconducting Bi-rich BiSb random alloys may potentially be used as an n-type thermoelement operating around 80 K. A ZT of 1.2 at 80 K in a magnetic field of 0.13 T has been reported. [8]

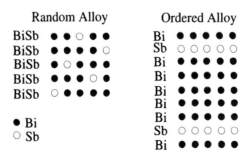

FIG. 1 The atomic arrangement of a random alloy and an artificially ordered alloy. Whereas the Sb atoms are randomly positioned in the random alloy, they are confined to specific planes in the ordered superlattice alloy.

InSb/Bi has been chosen as another example of a system to apply the superlattice alloy scheme. InSb is a material with a small band-gap (0.165 eV at room temp.), a high mobility and a large thermal conductivity which prevents InSb from being considered as a good thermoelectric material. It is known that InSb can be alloyed with Bi, $InSb_{1-x}B_x$, up to 12 at. % Bi ($x = 0.12$), with a small change of band-gap. [16,17] The InSb/Bi superlattice alloy system is a good thermoelement candidate since the reduction of the lattice thermal conductivity, by increasing the phonon scattering at interfaces, may be remarkable.

EXPERIMENT

Bi/Sb superlattices were grown on semi-insulating CdTe(111)B substrates in a custom-built MBE system similar in design to a Varian Model 360, as reported elsewhere. [18] The base pressure of the growth chamber was in the range of 10^{-10} Torr. We first deposited a 3000 Å CdTe buffer layer on the CdTe substrate at 250 °C, followed by the deposition of the Bi and Sb layers at a rate of about 0.4 Å/s and at a growth temperature of 100 °C. Growth was started with Bi on CdTe, followed by Sb and repeated to form a superlattice. The growth direction of the Bi and Sb layers on CdTe(111)B was parallel to the trigonal axis. RHEED was used to examine the specific surface reconstruction, the growth mode and the growth orientation of the deposited layers. The in-plane lattice constants of Bi and Sb in the hexagonal representation are 4.546 and 4.308 Å, respectively. Consequently, the lattice mismatch of Bi and Sb with CdTe(111)B (4.58 Å) are 0.7 % and 6 %, respectively. A two-second interruption (or dwell) time was introduced between the deposition of the Bi and Sb layers to enhance surface migration of the absorbed atoms during the growth. We prepared BiSb superlattice alloys with six different modulation periods: 7.7, 11, 12, 16.5, 18, 27, 40, and 55 Å, while maintaining the Bi to Sb ratio in the range 12-17 %. The total thickness of the alloys was 1.2 μm.

InSb/Bi superlattices were grown on semi-insulating CdTe(111)B substrates in a similar manner as the Bi/Sb superlattice case. The growth temperature was in the range of 300-350 °C. The in-plane lattice constant of InSb is 4.581 Å and the lattice mismatch of InSb with CdTe(111)B are 0.04 %. We prepared 18 Å InSb and 2.4 Å Bi superlattice alloys with several different growth temperatures.

RESULTS AND DISCUSSION

A. Bi/Sb Superlattice alloy

We observed a streaky RHEED pattern with Kikuchi lines when the first Bi layer was deposited on CdTe(111)B, as reported elsewhere. [15,19] The RHEED patterns of Bi(Sb) on Sb(Bi) also showed streaks, representing the 2D layer-by-layer growth of Bi on Sb or Sb on Bi and sharp interfaces. These growth properties are important for achieving superlattices because 3D nucleation introduces height variations and a number of defects where the 3D islands coalesce. There was no surface reconstructions of Bi(Sb) on Sb(Bi).

Standard θ-2θ XRD patterns for the superlattice alloys with different modulation periods are shown in Fig. 2. A well resolved pattern of symmetric satellite peaks is observed. Besides the fundamental Bragg diffraction peak, the first-, second-, and third-order superlattice reflections can be clearly seen for larger modulation periods, indicating relatively abrupt interfaces between the layers. Remarkably, in the superlattice alloys with monolayer and even submonolayer Sb layers ((10 Å Bi/1 Å Sb) and (15 Å/1.5 Å) samples), we see distinct superlattice satellites. The modulation period, λ, of a superlattice can be calculated from the angular position of the satellite peaks, as indicated on the right side in Fig. 2, and is consistent with the value determined by the quartz crystal balance.

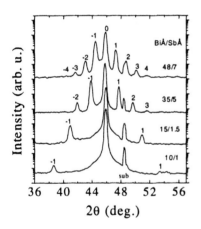
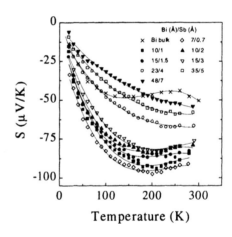

FIG. 2 θ-2θ X-ray scans around the (00.4) reflection for ordered Bi/Sb superlattice alloys with different modulation periods.

FIG. 3 Temperature-dependent thermopower of the superlattice alloys.

Figure 3 shows temperature-dependent thermopowers of Bi/Sb superlattice alloys with different modulation periods. All samples show negative thermopowers because they are either n-type or intrinsic, and the electrons have higher mobilities than holes. As the period decreases the magnitude of the negative thermopower increases. The short period superlattice alloys show slight thermopower increases above 200 K, characteristic of an intrinsic semiconductor. Other samples show semimetallic behavior.

Figure 4 shows the temperature-dependent electrical resistivities of Bi/Sb alloys with different modulation parameters and 1.2μm total thickness. The samples with long periods show semimetallic behavior. However, as the modulation period decreases, the electrical resistivity rapidly increases at low temperature. These data indicate that a semimetal-semiconductor transition occurs as the modulation period decreases. Effective thermal band-gaps (E_g) for the superlattice alloys were determined using the relation $\rho = \rho_0 \exp(E_g/2k_BT)$ as follows: 29, 29, 25, 24, and 22 meV for the samples with the modulation periods of 7.7, 11, 12, 16.5, and 18 Å, respectively.

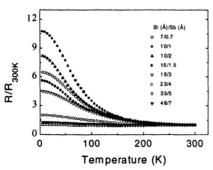

FIG. 4 Temperature dependent electrical resistivities of superlattice alloys.

L_{Bi} (Å)	L_{Sb} (Å)	x_{av}	E_g (meV)	n_0 (cm^{-3})	μ_n (cm^2/Vs)	p_0 (cm^{-3})
$Bi_{0.91}Sb_{0.09}$		0.09	35	3.8×10^{16}	2.6×10^4	$\ll n_0$
7	0.7	0.09	5	1.2×10^{17}	1.9×10^4	$\ll n_0$
10	1	0.09	0	1.1×10^{17}	1.9×10^4	$\ll n_0$
15	1.5	0.09	-10	2.2×10^{17}	1.9×10^4	7.0×10^{16}
15	3	0.17	-15	3.7×10^{17}	9.7×10^3	$\ll n_0$
10	2	0.17	-20	3.8×10^{17}	7.8×10^3	$\ll n_0$
Bi		0	-25	2.1×10^{17}	6.0×10^4	$\approx n_0$
23	4	0.15	-50	1.3×10^{18}	1.6×10^4	$\approx n_0$
35	5	0.13	-50	1.8×10^{18}	1.8×10^4	$\approx n_0$
48	7	0.13	-50	1.5×10^{18}	1.4×10^4	$\approx n_0$

Table. 1 Some electronic characteristics derived from a QMSA of the magneto-transport measurements on Bi/Sb random and superlattice alloys.

To investigate this phenomenon in detail, we determined electron and hole densities and mobilities using a quantitative mobility spectrum analysis (QMSA) [20,21] of the temperature- and field-dependent Hall and resistivity measurements, as shown in Table 1. In the low-temperature limit, all of the samples were either n-type or intrinsic ($n = p$), with electron concentrations in the 1.1-18×10^{17}cm^{-3} range and electron mobilities ranging from 7800-19000 cm^2/Vs.

The band-gaps were also derived in the low-temperature limit, although they are only semi-quantitative in character since the masses and their temperature behavior are complex and not known precisely and because the band-gap is very small and the nonparabolicity is strong. However, we can see that the relative trends of the derived band-gaps show a consistent behaviors in Table 1. All of the thicker-Sb-layer samples (23/4, 35/5, 48/7) have a much larger band overlap than pure Bi. However, shorter-period samples have a smaller overlap or became semiconducting, even when the average Sb concentration was held fixed. This confirms that the ordered superlattice alloys have a distinct band structure, which depends on layer thickness, rather than reverting to the properties of the equivalent random alloy. The differences in the electronic energy spectra between superlattice alloys and random alloys of similar average composition clearly show that the ordered structures correspond to a different (semiconducting) material.

B. InSb/Bi Superlattice alloy

Typical θ-2θ XRD patterns for InSb/Bi superlattice alloys and an InSb thin film on CdTe(111)B are shown in Fig. 5. In the superlattice alloy XRD pattern, several peaks, similar to satellite peaks but not equally spaced, are observed near the fundamental Bragg diffraction peak. The modulation periods calculated from the angular position of the satellite-like peaks do not correspond to som specific period within a reasonable margin. We tried to match these satellite-like peaks with known XRD peak data of InSb, InBi, and InSbBi random alloy but failed. We conclude, provisionally, that these superlattice alloys are not a conventional superlattice but involve some new compounds formed by reaction between the ingredients during the deposition. More XRD studies are in progress to interpret the structure of these superlattice alloys.

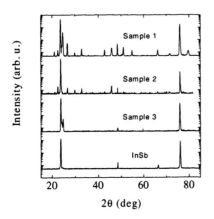
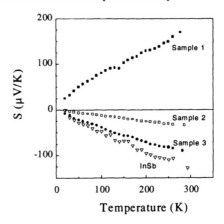

FIG. 5 θ-2θ X-ray scans for InSb/Bi superlattice alloys and a InSb thin film.

FIG. 6 Temperature-dependent thermopower of the superlattice alloys and the InSb film.

Figure 6 shows a plot of the temperature-dependent thermopowers measured for InSb/Bi superlattice alloys. One of the samples shows positive values. This is quite interesting because intrinsic InSbBi, InSb, and Bi all show n-type conduction, due to the high electron mobilities. [16,17] The measured positive thermopower values seem to be above those achievable by doping, which is another reason to think these superlattice alloys may be regarded as a new compound, and possibly large enough for thermoelectric applications. With the help of the expected reduction in the lattice thermal conductivity of these superlattice alloys (by the phonon scattering at the interfaces), these InSb/Bi superlattice alloy structures might open a possibility to make both n- and p-type thermoelectric elements from the same starting materials without any other special treatments.

SUMMARY

By alternating thin layers of Bi and Sb in a periodic geometry, we achieved artificially atomic-scale ordered Bi/Sb superlattice alloys; i.e., we can engineer the electronic band structure by changing the thicknesses of the Bi and Sb layers. While long-period samples showed the expected semimetallic behavior in the electrical resistivity, thermopower, and field-dependent magneto-transport, a semimetal-semiconductor transition was observed with decreasing period. The power factor is comparable to that of the random alloy. We expect a lower thermal conductivity due to an increase in phonon scattering from the layers, which may result in a higher ZT over a random alloy. InSb/Bi superlattice alloys have also been prepared and studied using XRD and thermopower measurements, which show that their thermoelectric transport properties can be modified in accordance with their structural changes. In conclusion, the superlattice alloy scheme gives one more tool to control and tune the electronic structure and consequently the thermoelectric properties.

ACKNOWLEDGMENTS

This work was supported by DARPA under Grant No. DAAG55-97-1-0130. Use was made of MRL Central Facilities supported by the National Science Foundation, at the Materials Research Center of Northwestern University, under Award No. DMR-9120521.

REFERENCES

1. L. D. Hicks and M. S. Dresselhaus, Phys. Rev. B **47**, 12727 (1993).
2. L. D. Hicks, T. C. Harman, and M. S. Dresselhaus, Appl. Phys. Lett. **63**, 3230 (1993).
3. T. C. Harman, D. L. Spears, and M. J. Manfra, J. Electron. Mat. **25**, 1121 (1996).
4. A. L. Jain, Phys. Rev. **114**, 1518 (1959).
5. S. Golin, Phys. Rev. **176**, 830 (1968).
6. E. J. Tichovolski and J. G. Mavroides, Solid State Commun. **7**, 927 (1969).
7. G. Oelgart, G. Schneider, W. Kraak and R. Herrmann, Phys. Stat. Sol. (b) **74**, K75 (1976).
8. W. M. Yim and A. Amith, Solid-State Electron. **15**, 1141 (1972).
9. B. Lenoir, M. Cassart, J.-P. Michenaud, H. Scherrer and S. Scherrer, J. Phys. Chem. Solids **57**, 89 (1996).
10. D. M. Brown and S. J. Silverman, Phys. Rev. 136, A290 (1964).
11. E. E. Mendez, A. Misu, and M. S. Dresselhaus, Phys. Rev. B **24**, 639 (1981).
12. M. Lu, R. J. Zieve, A. van Hulst, H. M. Jaeger, T. F. Rosenbaum, and S. Radelaar, Phys. Rev. B **53**, 1609 (1996).
13. D. T. Morelli, D. L. Partin and J. Heremans, Semicon. Sci. Technol. **5**, S257 (1990).
14. D. M. Brown and S. J. Silverman, Phys. Rev. **136**, A290(1964).
15. S. Cho, A. DiVenere, G. K. Wong, J. B. Ketterson, and J. R. Meyer, Phys. Rev. B**59**, 10691 (1999).
16. J. L. Zilko and J. E. Greene, J. Appl. Phys. **51**, 1549 (1980).
17. J. J. Lee, J. D. Kim, and M. Razeghi, Appl. Phys. Lett. **70**, 3266 (1997).
18. A. DiVenere, X. J. Yi, C. L. Hou, H. C. Wang, J. B. Ketterson, G. K. Wong, and I. K. Sou, Appl. Phys. Lett. **62**, 2640 (1993).
19. S. Cho, A. DiVenere, G. K. Wong, J. B. Ketterson, J. R. Meyer, J. I. Hong, Phys. Rev. B **54**, 2324 (1998).
20. I. Vurgaftman, J. R. Meyer, C. A. Hoffman, S. Cho, J. B. Ketterson, L. Faraone, J. Antoszewski and J. R. Lindemuth, J. Electron. Mat. **28**, 548 (1999).
21. Vurgaftman, J. R. Meyer, C. A. Hoffman, S. Cho, A. DiVenere, G. K. Wong, and J. B. Ketterson, J. Phys: Condens. Matter **11**, 5157 (1999).

Thermoelectric Properties of PbSr(Se,Te)-Based Low Dimensional Structures

Harald Beyer[1], Joachim Nurnus[1], Harald Böttner[1], Armin Lambrecht[1], Lothar Schmitt[2], Friedemann Völklein[2]
[1]Fraunhofer Institute of Physical Measurement Techniques (IPM),
Heidenhofstr. 8, D-79110 Freiburg i. Br., Germany
[2]FB Physikalische Technik, Fachhochschule Wiesbaden,
Am Brückweg 26, D-65428 Rüsselsheim, Germany

ABSTRACT

Thermoelectric properties of low dimensional structures based on PbTe/PbSrTe-multiple quantum-well (MQW)-structures with regard to the structural dimensions, doping profiles and levels are presented. Interband transition energies and barrier band-gap are determined from IR-transmission spectra and compared with Kronig-Penney calculations. The influence of the data evaluation method to obtain the 2D power factor will be discussed. The thermoelectrical data of our layers show a more modest enhancement in the power factor σS^2 compared with former publications and are in good agreement with calculated data from Broido et al. [5]. The maximum allowed doping level for modulation doped MQW structures is determined. Thermal conductivity measurements show that a ZT enhancement can be achieved by reducing the thermal conductivity due to interface scattering. Additionally promising lead chalcogenide based superlattices for an increased 3D figure of merit are presented.

INTRODUCTION

Model calculations predict an enhancement of the thermoelectric figure of merit ZT in MQW structures compared with bulk material [1]. This large increase in ZT will result mainly from the two dimensional electron density of states causing a strong thermopower enhancement. Additionally the thermal conductivity is reduced by interface scattering in the boundary regions between wells and barriers [2]. Other authors predict a more modest enhancement in ZT because of electron tunneling [3,4] and enhanced carrier-phonon scattering for decreasing well-widths [5]. A first experimental evidence for an increasing figure of merit in n-type PbTe/PbEuTe-MQW is given by Harman et al. [6]. MQW structures with well-widths less than 3 nm show a strong enhancement in the power factor σS^2. A detailed analysis of electrical and optical properties of this system is given by Yuan et al. [7]. Recently first results on PbSe/PbTe-quantum dot-superlattices show an enhancement of the 3D figure of merit [8] allowing this material system for the use in highly efficient thermoelectrical applications.

However some problems arise from the evaluation method for the thermoelectrical data of MQW structures given in [6]. A complete carrier transfer from barriers into the wells is assumed approximating the barrier layers as ideal insulator. Based on this idea the well thermopower is given by the thermopower of the whole structure. This might cause problems in modulation doped heterostructures.

In the present study we show the thermoelectric properties of well and barrier doped PbTe/PbSrTe-MQW's with different well widths and doping levels. The thermoelectrical data were evaluated like in the Hicks-model [6]. To study the limitations of this model further MQW's with different doping levels at the same well and barrier width were investigated.

EXPERIMENTAL DETAILS

PbTe/PbSrTe-MQW's were grown by molecular beam epitaxy (MBE) on freshly cleaved $BaF_2(111)$ at a substrate temperature of 370°C. PbTe was evaporated from a PbTe compound source achieving a growth rate of about 1.5 µm/h. 10 % additional Te-flux was used to stabilize the stoichiometry. N-type doping was performed by elementary bismuth. To increase the barrier band gap up to 0.63 eV elementary strontium was coevaporated during PbTe-growth. Two well doped MQW series with well-width varying between 1 and 10 nm at a constant barrier-width of 14 nm were made. The well-carrier concentration n_{well} was $4.5*10^{18}$ cm^{-3} and $1.2*10^{19}$ cm^{-3} respectively. Barrier doped MQW's with a barrier thickness of 15 nm also varying the well width in a range between 1 and 7 nm had a barrier doping-level $n_{Barr} = 8*10^{17}$ cm^{-3} and $2.5*10^{18}$ cm^{-3}. To investigate the carrier transfer between barriers and wells a series with varying doping levels at $d_{Well} = 2$ nm and $d_{Barr} = 40$ nm has been grown. Electrical conductivity was measured in Van der Pauw-geometry. Infrared spectroscopy was performed by FT-IR in a wavenumber range of 500...6000 cm^{-1}. Thermal conductivity was measured after removing the substrate by the method described in [9]. All measurements of transport properties were performed at 300 K.

DISCUSSION

Infrared transmission and photoconductivity of a PbTe/PbSrTe-MQW structure with d_{Well} = 4.69 nm and d_{Barr} = 67 nm are shown in fig. 1a. The step-like behavior of IR-transmission and photoconductivity is corresponding to interband transitions. The sharp steps and the transition energies indicate a type I superlattice structure. The temperature dependence of interband transition energies, well and barrier band-gap is shown in fig. 1b. Transition energies and barrier band-gap show a nearly linear temperature dependency and are in good agreement with calculations in a Kronig-Penney model for superlattice structures [10] taking the non-

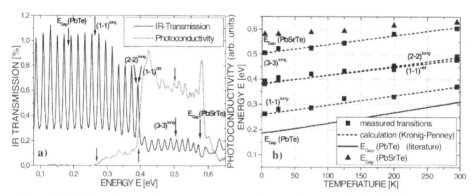

Figure 1. *a) Infrared transmission spectrum and photoconductivity of a PbTe/PbSrTe-MQW structure with d_{Well} = 4.69 nm and d_{Barr} =67 nm at T = 4.8 K. The step-like character corresponds to interband transitions. b) T-dependence of interband transitions: The measured transition energies are in good agreement with calculations in a Kronig-Penney model.*

parabolicity of the band structure (Kane model) into account. A conduction band offset $\Delta E_C/\Delta E_{Gap} = 0.55$ according to [7] was used. Over the whole temperature range between 4.8 K and 300 K the MQW structure shows type I band alignment.

Thermoelectrical properties of all MQW structures were evaluated according to [6]. There any electrical transport in the barrier layers is neglected. A strong enhancement of nS^2_{2D} for small well-widths can be observed compared with nS^2 for bulk PbTe with the same doping level. This behaviour agrees well with calculations made by Broido et al. [5] taking the influence of surface roughness or carrier-phonon scattering on the carrier mobility into account (see fig. 2). Consequently these results are a good evidence of the influence of quantum confinement on the thermoelectric properties and would result in a significantly increased 2D power factor by using the bulk carrier mobility of 1400 cm²/V*s in MQW structures. However with regard to scattering mechanisms the carrier mobility in our MQW structures is reduced: $\mu \approx 1000$ cm²/V*s for d_{well} = 9 nm; $\mu \approx 600$ cm²/V*s for d_{well} = 2 nm (see table I). Consequently the 2D power factor is reduced compared with Hicks [1]. As shown in [11] for 2D carrier concentrations up to n_{well} = $1,2*10^{19}$ cm^{-3} in well doped MQW's the 2D power factor has a minimum at well widths between 3 and 5 nm. For smaller well widths an increasing power factor can be observed due to thermopower enhancement, but it is not exceeding bulk values because of mobility reduction. The mobility reduction can be slightly reduced by barrier doping due to separation of carriers and dopants.

A ZT enhancement might be possible by high doping levels in barrier doped MQW structures or a reduction in the thermal conductivity due to interface scattering. The first way means reaching a 2D carrier concentration $n_{2D} > 1*10^{19}$ cm^{-3}. On one hand at high carrier concentrations the carrier mobility of bulk material is reduced due to scattering at the dopants. Thus the mobility difference between bulk PbTe and MQW structures is reduced. On the other hand a high doping level results in occupation of continuum or barrier states. Therefore we investigated barrier doped MQW's with $d_{Well} = 2$ nm and $d_{Barr} = 40$ nm in a range of $n_{2D} = 4*10^{18}...5*10^{19}$ cm^{-3}. The results of conductivity and thermopower measurements are shown in

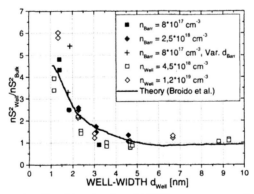

Figure 2. nS^2_{2D}/nS^2_{Bulk} of well and barrier doped MQW structures as a function of d_{well} compared with calculations of Broido et al. [5] neglecting any barrier transport.

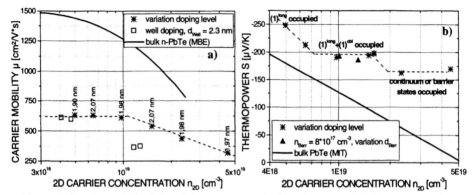

Figure 3. Carrier mobility (a) and thermopower (b) of barrier doped MQW structures (d_{Well} = 2 nm, $d_{Barr} \approx 40$ nm) depending on n_{2D} evaluated as in [6]. The mobility decreases for $n_{2D} > 10^{19}$ cm^{-3}. A step-like behavior of thermopower can be observed.

fig. 3. As already described the carrier mobility is reduced compared with bulk PbTe. Up to $n_{2D} = 10^{19}$ cm^{-3} the mobility is nearly constant at $\mu \approx 600$ cm^2/V*s. For higher n_{2D} a decreasing carrier mobility was observed. Data obtained for well doped samples are significantly lower. Additionally a step-like behaviour in the thermopower can be found showing plateaus with constant thermopower for $n_{2D} = 7*10^{18}...1.5*10^{19}$ cm^{-3} and $n_{2D} > 2*10^{19}$ cm^{-3}. No influence of barrier width variation on the thermopower was found. This behaviour can be interpreted by the occupation of subbands and continuum or barrier states respectively. For $n_{2D} < 7*10^{18}$ cm^{-3} only one subband is significantly occupied. At carrier concentrations above $7*10^{18}$ cm^{-3} the second subband has a strong contribution to electrical conductivity and thermopower. For $n_{2D} > 1.5*10^{19}$ cm^{-3} the contribution of continuum or barrier states to the thermopower can't be neglected. These results show an upper limit of n_{2D} between $1*10^{19}$ and $1.5*10^{19}$ cm^{-3} in modulation doped MQW structures. By this way in single structures a maximum value of the 2D power factor $\sigma S^2_{2D} = 40...50$ µW/cm*K^2 can be reached, but with the uncertainty of electrical transport in the barriers.

Recently, a reduction of the thermal conductivity λ was observed for MQW structures. Two MQW's with d_{Well} = 2.38 nm and 9.26 nm and d_{Barr} = 13.5 nm are compared to a n-PbTe layer, the maximum literature values of n-PbTe and a PbSrTe layer (table I). The measured values for PbTe and PbSrTe are 4.15 and 4.5 W/m*K. These values are higher than the literature value of 3.1 W/m*K [7]. With 0.60 W/m*K for d_{Well} = 2.38 nm and 1.25 for d_{Well} = 9.26 nm the MQW structures have a strongly reduced thermal conductivity compared with bulk material.

According to these results a clear evidence for thermal conductivity reduction in multi-layered structures due to interface scattering is shown, which has also influence on parallel transport. Additionally the different thermal conductivities of the two MQW's show a good consistence with this assumption. The higher periodicity and the enhanced electrical conductivity lead to higher thermal conductivity because of a reduced number of interfaces and higher electronic contribution.

Based on these results the 2D figure of merit ZT_{2D} can be calculated. Assuming the same thermal conductivity for well and barrier layers (in the wells the lattice part of thermal conductivity might be smaller due to higher interface influence, but the electronic part may be larger

Table I. Thermoelectrical data of PbTe, PbSrTe and two PbTe/PbSrTe-MQW structures at room tempeature including thermal conductivity measurements and resulting figure of merit (3D values given for bulk layers, 2D values given for MQW structures).

	d_{Well} [nm]	μ [cm²/Vs]	$\sigma_{3D/2D}$ [1/(Ω cm)]	S [μV/K]	$\sigma S^2_{3D/2D}$ [μW/cmK²]	λ_{total} [W/m*K]	$\lambda_{el,\,2D/3D}$ [W/mK]	$ZT_{3D/2D}$
epi-PbTe		1250	1183	-169.6	31.6	4.15	0.87	0.23
PbTe (lit.)		1400	1350	-165.0	36.7	3.1	0.99	0.36
PbSrTe		118	2.4	-530.0	0.7	4.5	0.0018	0.005
MQW 1	2.38	605	426	-230.2	21.8	0.65	0.30	1.01
MQW 2	9.26	1025	978	-171.8	29.0	1.25	0.69	0.70

due to carrier confinement) the MQW structures have room temperature figures of merit of 1.01 and 0.70 respectively. Compared with literature values of bulk PbTe this results in an ZT enhancement by a factor > 2.

Consequently other kinds of lead salt-based multi-layered structures can be made to increase the 3D figure of merit ZT_{3D}. A significant reduction of the thermal conductivity must be reached under conservation of a high 3D power factor σS^2 e.g. by quantum dot superlattices [10]. In addition to QD-SL doping SL, PbTe/PbSrTe- and PbTe/PbSeTe-SL were prepared. The power factors depending on the carrier concentration of these structures are shown in fig. 4.

Most promising results are achieved using doping superlattices. Interfaces are introduced by the high dopant concentration (~ 1 at%) in the small doped parts of the layers. In the other material systems the power factor is reduced due to carrier mobility reduction because of electron scattering at the interfaces or within the low mobility material (e.g. PbSrTe). By quantum dot superlattices the high power factors shown in [8] were not reached, but for high carrier concentrations ($n > 10^{19}$ cm^{-3}) the values of bulk PbTe can be approached. QD superlattices can be grown with a periodicity down to $\approx$ 20 nm. Additionally PbTe/PbSeTe and PbTe/PbSrTe superlattices (wide well, small barrier) are promising material systems. The periodicity can be reduced to 10 nm and even less. The layers shown in figure 1 have a periodicity of 12.5 nm. In the case of the PbSeTe superlattices a power factor of 26 μW/cm*K², with PbTe/PbSrTe of 21 μW/cm*K² could be reached. If a similar thermal conductivity reduction as with MQW samples could be found, this would lead in the PbTe/PbSrTe or in the related PbTe/PbSeTe material system to a significantly enhanced 3D figure of merit.

CONCLUSIONS

A systematical study of the thermoelectric properties MQW structures with different structural properties and and doping profiles was performed. The 2D power factor has a minimum for d_{well} = 3...5 nm and is increasing for smaller well widths due to a strong thermopower enhancement. This is a good evidence of the influence of quantum confinement on the thermoelectric properties. However the maximum bulk values of PbTe can't be surpassed because of carrier mobility reduction at small well widths. These results are in good agreement with model calculations performed by Broido and Reinecke [5]. Additionally a power factor enhancement by high 2D carrier concentrations is limited by the occupation of continuum or barrier states. A maximum 2D power factor between 40 and 50 μW/cm*K² may be reached, but

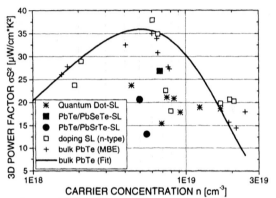

Figure 4. *3D power factor of different n-type superlattice structures compared with PbTe bulk layers.*

with the uncertainty of electrical transport in the barrier layers. The reduction in the thermal conductivity for superlattice structures is more promising. In MQW's with a period of 16 nm the thermal conductivity is reduced to 0.65 resulting in 2D $ZT_{300 K} = 1.01$. Based on these results lead salt based superlattices are promising candidates for a significant enhancement in the 3D figure of merit opening the way to highly efficient thermoelectric devices.

ACKNOWLEDGMENTS

This work was supported by the German Ministry for Education and Research (BMBF), grant No. 03N2014A. We thank G. Springholz and G. Bauer of university of Linz for fruitful discussions.

REFERENCES

1. L. D. Hicks, M. S. Dresselhaus, *Phys. Rev. B*, **47**(19), 12727-12731 (1993)
2. R. Venkatasubramanian, *Mat. Res. Soc. Symp. Proc.*, **478**, 73-84 (1997)
3. J. O. Sofo, G. D. Mahan, *Appl. Phys. Lett.*, **65**(21), 2690-2692 (1994)
4. P. J. Lin-Chung, T. L. Reinecke, *Phys. Rev. B*, **51**(19), 13244-13248 (1995)
5. D. A. Broido, T. L. Reinecke, *Appl. Phys. Lett.*, **70**(21), 2834-2836 (1997)
6. T. C. Harman, D. L. Spears, M. J. Manfra, *J. Electr. Mater.*, **25**(7), 1121-1127 (1996)
7. S. Yuan, G. Springholz, G. Bauer, *Phys. Rev. B*, **49**(8), 5476-5489 (1994)
8. T. C. Harman, P. J. Taylor, D. L. Spears, M. P. Walsh, *Proc. 18th International Conference on Thermoelectrics*, Baltimore, USA, 1999, in press
9. F. Völklein, T. Stärz, *Proc. 16th International Conference on Thermoelectrics*, Dresden, Germany, 1997, 711-718
10. G. Bastard, Wave Mechanics Applied to Semiconductor Heterostructures, *Les éditions de physique*, Les Ulis, 1992, ISBN 202645, 18-26
11. H. Beyer, A. Lambrecht, J. Nurnus, H. Böttner, H. Griessmann, A. Heinrich, L. Schmitt, M. Blumers, F. Völklein, *Proc. 18th International Conference on Thermoelectrics*, Baltimore, USA, 1999, in press

New Materials I

Thermoelectric Figure of Merit, ZT, of Single Crystal Pentatellurides ($MTe_{5-x}Se_x$: M = Hf, Zr and x = 0, 0.25)

R. T. Littleton IV[1], Terry M. Tritt [1,2], B. Zawilski[2],
J. W. Kolis[1,3], D. R. Ketchum[3] and M. Brooks Derrick[3]
1.) Materials Science and Engineering Department
2.) Department of Physics and Astronomy
3.) Department of Chemistry
Clemson University, Clemson, SC 29634 USA

ABSTRACT

The thermoelectric figure of merit, $ZT = \alpha^2\sigma T/\lambda$, has been measured for pentatelluride single crystals of $HfTe_5$, $ZrTe_5$, as well as Se substituted pentatellurides. The parent materials, $HfTe_5$ and $ZrTe_5$, exhibit relatively large p- and n- type thermopower, $|\alpha| \geq 125$ μV/K, and low resistivity, $\rho \leq 1$ mΩ•cm. These values lead to a large power factor ($\alpha^2\sigma T$) which is substantially increased with proper Se substitution on the Te sites. The thermal conductivity of these needle-like crystals has also been measured as a function of temperature from $10 \text{ K} \leq T \leq 300 \text{ K}$. The room temperature figure of merit for these materials varies from $ZT \approx 0.1$ for the parent materials to $ZT \approx 0.25$ for Se substituted samples. These results as well as experimental procedures will be presented and discussed.

INTRODUCTION

There have been extensive efforts recently to discover and optimize thermoelectric (TE) materials especially below room temperature.[1,2] Efficient TE materials at lower temperatures could eventually lead to thermoelectric modules that would greatly aid in localized cooling of many electronic devices.[3] The efficiency of a thermoelectric material is commonly determined by the dimensionless figure of merit, ZT ($\alpha^2\sigma T/\lambda$ or $\alpha^2 T/\rho\lambda$), where α is the Seebeck coefficient (or thermopower), σ is the electrical conductivity, ρ (1/σ) is the electrical resistivity, T is the absolute temperature, and λ is the thermal conductivity. The thermopower should be maximized so a large Peltier effect is observed. Electrical resistivity should be small to minimize the Joule heating (I^2R) contribution. The thermal conductivity in a good thermoelectric material should be relatively low so that a temperature difference, ΔT, may be established and maintained across the material. An ideal thermoelectric material should exhibit a large amount of phonon scattering to minimize thermal conduction and a small amount of electron scattering to maximize electrical conduction. Low temperature thermoelectric materials are even more difficult to achieve, since the absolute value of the thermopower typically decreases with decreasing temperature.

The numerator or power factor, $\alpha^2\sigma T$, can typically be tuned through chemical doping and substitution. The total thermal conductivity is composed of a lattice and electronic part ($\lambda_{TOT} = \lambda_L + \lambda_E$, respectively). The electronic thermal conductivity, in many materials, may be related to σ through the Wiedemann-Franz law [$\lambda_E = L_0 \sigma T$ where L_0 is the Lorentz number ($L_0 = 2.45 \times 10^{-8}$ W-Ω/K^2)]. Consequently, changes in the electrical conductivity will affect the electronic thermal conductivity (λ_E). These influences (such as alloying) that affect σ can also affect the lattice thermal conductivity (λ_L). A good thermoelectric material has been described by Glen Slack to be a "phonon-glass, electron-crystal".[4] The electrons in these materials would behave as if the material were a crystal (low electron scattering) while the phonons would behave as if the material were a glass (high phonon scattering).

Achieving high thermopower at low temperatures is one of the important issues for discovering or optimizing thermoelectric materials for low temperature applications. Heavy fermion materials, Kondo systems, as well as quasi-one-dimensional materials are a few of a number of systems that are candidates for low temperature thermoelectric materials. Low dimensional systems are known to be susceptible to van Hove singularities (or cusps) in their density of states, g(E), electronic phase transitions, and exotic transport properties, which can add structure in g(E) near E_F. Doping can produce very substantial effects in these types of materials and can drastically change their electronic transport. We have recently been investigating a class of materials known as pentatellurides, with "parent" materials HfTe$_5$ and ZrTe$_5$.[5,6] The resistivity and thermopower as a function of temperature for the parent compounds are shown in Figure 1(a) and 1(b). Both parent materials (HfTe$_5$ and ZrTe$_5$) exhibit a resistive anomaly peak, $T_p \approx 80$ K for HfTe$_5$ and $T_p \approx 145$ K for ZrTe$_5$. As shown in Figure 1(b), each display a large positive (p-type) thermopower ($\alpha \geq +125$ µV/K) around room temperature, which undergoes a change to a large negative (n-type) thermopower ($\alpha \leq -125$ µV/K) below the peak temperature. The temperature at which the thermopower crosses zero, T_0, correlates well with the peak temperature of the resistivity, T_p. These materials have shown promise as potential low temperature thermoelectric materials[7] and exhibit interesting magnetic properties as well.[8]

EXPERIMENTAL PROCEDURE

Single crystals of MTe$_{5-x}$Se$_x$ (M = Hf, Zr and x = 0, 0.25) transition metal pentatellurides were grown in similar conditions to previously reported methods.[9] Stoichiometric amounts of the starting materials were sealed in a fused silica tube with iodine ($\approx$ 5 mg/mL) and placed in a tube furnace. Initial materials were at the center of the furnace with the other end of the reaction vessel near the end of the furnace to provide a temperature gradient. Ribbon-like crystals were obtained in excess of 1.5 mm long and 100 µm in diameter with the preferred direction of growth along the *a*-axis, as determined by face indexing. The pentatelluride crystals are long chain systems with an orthorhombic crystal structure.[10] They exhibit slightly anisotropic transport properties with the high conductivity axis along the growth, *a*, axis.

Seebeck, resistance, and thermal conductivity measurements as a function of temperature (10 K ≤ T ≤ 300 K), were performed in APD closed cycle cryocoolers. Single crystals were mounted across two copper blocks in a standard four-wire resistivity and thermopower measurement along the preferred axis. Thermal conductivity was measured using a Parallel Thermal Conductance (PTC) method which is described in detail elsewhere.[11] The PTC method requires the thermal conductance of the sample holder to be known (measured) and subtracted from the measured thermal conductance of the sample holder together with the mounted sample. Sample conductance is measurable down to ~ 10^{-6} W/K.

Sample dimensions for each measurement are taken using an Olympus 500X metallurgical microscope with digital imaging. Nevertheless, the measured cross-sectional area, A, contains a significant amount of error (≈ 10% to 50%), which makes accurate resistivity and thermal conductivity determinations difficult. However, this error can be eliminated when determining the sample ZT. This can be accomplished by measuring the same sample in the resistance and thermopower system and in the PTC system while accurately measuring the respective length l_R and l_K, which vary only slightly. The ZT calculation is then:

$$ZT = \frac{\alpha^2 T}{\rho \lambda} = \alpha^2 T \left(\frac{l_R}{R \bullet A}\right)\left(\frac{A}{K \bullet l_K}\right) = \frac{\alpha^2 T}{R \bullet K} \frac{l_R}{l_K}$$

Where R is the electrical resistance and K is the thermal conductance. In this manner the measured cross-area along with the uncertainty in the sample cross-area is canceled.

RESULTS AND DISCUSSION

The measured resistivity, thermopower, thermal conductivity, and ZT values as a function of temperature (10 K ≤ T ≤ 300 K), are shown in Figure 1, for single crystals of nominal concentrations: $HfTe_5$, $ZrTe_5$, $HfTe_{4.75}Se_{0.25}$, and $ZrTe_{4.75}Se_{0.25}$. As shown in Figure 1(a), each sample exhibits a broad resistivity versus temperature anomaly as discussed earlier. Samples with a 5% nominal Se substitution for Te display a small reduction (≈ 10 – 15 K) in the resistive peak temperature. The magnitude of the resistivity is also reduced approximately 25% from the parent compounds. Figure 1(b) shows the thermopower as a function of temperature for the same nominal concentrations. Again the temperature dependence of the thermopower is shifted slightly while the magnitude is increased nearly 20% relative to that of the parent compounds. The reduction in resistivity and enhancement of the thermopower results in an enhancement of the power factor (related to the electronic properties) which is nearly double that of the parent materials.

Thermal conductivity is shown as a function of temperature in Figure 1(c). It should be noted that the magnitude of the thermal conductivity can vary greatly from sample to sample due to sample quality and dimensional error.[12] Nevertheless, the magnitude of the thermal conductivity shown is a good approximation, especially given the challenges due to the sample dimensions. The temperature behavior of the thermal conductivity, λ(T), for the pentatellurides is representative of typical crystalline behavior. Near room temperature, λ(300 K), the thermal conductivity ranges from 4 to 8 W/m•K. At lower temperatures the thermal conductivity increases with an approximate 1/T dependence to a maximum value, λ_{max}, of ≈ 15 to 30 W/m•K around 20 to 30 K. Below the peak temperature, λ decreases sharply to the lowest temperatures measured. The peak in the thermal conductivity is characteristic of single crystal data showing depletion of Umklapp scattering of the phonons, where the magnitude of the peak is related to the "perfection" or lack of defects within the sample.

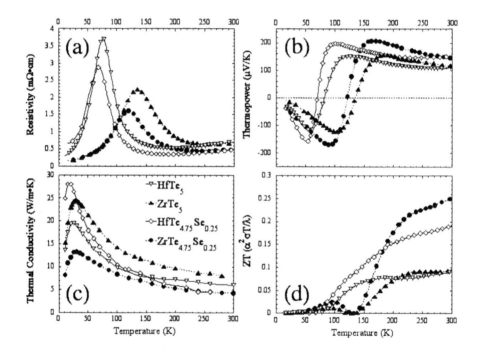

Figure 1. Resistivity (a), thermopower (b), thermal conductivity (c), and ZT (d) as a function of temperature for pentatelluride single crystals of nominal concentrations $HfTe_5$, $ZrTe_5$, $HfTe_{4.75}Se_{0.25}$, and $ZrTe_{4.75}Se_{0.25}$.

The figure of merit, ZT, is shown in Figure 1(d). ZT of $HfTe_5$, $ZrTe_5$, $HfTe_{4.75}Se_{0.25}$, and $ZrTe_{4.75}Se_{0.25}$ were calculated using the measured values of $\rho(T)$, $\alpha(T)$, and $\lambda(T)$ from the same sample as discussed earlier in the Experimental Procedure. Below 50 K the figure of merit for each sample is nearly insignificant due to the high thermal conductivity and low temperature values. Above 50 K the ZT of $HfTe_5$ and $HfTe_{4.75}Se_{0.25}$ begin to increase with temperature while $ZrTe_5$ and $ZrTe_{4.75}Se_{0.25}$ exhibit a small peak in ZT around 100 K due to the large contribution of the n-type thermopower. At higher temperatures, ZT for each of the parent materials slowly increases with temperature reaching a room temperature value of ZT (300 K) $\approx$ 0.1 for both $HfTe_5$ and $ZrTe_5$. The Se doped samples exhibit a much higher figure of merit resulting from a lower resistivity and higher thermopower. The ZT (300K) reaches $\approx$ 0.20 and 0.25 for $HfTe_{4.75}Se_{0.25}$ and $ZrTe_{4.75}Se_{0.25}$ respectively. Unfortunately, these values of the figure of merit are not competitive with existing materials and further investigations are in progress. However, these results further illustrate the importance of measuring all the properties on the same sample. The thermal conductivity of the single crystals reported here is a factor of 4 to 5 times larger than that of a pressed pellet of pentatelluride powders, which will average the anisotropy of these materials.

Transition metal pentatellurides feature exotic quasi-one dimensional transport properties that exhibit relatively low electrical resistivity, high thermoelectric power, and moderate thermal conductivity. These materials, like many low dimensional systems, are sensitive to chemical doping that can greatly alter their electrical and thermal transport properties. Further doping studies are underway to continue optimizing the thermoelectric properties of this system. Given that the thermal conductance of single crystals can now accurately be measured and determined, the primary focus of our research is on chemical compositions that will lower the thermal conductivity without compromising the electrical properties. The optimum combination of substitutions and doping should enhance ZT of the pentatellurides even further.

SUMMARY

Resistivity, thermopower, thermal conductivity, and ZT for single crystals of nominal concentrations $HfTe_5$, $ZrTe_5$, $HfTe_{4.75}Se_{0.25}$, and $ZrTe_{4.75}Se_{0.25}$ have been measured as a function of temperature (10K $\leq$ T $\leq$ 300K). Each ZT calculation was made using $\rho(T)$, $\alpha(T)$, and $\lambda(T)$ measured values from the same sample to minimize any differences in the samples or sample configuration. These results further illustrate the importance of such measurements on the same sample and certainly the same type of crystal. The parent materials, $HfTe_5$ and $ZrTe_5$, exhibit relatively large p and n- type thermopower and low resistivity. These values lead to a large power factor ($\alpha^2\sigma T$) which is significantly increased with a 5% nominal Se substitution. The room temperature ZT for these materials range from $\approx$ 0.1 for $HfTe_5$ and $ZrTe_5$ to $\approx$ 0.25 for $ZrTe_{4.75}Se_{0.25}$. The electrical and thermal transport properties of transition metal pentatellurides can be substantially altered with substitutional doping. This allows for a broad range of manipulation in which to search for the optimal thermoelectric properties of these materials for potential low temperature applications.

ACKNOWLEDGEMENTS

The authors would also like to acknowledge support from ARO/DARPA (grant #DAAG55-97-1-0267) for the research funds provided for this work. One of the authors (RTL) is the recipient of the 1999 *H. J. Goldsmid Award*, a graduate student award for outstanding accomplishments in thermoelectric materials research.

REFERENCES

1. F. J. DiSalvo, Science, **285**, 703-6 (1999).
2. D. Y. Chung, T. Hogan, P. Brazis, M. Rocci-Lane, C. Kannewurf, M. Bastea, C. Uher, and M. G. Kanatzidis, Science, **287**, 1024-7 (2000.)
3. T. M. Tritt, Science, **283**, 804-5 (1999).
4. *New Materials and Performance Limits for Thermoelectric Cooling*, G. A. Slack, in CRC Handbook of Thermoelectrics, Rowe ed. p 407- 4 (1995).
5. R. T. Littleton, J. W. Kolis, C. R. Feger and T.M. Tritt, MRS 1998: *Thermoelectric Materials 1998.* Edited by T. M. Tritt et. al. Vol. **545,** p 381-396 (1998).
6. R. T. Littleton, T. M. Tritt, C. R. Feger, J. Kolis, M. L. Wilson, M. Marone, J. Payne, D. Verebeli, and F. Levy, Appl. Phys. Lett., **72**, 2056-8 (1998).
7. R. T. Littleton, T. M. Tritt, J. W. Kolis, and D. Ketchum, Phys. Rev. B., **60**, 13453-7 (1999).
8. T. M. Tritt, N. D. Lowhorn, R. T. Littleton, A. Pope, C. R. Feger, and J. W. Kolis, Phys. Rev. B., **60**, 7816-9 (1999).
9. F. Levy and H. Berger, J. Cryst. Growth, **61**, 61-8 (1983).
10. S. Furuseth, L. Brattas, and A. Kjekshus, Acta. Chem. Scand., **27**, 2367-2374 (1973).
11. B. Zawilski, R. T. Littleton, and T. M. Tritt, Rev. Sci. Instrum., in progress (2000).
12. B. Zawilski, R. T. Littleton, and T. M. Tritt, MRS, this proceeding, (2000).

THERMOELECTRIC PROPERTIES OF SELENIDE SPINELS

G. Jeffrey Snyder*, T. Caillat, and J. -P. Fleurial
Jet Propulsion Laboratory/California Institute of Technology
4800, Oak Grove Drive, MS 277-207, Pasadena, CA 91109
*jeff.snyder@jpl.nasa.gov

ABSTRACT

Many compounds with the spinel structure type have been analyzed for their thermoelectric properties. Published data was used to augment experimental results presented here and to select promising thermoelectric spinels. Compounds studied here include $Cu_{0.5}Al_{0.5}Cr_2Se_4$, $Cu_{0.5}Co_{0.5}Cr_2Se_4$, $Cu_{0.5}In_{0.5}Cr_2Se_4$, and $CuIr_2Se_4$. Many exhibit low lattice thermal conductivity of about 20 mW/cmK, independent of temperature. Preliminary results are given for two series of compounds that were selected for further study: $Ga_xCu_{1-x}Cr_2Se_4$ and $Zn_xCu_{1-x}Cr_2Se_4$.

INTRODUCTION

The growth of commercial applications of thermoelectric devices depends primarily on increasing the figure of merit, ZT, for thermoelectric materials. The figure of merit is defined as $ZT = \alpha^2 \sigma T/\lambda$, where α is the Seebeck coefficient, σ the electrical conductivity, λ the thermal conductivity, and T is the absolute temperature. Materials with a large $\alpha^2\sigma$ value, or power factor, are usually heavily doped semiconductors, such as Bi_2Te_3. The thermal conductivity of semiconductors is usually dominated by phonon or lattice thermal conductivity. Thus, one method for finding new, advanced thermoelectric materials is to search for semiconductors with low lattice thermal conductivity.

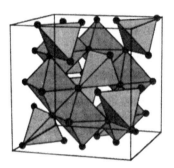

Figure 1. Illustration of the Spinel unit cell (e.g. $ZnCr_2Se_4$) showing Se atoms as spheres Cr atoms (not shown) at the center of the shaded octahedra and Zn atoms (not shown) at the center of the shaded tetrahedra. The cubic unit cell is indicated.

In this paper we evaluate compounds based on the Spinel structure with general composition $A_1B_2X_4$ where A and B are transition metals and X is a chalcogen, primarily Se. Previous work on such compounds [1] have shown that a range of metals and insulators exist with this structure type.

The structure of Spinel (Figure 1) consists of cubic close packed chalcogen atoms with metal B atoms in half the octahedral holes and metal A atoms in 1/8 of the tetrahedral holes. There can be significant mixing of the different metal atoms on the two metal sites. As suggested by Spitzer [2] the relatively high coordination number of the B atoms in this structure may favor low lattice thermal conductivity. The large cubic unit cells (about 10Å) full of vacant octahedral holes should reduce the lattice thermal conductivity by increasing the scattering of phonons. The multi-valley electronic structure associated with such cubic compounds can be expected to enhance the thermopower.

There are approximately 300 known Spinels with X = Se or S. Many of these compounds have X = Se compounds with 3-d transition metals for A and B atoms (Figure 2), and constitute most of the samples in this investigation. Many oxide spinels have been studied for their magnetic properties but are not suitable for thermoelectric applications because they are too insulating. Only a few of the known spinel sulfides are metals or heavily doped semiconductors; such compounds based on the sulfides of V, Co, Fe or Ni, are somewhat unstable in the spinel structure, preferring the Cr_3S_4-type at high temperature and pressure[3].

Several of the known AB_2S_4 sulfides with the Cr_3S_4-type structure are high temperature/pressure polymorphs of compounds with the Spinel structure at room temperature and pressure. Known compounds with X = Te are metals with low thermopower (α) (Table 1).

The Cr_3S_4-type selenides are attractive for thermoelectric applications not only because they may have low thermal conductivity as suggested above, but they also exhibit a range of electronic properties – from metals to semiconductors. Precise, heavy doping of the semiconductor is critical to obtain optimal power factor for both n- and p-type materials. Proven thermoelectric materials such as filled Skutterudites and Zn_4Sb_3 are often difficult to dope to the optimal n- or p-type carrier concentration. An advantage of the $A_xB_{3-x}X_4$ compounds is the chemical versatility of the structure, allowing continuous doping from metal to n- and p- type semiconductor.

Experimental

The thermoelectric properties of many sulfur and selenide spinel compounds reported in the literature [1] were used to narrow the search. For insulating compositions, which would need to be heavily doped for thermoelectric applications, the most useful of this information for thermoelectric considerations is the apparent band gap and carrier mobility. These data are summarized in Table 1.

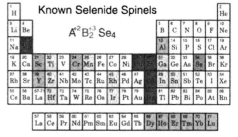

Figure 2. Elements known to make $A_1B_2Se_4$ compounds with the spinel structure type, where A elements are horizontally hatched and B elements vertically hatched.

The metallic spinels, such as $CuCr_2S_4$, can be used to dope or alloy with related semiconducting spinels. The solid solution $Fe_xCu_{1-x}Cr_2S_4$ changes from a p-type semimetal to n-type, and then back to p-type as x is increased[4]. The maximum room temperature power factor for this system is about 1 $\mu W/cmK^2$.

Most of the known selenide spinels contain Cr as the B atom, and a +2 element such as Zn or Cd as the A atom. The other elements that can go on the B site are mostly rare earth group elements and Al which will make less chemically stable spinels with a strongly ionic character (large band gap insulators). $CuIr_2Se_4$ is both a metal and a pressure induced semiconductor, and as such may have interesting thermoelectric properties [5]. Various doped chromium selenide spinel semiconductors have been made by doping with about one percent of a +1 or +3 element on the A^{+2} site, such as with $CdCr_2Se_4$ [6, 7] or $HgCr_2Se_4$ [8] where a RT power factor of about 1 $\mu W/cmK^2$ was found [9]. $CuCr_2Se_4$ like its S counterpart is metallic and can be used not only to dope but also alloy with the insulating chromium spinels such as $Cd_xCu_{1-x}Cr_2Se_4$[10] (maximum RT power factor 10^{-4} $\mu W/cmK^2$) and $Ga_xCu_{1-x}Cr_2Se_4$[11]. Such solid solutions are often not completely miscible, forming two spinel phases with different crystallographic cells [12, 13] particularly when the two types of A ions differ in size.

Table 1. Room Temperature thermoelectric properties of S or Se spinel compounds reported in the literature [1, 5, 14, 15]. I indicates insulating.

Composition	Resistivity mΩcm	Seebeck μV/K	$\Delta E_{optical}$ eV	$\Delta E_{resistivity}$ eV	Mobility cm^2/Vs
FeCr$_2$S$_4$	I	-		0.02 - 0.2	0.3
CoCr$_2$S$_4$	I	-		0.01 - 0.3	0.2
MnCr$_2$S$_4$	I	-		0.1 - 0.3	
ZnCr$_2$S$_4$	I	-		0.6	
CdCr$_2$S$_4$	I	-	1.6	0.2-0.6	
HgCr$_2$S$_4$	I	-	1.4	0.4 - 2	
Ni$_{0.5}$Co$_{0.5}$Cr$_2$S$_4$	200	60		0 - 0.12	1
CuCr$_2$S$_4$	0.9	16		0.03	
CuV$_2$S$_4$	0.6	5			
CuTi$_2$S$_4$	0.4	-12			
CuCo$_2$S$_4$	0.4	13			
NiCo$_2$S$_4$	0.8	-18			
CoNi$_2$S$_4$	0.4	-2			
Co$_3$S$_4$	0.3	5			
ZnCr$_2$Se$_4$	I	-	1.3	0.3	5
CdCr$_2$Se$_4$	I	-	1.3	0.2 - 0.6	50
HgCr$_2$Se$_4$	I	-	0.84	0.4 - 2	30
CuCr$_2$Se$_4$	0.1	16			<10
CuIr$_2$Se$_4$	5				

Using these previous results, we prepared compositions that looked promising for thermoelectric applications and/or where thermoelectric data was missing.

Polycrystalline samples were prepared by mixing and reacting elemental powders in evacuated silica ampoules for several days at 700° - 800° C. The samples were analyzed by x-ray diffractometry to confirm the crystalline structure. The powders were then hot-pressed in graphite dies into dense samples, 3 mm long and 12 mm in diameter. The hot-pressing was conducted at a pressure of 1400 kg/cm^2 and at a temperature of 700° - 800° C for about 2 hours under argon atmosphere. The density of the samples was calculated from the measured weight and dimensions and was found to be greater than 90% of the theoretical density for all samples.

The samples were also characterized by microprobe analysis which was performed using a JEOL JXA-733 electron superprobe. The Al and In concentration in the samples of Cu$_{0.5}$In$_{0.5}$Cr$_2$Se$_4$ and Cu$_{0.5}$Al$_{0.5}$Cr$_2$Se$_4$ were not uniform. The CuIr$_2$Se$_4$ contained some (~10%) IrSe$_2$ secondary phase. The elemental concentrations determined from microprobe analysis for Ga$_x$Cu$_{1-x}$Cr$_2$Se$_4$ were within a few atomic percent of the expected values.

Table 2. Room temperature thermoelectric properties from this study. I indicates insulating. FM indicates Ferromagnetic.

Composition	Resistivity mΩcm	Seebeck μV/K	Thermal Conductivity mW/cmK	Mobility cm²/Vs
$FeCr_2S_4$	2×10^4	400	28	0.1
$ZnCr_2Se_4$	I	-	31	2
$CdCr_2Se_4$	I	-	30	<5
$Cu_{0.5}Al_{0.5}Cr_2Se_4$	30	100	16	6
$Cu_{0.5}In_{0.5}Cr_2Se_4$	30	300	20	16
$Cu_{0.5}Co_{0.5}Cr_2Se_4$	0.6	30	25	10
$CuSnCrSe_4$	10	100	20	5
$CuCr_2Se_4$	0.28	25	37	FM
$CuIr_2Se_4$	0.3	3		-2

Samples in the form of disks (typically a 1.0 mm thick, 12 mm diameter slice) were cut from the cylinders using a diamond saw for electrical and thermal transport property measurements. Temperature dependence of electrical resistivity, Hall effect, Seebeck coefficient, thermal diffusivity and heat capacity measurements were conducted on selected samples between 80 and 800K. The resistivity and Hall effect were measured using the method of Van der Pauw [16]. The carrier density was calculated from the Hall coefficient, assuming a scattering factor of 1.0 in a single carrier scheme, by $n = 1/R_H e$, where n is the density of holes or electrons, and e is the electron charge. The Hall mobility (μ_H) was calculated from the Hall coefficient and the resistivity values by $\mu_H = R_H/\rho$. The normal Hall effect, however, is often compounded by the anomalous Hall effect because of ferromagnetism in many of the compounds. Therefore it is often difficult to estimate the Hall mobility or carrier concentration. The Seebeck coefficient (α) was measured with a high temperature light pulse technique [17]. Room temperature thermal conductivity was measured using the comparison method [18]. High temperature heat capacity and thermal diffusivity were measured using a flash diffusivity technique [19]. The thermal conductivity (λ) was calculated from the experimental density, heat capacity, and thermal diffusivity values.

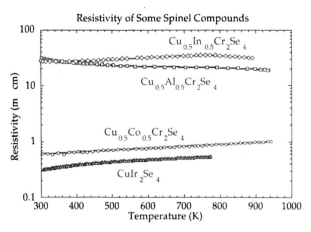

Figure 3. Electrical Resistivity of some Spinel compounds.

Results and Discussion

The Spinel selenides and sulfides exhibit a wide variety of electronic properties (Tables 1 and 2). Some, $CuIr_2Se_4$ and $Cu_{0.5}Co_{0.5}Cr_2Se_4$ for example, are metals having room temperature resistivity (ρ) from about 10^{-4} to 10^{-3} Ωcm. Others, such as the alloys $Cu_{0.5}In_{0.5}Cr_2Se_4$ and $Cu_{0.5}Al_{0.5}Cr_2Se_4$, are semimetals or very low band gap semiconductors with high carrier concentrations. Spinels having a stable +2 A atom, are usually insulators or high band gap (~1 eV optical gap) semiconductors. These semiconductors can then be doped n- or p-type [6-10]. Measurements of the resistivity and thermopower as a function of temperature frequently contain discontinuities and hysteresis (hysteretic samples not shown), which may be due to magnetic or structural changes, or even loss of S or Se at high temperatures.

Exchange split Cr^{+3} in octahedral coordination will have 3 electrons to completely fill the majority spin T_{2g} orbital (or subsequent band); thus Cr^{+3} may not provide metallic carriers. This is certainly the case for the insulating chromium spinels in Tables 1 and 2. The doping or alloying on the A site should produce doped semiconductors or metals. However, significant alloying on the A site may result in a polaron semiconductor instead of a metal due to the localization effects of the distant dopant atoms. Such behavior is more clearly demonstrated in the related defect NiAs-type ACr_2Se_4 chromium selenides [3, 20].

The resistivity due to small polaron hopping conduction has only a slightly different temperature dependence ($T\exp(E_a/kT)$) than that expected of a semiconductor ($\mu^{-1}\exp(E_a/kT)$ where the mobility μ is proportional to $T^{-3/2}$ for many semiconductors). Both forms are dominated by an exponential with characteristic energy E_a. For the materials described here the resistivity data is not sufficiently well described by either of the exact forms to determine the transport mechanism. Nevertheless, band semiconductor transport characteristically has carriers with high mobility and low concentration while small polarons have high concentration and low mobility. Thus, the data suggests that these materials have polaron conductivity.

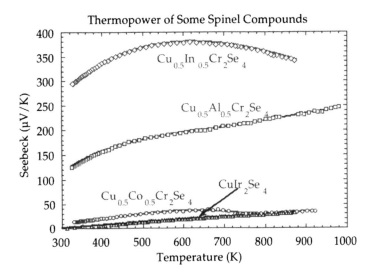

Figure 4. Thermopower of some Spinel compounds.

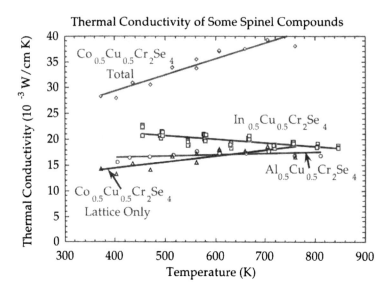

Figure 5. Thermal Conductivity of some Spinel compounds.

The thermopower of some of the spinel compounds studied is shown in Figure 4. The low resistivity samples have metallic like thermopowers that are small and linear. The higher resistivity samples can have large thermopowers like conventional semiconductors, even though transport is probably by small polarons.

The thermal conductivity λ is given by the sum of the electronic λ_E and lattice contribution λ_L. λ_E is directly related to the electronic conductivity: $\lambda_E = L\sigma T$, where L is the Lorenz factor. The Lorenz factor used is that typical for metals (2.4×10^{-8} J^2/K^2C^2). The measured thermal conductivity of some spinel compounds is shown in Figure 5. The electronic contribution of the high resistivity samples ($Cu_{0.5}In_{0.5}Cr_2Se_4$ and $Cu_{0.5}Al_{0.5}Cr_2Se_4$) is not significant (< 1 mW/cm K) so the measured values are due to the lattice contribution.

The lattice thermal conductivity is relatively independent of temperature (Figure 7) indicating multiple scattering processes. Low, temperature independent thermal conductivity is found in complex structures such as glasses. Common crystalline materials have large lattice thermal conductivity that is proportional to $1/T$ at high temperatures [21]. The quaternary spinel compounds should have lower lattice thermal conductivity because of additional alloy scattering.

The power factor and therefore figure of merit is relatively low for most of these compounds. These materials should have carriers with high effective masses ($m^*/m \sim 4$), that improve the thermoelectric properties, but the very low mobility of these carriers cancels any improvement. Typical thermoelectric materials have Hall mobilities greater than 10 cm^2/Vs, whereas the materials in Table 2 have mobilities at least 10 times less. This may be due to the hopping method of transport, the increased electron scattering from the transition metal magnetic moments (magnon scattering) or due to the lower covalency of these materials as compared to conventional thermoelectric semiconductors.

The spinels with the most promising thermoelectric properties are derivatives of $CuCr_2Se_4$. Our next task is to select a representative series of compounds to study that may give good thermoelectric properties with optimal doping or alloying. The series of compounds $Fe_xCu_{1-x}Cr_2S_4$ [4] and $Cd_xCu_{1-x}Cr_2Se_4$[10] have been studied in the composition ranges appropriate for thermoelectric materials with some promising results. The compounds $Ga_xCu_{1-x}Cr_2Se_4$[11] should be semiconducting at $x= 1/2$ assuming Ga is +3 and Cu is +1. The alloys should produce both p- and n-type compositions, depending on whether $x < 1/2$ or $x > 1/2$ respectively. We found that the related series $In_xCu_{1-x}Cr_2Se_4$ makes two phase samples with the major phase being $Cu_{0.5}In_{0.5}Cr_2Se_4$. A previous study of $Ga_xCu_{1-x}Cr_2Se_4$[11] did not investigate bulk samples in the composition range useful for thermoelectrics. Thus we chose to examine the series $Ga_xCu_{1-x}Cr_2Se_4$.

The other series of samples being studied is $Zn_xCu_{1-x}Cr_2Se_4$. The parent compound $ZnCr_2Se_4$ undergoes magnetic ordering at low temperature. From band structure calculations [22] it was suggested that in the ordered state the magnon scattering will no longer hinder the mobility, but will still have the high effective mass carriers with large thermopower.

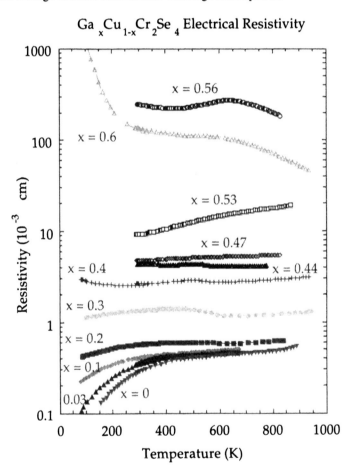

Figure 6. Electrical resistivity of $Ga_xCu_{1-x}Cr_2Se_4$ for various x.

The resistivity of $Ga_xCu_{1-x}Cr_2Se_4$ for various x is shown in Figure 6. There is a change in slope at the ferromagnetic curie temperature in samples with small x. This transition temperature decreases from 435K for $x = 0$ to about 375K for $x = 0.1$. There is another transition observable in the resistivity between 475K and 525K for $0.2 < x < 0.5$. This other transition appears to broaden and move to even higher temperature for $x > 0.5$.

The thermopower of $Ga_xCu_{1-x}Cr_2Se_4$ for various x is shown in Figure 7. As expected, the p-type range extends up to approximately $x < 0.5$, gradually changing from a p-type metal with linear Seebeck coefficient to a p-type semiconductor with a peaked Seebeck coefficient. The n-type region, however, is small because the spinel structure is not stable for x above about 0.6. The room temperature thermoelectric properties are shown in Figure 8, highlighting the metal-insulator transition and p-type to n-type transition at around $x = 0.55$.

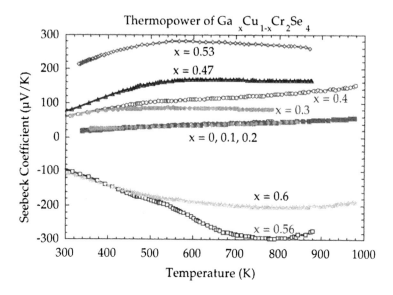

Figure 7. Thermopower of $Ga_xCu_{1-x}Cr_2Se_4$ for various x.

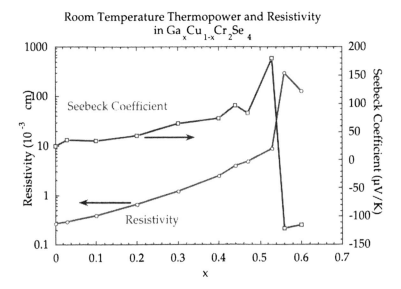

Figure 8. Room temperature values of electrical resistivity and thermopower for $Ga_xCu_{1-x}Cr_2Se_4$.

The room temperature values are also plotted in Figure 9 along with other $A_xCu_{1-x}Cr_2Se_4$ compounds for comparison. The room temperature values for p-type $Ga_xCu_{1-x}Cr_2Se_4$ fall near the line of constant 3 $\mu W/K^2$ power factor, indicating that these compounds have about the same thermoelectric figure of merit at this temperature. This power factor is significantly higher than that previously reported for $Ga_xCu_{1-x}Cr_2Se_4$ [11] on single crystals and $Cd_xCu_{1-x}Cr_2Se_4$ [13], but comparable to our preliminary results on $Zn_xCu_{1-x}Cr_2Se_4$. The maximum figure of merit, shown in Figure 10, is about 0.1 for the p-type compounds, which is comparable to that found in the defect NiAs - type series of compounds $Fe_xCr_{3-x}Se_4$ [20].

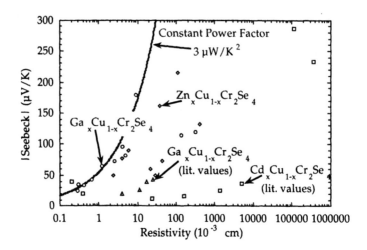

Figure 9. Room temperature values of electrical resistivity and thermopower for various $A_xCu_{1-x}Cr_2Se_4$. Samples with a high power factor ($\alpha^2\sigma$) are to the upper left, those with a low power factor to the lower right. A iso-line of constant power factor is shown. The literature values are obtained from [11, 13].

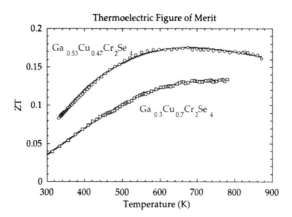

Figure 10. Thermoelectric figure of merit, ZT, for $x = 0.3, 0.53$ in $Ga_xCu_{1-x}Cr_2Se_4$.

Summary

A variety of spinel sulfides and selenides were examined for high thermoelectric figure of merit. Reported thermoelectric properties and the existence of known compounds helped guide the selection of materials to reexamine. Many showed low, glass-like thermal conductivities. Most compounds not containing chromium were eliminated due to their strongly ionic character (sulfides and rare earth selenides). Others show a variety of electronic properties from metals to small polaron semiconductors, with low carrier mobilities. A representative series of compounds $Ga_xCu_{1-x}Cr_2Se_4$ has been studied to examine the doping region of interest for thermoelectric applications. Both p- and n-type compounds have been found; the maximum figure of merit ZT is of order 0.1.

We would like to thank A. Zoltan, L. D. Zoltan, S. Chung, and A. Borshchevsky for their help on this project. This work was carried out at the Jet Propulsion Laboratory-California Institute of Technology, under contract with NASA and supported by the U. S. Defense Advanced Research Projects Agency, Grant No. E407.

REFERENCES

[1] *Landolt-Börnstein* Springer-Verlag, Berlin, Vol. NS III/17h;NS III/12b;NS III/4b.
[2] D. P. Spitzer, J. Phys. Chem. Solids **31**, 19 (1970).
[3] G. J. Snyder, T. Caillat, and J.-P. Fleurial, Mat. Res. Soc. Symp. Proc. **545**, 333 (1999).
[4] F. K. Lotgering, R. P. vanStapele, G. H. A. M. vanderSteen, *et al.*, J. Phys. Chem. Solids **30**, 799 (1969).
[5] T. Furubayashi, T. Kosaka, J. Tang, *et al.*, J. Phys. Soc. Jpn. **66**, 1563 (1997).
[6] H. W. Lehmann, Phys. Rev. **163**, 488 (1967).
[7] A. Amith and G. L. Gunsalus, J. Appl. Phys. **40**, 1020 (1969).
[8] M. R. Chaves, J. L. Ribeiro, A. Selmi, *et al.*, Phys. Stat. Sol. **92**, 263 (1985).
[9] K. Minematsu, K. Miyatani, and T. Takahashi, J. Phys. Soc. Jpn. **31**, 123 (1971).
[10] H. Duda, T. Gron, and J. Warczewski, J. Magn. Magn. Mater. **88**, 55 (1990).
[11] T. Gron, K. Baerner, C. Kleeberg, *et al.*, Physica B **225**, 191 (1996).
[12] H. D. Lutz, U. Koch, and I. Okonska-Kozlowska, J. Solid State Chem. **51**, 69 (1984).
[13] J. Krok-Kowalski, H. Rej, T. Gron, *et al.*, J. Magn. Magn. Mater. **137**, 329 (1994).
[14] R. J. Bouchard, P. A. Russo, and A. Wold, Inorg. Chem. **4**, 685 (1965).
[15] A. A. Abdurragimov, Z. M. Namazov, L. M. Valiev, *et al.*, Inorg. Mater. **17**, 1113 (1981).
[16] L. J. van der Pauw, Philips Res. Repts. **13**, 1 (1958).
[17] C. Wood, L. D. Zoltan, and G. Stapfer, Rev. Sci Instrum. **56**, 719 (1985).
[18] D. M. Rowe, *Thermoelectric Handbook* (CRC, Boca Raton, 1995).
[19] J. W. Vandersande, C. Wood, A. Zoltan, *et al.*, in *Thermal Conductivity* (Plenum, New York, 1988), p. 445.
[20] G. J. Snyder, PRB submitted (2000).
[21] C. M. Bhandari and D. M. Rowe, *Thermal Conduction in Semiconductors* (Wiley Eastern Limited, New Delhi, 1988).
[22] D. Singh, Private Communication.

Thermoelectric properties of Tl₉BiTe₆ / Tl₉BiSe₆ solid solutions

Bernd Wölfing[1], Christian Kloc[1] and Ernst Bucher[1,2]
[1]Department of Materials Physics Research, Bell Laboratories, Lucent Technologies, Murray Hill, NJ 07974, U.S.A.
[2]Lehrstuhl für angewandte Festkörperphysik, Universität Konstanz, Konstanz, Germany

ABSTRACT

The compounds Tl₉BiTe₆ (TBT) and Tl₉BiSe₆ (TBS) crystallize in the tetragonal space group I4/mcm. Tl₉BiTe₆ has a thermopower of 185 µV/K and an electrical resistivity of 5.5 mΩcm at 300K, resulting in a power factor of $S^2/\rho = 0.6$ mW/mK². Compared to Bi₂Te₃ which is the state of the art material at this temperature this is about a factor of 7 lower. At 300 K TBS has a thermopower of 750 µV/K but a high resistivity of 130 Ωcm. To optimize the thermoelectric properties of TBT solid solutions have been formed with TBS. The resistivities and have been measured on Tl₉BiTe₁₋ₓSeₓ with x = 0.05, 0.08, 0.2 and 0.5. In addition to the electrical properties the lattice constants have been measured by X-ray diffraction. The dependence of the lattice constants on the Te/Se ratio clearly deviates from Vegard's law. Different affinities of Te and Se towards the two chalcogenide sites in the crystal can explain this behavior.

INTRODUCTION

TBT and TBS belong to a larger family of ternary compounds. This family contains all combinations of Tl₉M^V Q₆ (M^V = Sb, Bi; Q = Se, Te) and Tl₈M^IV₂Q₆ (M^IV = Sn, Pb; Q = Se, Te) except Tl₈Sn₂Se₆. The 7 compounds in this family are completely miscible. This group of compounds is derived from the binary Tl₅Te₃. Assuming that Tl₅Te₃ is a semi-metal, the Tl atoms exhibit two different valence states: Tl^I and Tl^III [1]. The expanded formula can thus be written as Tl^I₈(Tl^I, Tl^III)Te₆ where the atoms in brackets are on the crystallographic 4c site. Tl^III can be replaced by trivalent metals Bi and Sb resulting in compounds with a disordered 4c site. Alternatively both atoms on this site can be substituted by Pb or Sn. In all but the Sn-compound Te can be substituted by Se. The existence of these compounds suggest that this simple valence model is accurate to a certain extent.

The quality of the electrical properties of a thermoelectric material is described by the power factor S^2/ρ where S is the thermopower and ρ the electrical resistivity. In comibination with the thermal conductivity κ the figure of merit $Z = S^2/\rho\kappa$ is obtained which describes the overall quality of a thermoelectric material.

Of all compounds in this family TBT has the highest power factor while TBS exhibits the highest thermopower. By forming solid solutions between these two compounds we aimed at increasing the thermopower without sacrificing too much electrical to obtain a solid solution with a power factor exceeding that of pure TBT.

EXPERIMENTAL

TBT and TBS are reported to melt congruently at temperatures of 540°C and 519°C [2,3]. Stoichiometric amounts of the elements were sealed in quartz tubes and heated to 600°C for several hours. Then the melt was furnace-cooled to room temperature. X-ray powder diffraction reveals the phase purity of the obtained compounds. The solid solutions were formed from master compounds of TBT and TBS. All samples were obtained as polycrystalline ingots.

The obtained material has been cut to samples of approximately 1x1x4 mm^3 for Seebeck and resistivity measurements. The Seebeck was measured with a temperature difference of 0.5K across the sample. Resistivities were measured by a standard 4-point AC method.

DISCUSSION

Pure TBT and TBS

In the following the properties of the pure compounds TBT and TBS are discussed before the effects of forming solid solutions are described.

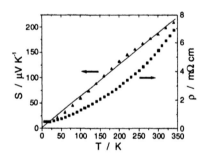 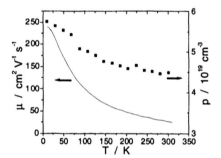

Fig. 1: Thermopower and resistivity of TBT as function of temperature.

Fig. 2: Hole concentration and mobility as determined from Hall measurements

The electrical properties of Tl_9BiTe_6 are shown in Fig. 1 and Fig. 2. The thermopower (Fig. 1) rises linearly with temperature over the entire temperature range with a slope of 0.6 µV/K^2. This results in a room temperature (300 K) value of +185 µV/K. The positive sign of the thermopower indicates hole-type conductivity. The electrical resistivity also rises monotonically with superlinear temperature dependence. At 300 K the resistivity is 5.5 mΩ. The power factor S^2/ρ is 0.6 mW/mK^2. Compared to Bi_2Te_3 which is the state of the art material at this temperature this is about a factor of 7 lower.

The hole concentration and their mobility as functions of temperature are shown in Fig. 2. The hole concentration is nearly constant over a temperature range from 20 K to 300 K with a value of approximately $5 \cdot 10^{19}$ cm^{-3}. The mobility at 20 K is almost 250 cm^2/Vs but it falls off by an order of magnitude to a room temperature value of 25 cm^2/Vs.

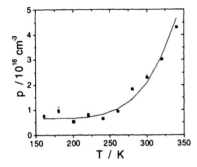
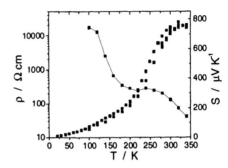

Fig. 3: *Temperature dependence of the hole concentration of TBS and fit to the data as described in the text.*

Fig. 4: *Resistivity and thermopower of TBS as function of temperature.*

In contrast to TBT which shows a "metallic" behavior, TBS is clearly semiconducting. Hall measurements have been carried out from 160 K to 340 K (Fig. 3). The temperature dependence of the hole concentration was fitted to the expression $p = p_0 + P \cdot exp(-E_g / 2kT)$ where E_g is the band gap energy and it is assumed that there is a concentration p_0 of acceptors completely ionized above 160 K. The factor P depends on the density of states in the bands. A good agreement with the experiment is achieved with $E_g = 0.45$ eV and $p_0 = 6.5 \cdot 10^{15}$ cm^{-3}. The temperature dependence of the carrier concentration is mirrored in the resistivity (Fig. 4). Above 250 K where carriers start to be excited over the band gap the slope of the thermopower begins to decrease. Below 150 K the "freeze out" of the acceptors can be observed in the resistivity.

An estimation of the band gap from the thermopower is given by Goldsmid and Sharp [4]. They show that $E_g = 2eS_{max}T_{max}$ where e is the elementary charge, S_{max} is the maximum value of the thermopower and T_{max} the temperature at which S_{max} occurs. Although the decline of the thermopower at high temperatures cannot be seen due to the limited temperature range of the experiment the maximum in the curve is clearly identifiable. With $S_{max} = 750$ µV/K at $T_{max} = 330$ K the band gap is estimated to be $E_g = 0.5$ eV. This is in very good agreement with the value determined from the hole concentration.

Solid Solutions TBT / TBS

The stoichiometry dependence of the lattice constant a and c is shown in Fig. 5. According to a "law" proposed by Vegard [5] the lattice spacing of solid solutions should vary linearly in proportion to the lattice spacing of the components. The solid solution system of TBT and TBS clearly deviates from this "law". The lattice constant c of TBT and TBS along the unique axis is 13.06Å and 12.69Å, respectively. Over a concentration range from 100% Te to 50% Te c does not change within the resolution of our X-ray diffraction

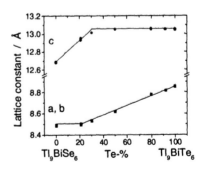

Fig. 5: *Lattice constants of TBS/TBT solid solutions.*

measurements. At a ratio of Te/Se = 30/70 the lattice constant c has decreased only by 0.04Å. From there it falls off linearly towards 12.69Å (TBS). The entire change in the lattice constant takes place over 30% of the substitution range.

For the lattice constant a (and b) the picture is similar but less dramatic. Starting from TBS no change in a is observed up to a substitution of 20% Te. From there the lattice constant increases linearily towards 13.06Å (TBT).

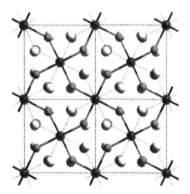

Fig.6: Structure of TBT and TBS as seen along the c-axis. The bonds between the chalcongenide atoms and the metal atoms on the 4c site are shown as sticks. Tl: light, Q: medium, (Tl, Bi): dark

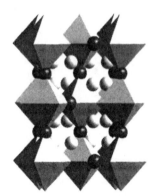

Fig. 7: Crystal structure perpendicular to the c-axis. The $(Tl,Bi)Q_6$ units are shown as octahedrons. Tl: light, Q: dark.

A closer look at the crystal structure and the bonds offers an explaination for this peculiarity. The Tl and Bi atoms on the *4c* site have 6 nearest neighbours (Te) which form octahedrons. The 4 Te atoms on the *8h* site share the *a, b* – plane with the central Tl or Bi atom. The coordination in the *a, b* – plane is shown in Fig. 6. For clarity reasons a ball and stick representation has been preferred to polyhedrons. The third axis of the octahedrons is parallel to the c-axis. These corners correspond to the *4a* site. Figure 7 shows that the $(Tl,Bi)Q_6$ units form chains along the c-axis. In the *a, b* – plane a corner sharing network is formed.

It is now comprehensible that the lattice constant c is mainly determined by the atoms on the *4a* site while the lattice constant a is determined by the *8h* site. Assuming that Te shows a greater affinity towards the *4a* site than Se the latter will start to occupy the *8h* site upon Se substitution in TBT. In a first order approximation the lattice constant a can be expected to decrease while c remains constant. By substituting two thirds of the Te atoms with Se the *8h* site will be completely occupied by Se while the *4a* site remains occupied by Te. This gives an ordered compound with a formula unit $Tl_9BiSe_4Te_2$. By further increasing the Se/Te ratio also the Te on the *4a* site will be replaced by Se. The lattice constant c will shrink to 12.69Å for pure TBS.

From this model we expect c to be constant from Te/Se 100/0 to Te/Se 67/33 and then decrease linearly. On the other hand a is expected to be constant from Se/Te 100/0 to Se/Te 33/67 and then increase linearly with the Te/Se ratio. Experimentally the critical concentrations for changes in the lattice constants are 23% Se for a and 32% for c. Since we have omitted half of the atoms in the crystal (Tl on *16l* site) from our argumentation this is a very good agreement to our model.

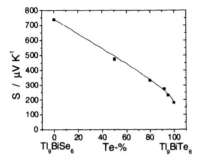

Fig. 8: *Thermopower of TBS/TBT solid solutions.*

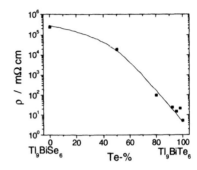

Fig. 9: *Electrical resistivity as function of TBS/TBT ratio.*

In the following we describe the dependencies of the electrical properties, resistivity and thermopower on the ratio of TBT/TBS in the solid solutions. The dependency of the thermopower on the Te/Se ratio is shown in Fig. 8. The thermopower changes approximately linearly from TBS to TBT. In contrast to the thermopower, which changes only by a factor of 8, from TBT to TBS the resistivity (Fig. 9) changes by more than five orders of magnitude. Especially in the Te-rich part of the solution diagram there seems to be an exponential dependency on the Se concentration. This very steep increase in resistivity is unfavorable for thermoelectric optimization because it clearly overcompensates the gain from increasing the thermopower. As shown in Fig. 10 the power factor falls off rapidly when the composition is change from TBT towards TBS. Since there is no intermediate maximum an optimization of the thermoelectric properties of TBT by forming solid solutions with TBS seems not possible. Even an imaginable decrease of the thermal conductivity by enhanced phonon scattering on the disordered Te/Se site will not make up for the loss in the power factor.

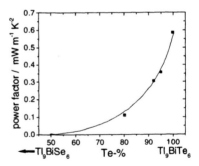

Fig.10: *Power factor of TBS / TBT solid solutions.*

CONCLUSIONS

In this study we have examined the potential of optimizing the thermoelectric properties of TBT by forming solid solutions with TBS. The two lattice constants, a and c, of the orthorombic cell show a peculiar dependence on the Te/Se ratio which clearly deviates from Vegard's law. This deviation can be explained by different affinities of Te and Se towards the two chalcogenide sites $4a$ and $8h$. Compared to Te, Se seems to favor the $8h$ site. Thus at a ratio of Se/Te = 2/1 we expect to have an ordered compound with the formula unit $Tl_9BiSe_4Te_2$. The behavior of the electrical properties upon forming solid solutions is unfavorable for thermoelectric optimization. Upon alloying TBS to TBT the increase in the resistivity is much more pronounced than the increase in the thermopower. This results in a rapid drop in the power factor which cannot be expected to be compensated by a possible reduction of the thermal conductivity. Since TBT belongs to a family of seven compounds other approaches to optimize the thermoelectrical properties will be pursued in future.

REFERENCES

1. A. A. Toure, G. Kra and R. Eholie, *Journal of Solid State Chemistry,* **87**. 229 (1990).
2. A. Pradel, J.-C. Tedenac, D. Coquillat and G. Brun, *Revue de Chemie Minerale,* **19**, 43 (1982).
3. I. E. Barchii, V. B. Lazarev, E. Yu. Peresh, Yu. V. Voroshilov, and V. I. Tkachenko, *Neorganicheskie Materialy,* **24,** 1791 (1988).
4. H.J. Goldsmid and J.W. Sharp, *Journal of Electronic Materials,* **28**, 869 (1999).
5. L. Vegard, *Z. Phys,* **5**, 17 (1921).

INVESTIGATIONS OF SOLID SOLUTIONS OF CsBi$_4$Te$_6$

Duck-Young Chung[1], Tim P. Hogan[2], Nishant Ghelani[2], Paul W. Brazis[3], Melissa A. Lane[3], Carl R. Kannewurf[3] and Mercouri G. Kanatzidis[1]
[1]Department of Chemistry, Michigan State University, East Lansing, MI 48824.
[2]Dept of Electrical Engineering, Michigan State University, East Lansing, MI 48824.
[2]Dept of Electrical and Computer Engineering, Northwestern University, Evanston, IL 60208.

ABSTRACT

Results on the synthesis and characterization of the solid solutions CsBi$_{4-x}$Sb$_x$Te$_6$, CsBi$_4$Te$_{6-y}$Se$_y$, as well as doping experiments on CsBi$_4$Te$_6$ are reported. We report X-ray structural investigations showing that the Sb or Se atoms in these compounds are not uniformly distributed in the lattice but show preferential occupation of specific crystallographic sites. Thermoelectric properties of selected systems are presented.

INTRODUCTION

Recently, we described the synthesis, structure and thermoelectric (TE) properties of CsBi$_4$Te$_6$, which, when doped with SbI$_3$, achieves a maximum ZT of ~0.8 at 225 K [1]. We believe additional improvements in TE performance of this material are possible. Thus we are pursuing further exploration of doping agents and the investigation of solid solutions such as CsBi$_{4-x}$Sb$_x$Te$_6$, CsBi$_4$Te$_{6-y}$Se$_y$ and Cs$_{1-z}$Rb$_z$Bi$_4$Te$_6$. The latter could result in substantially lower thermal conductivities. In addition, a complete TE cooling device needs both a p-type and n-type versions of a material to operate. Therefore we explored whether n-type doping is possible with various doping agents for CsBi$_4$Te$_6$. Here we report initial results on the synthesis of the solid solutions CsBi$_{4-x}$Sb$_x$Te$_6$, CsBi$_4$Te$_{6-y}$Se$_y$, and their characterization and doping studies of CsBi$_4$Te$_6$ with In$_2$Te$_3$ and BiI$_3$.

RESULTS AND DISCUSSION

CsBi$_4$Te$_6$ is composed of anionic [Bi$_4$Te$_6$] slabs alternating with layers of Cs$^+$ ions, Figure 1A. The reaction of cesium with two equivalents of Bi$_2$Te$_3$ does not produce a formal intercalation compound, but causes a complete reorganization of the bismuth telluride framework to produce a new structure type. The added electrons localize on the Bi atoms and form Bi-Bi bonds that are 3.238(1) Å long. The presence of these bonds seems to play a role in the charge transport properties of the material. The [Bi$_4$Te$_6$] layers are strongly anisotropic as they consist of one-dimensional [Bi$_4$Te$_6$] laths running parallel to the crystallographic b-axis. The width and height of these laths is 23 Å by 12 Å.

This structure type is not stable when Bi is substituted with Sb or when Te is substituted with Se. Therefore the entire range of solid solutions CsBi$_{4-x}$Sb$_x$Te$_6$ and CsBi$_4$Te$_{6-y}$Se$_y$ is not possible. In this work we probed the extent of x and y, and report the crystallographic refinement of CsBi$_{4-x}$Sb$_x$Te$_6$ and CsBi$_4$Te$_{6-y}$Se$_y$ (x = 0.3; y = 0.3). Interestingly, we found the Sb or Se atoms in these compounds to be distributed in a preferential manner as opposed to uniformly. This probably originates from the structural type and rigidity of the [Bi$_4$Te$_6$] framework associated with size and electronegativity difference between the atoms (r_{Bi} = 1.46 Å, EN = 2.02; r_{Sb} = 1.40 Å, EN = 2.05; r_{Te} = 1.36 Å, EN = 2.10)[2]. For example, certain sites contain greater concentrations of Sb or Se while others contain much less. As shown in Figure 1A and B, the Sb and Se atoms are preferentially located on the surface of [Bi$_4$Te$_6$] laths. From the refinement of the crystallographic data we can see that in the case of CsBi$_{4-x}$Sb$_x$Te$_6$ (x = 0.3) the Sb atoms prefer the metal positions lying at the outer layer of the [Bi$_4$Te$_6$] lath. So at the core of this lath the metal positions are purely Bi. This positional preference of Sb and Se atoms minimizes the structural tension of the rigid [Bi$_4$Te$_6$] slabs, which is responsible for the existence of compositional limits, x < 0.8 and y < 1.2 for CsBi$_{4-x}$Sb$_x$Te$_6$ and CsBi$_4$Te$_{6-y}$Se$_y$, respectively. The Bi/Sb and Bi/Te distribution in these materials will be investigated for other x and y values as well. The implications of this inhomogeneous Bi/Sb and Te/Se distribution in the charge transport properties of these materials are not entirely clear, nevertheless we suspect they would be significant by exerting an influence on carrier mobilities.

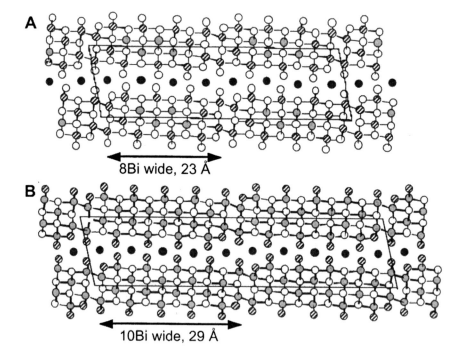

Figure 1. The structure of (A) CsBi$_{4-x}$Sb$_x$Te$_6$ (x = 0.3) and (B) CsBi$_4$Te$_{6-y}$Se$_y$ (y = 0.3) viewed down the b-axis. Open circles are Te atoms, gray ones Bi atoms, black ones Cs atoms, and striped ones Sb atoms (A) or Se atoms (B). The widths of the [Bi$_4$Te$_6$] slabs are shown.

It is also very interesting that in the case of $CsBi_4Te_{6-y}Se_y$ (y = 0.3) we discovered a closely related phase whose structure has wider $[Bi_4Te_6]$ laths. These laths are two Bi atom wider (~ 29 Å) than those in $CsBi_4Te_6$ (~ 23 Å) and the larger Bi_2Te_3 portion in this compound, compared to that in $CsBi_4Te_6$, causes a lower $Cs/[Bi_4Te_6]$ ratio, resulting in a less reduced compound with the formula $Cs_{0.8}Bi_4Te_{5.7}Se_{0.3}$. It is not clear whether this structural modification is due to the participation of Se atom in the framework or due to the synthetic conditions in which this compound was prepared. The discovery of a modified structure of $CsBi_4Te_6$ suggests that compounds possessing either wider or narrower $[Bi_4Te_6]$ slabs could possibly be formed. Because the $[Bi_4Te_6]$ slab is the major part participating in the electronic structure near Fermi level[3] this new structural modification may provide us with more chances to find additional interesting new TE materials.

Doping

As we reported recently, $CsBi_4Te_6$ is amenable to considerable doping manipulation, much like Bi_2Te_3, and thus higher ZT values may be obtained. We prepared doped samples of $CsBi_4Te_6$ using a procedure to be reported elsewhere [4]. We investigated SbI_3, BiI_3, and In_2Te_3 in amounts varying from 0.02 to 4 mol%. These dopants were chosen for the purpose of placing halide atoms in the Te sites (to produce n-type doping) and In atoms in the Bi sites (to produce p-type doping). Doping with these agents does occur, but we currently do not know what sites in the crystal structure are being occupied. Surprisingly, doping with SbI_3 and BiI_3 produces p-doped rather than n-doped samples. This observation is not consistent with iodine atoms occupying Te sites but instead suggests Sb and Bi atoms on Te sites. The Sb and Bi atoms, having only five electrons, introduce holes in the valence Te-based band.

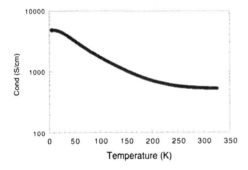

Figure 2. Electrical conductivity data for 1.5 mol% In_2Te_3-doped $CsBi_4Te_6$.

Electrical conductivity data for almost all doped samples exhibit a weak metallic dependence, with the conductivity decreasing as temperature increased from 4.2 to 340 K. Room temperature (295 K) electrical conductivities ranged from 420 S/cm (0.3% SbI_3) to 2100 S/cm (2.5% In_2Te_3).

Interestingly, depending on sample preparation conditions In_2Te_3 doping can produce n-type samples. Depending on the type and degree of doping, room-temperature thermopower values between -40 and -100 µV/K were observed. Room temperature (295 K) electrical conductivities ranged from 500 S/cm (0.3% In$_2$Te$_3$) to 2100 S/cm (2.5% In$_2$Te$_3$). Though optimum levels have not been reached yet with these dopants, Figures 2 and 3, show that n-type behavior is possible. It is interesting that maximum thermopower of almost –100 µV/K occurs at ~160 K, a temperature which has important implications for the development of even lower temperature TEs. Surprisingly, larger amounts (>3%) of In$_2$Te$_3$ produce p-type samples.

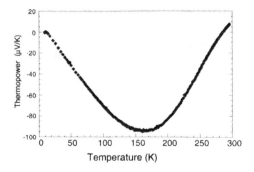

Figure 3. Thermopower data from a single crystal of 1.5 mol % In$_2$Te$_3$-doped CsBi$_4$Te$_6$.

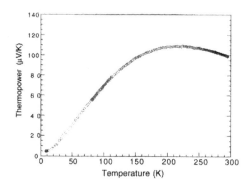

Figure 4. Thermopower data from a single crystal of 0.3 mol % BiI$_3$-doped CsBi$_4$Te$_6$.

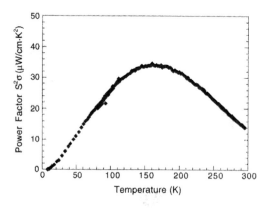

Figure 5. Power factor for a single crystal of 0.3 mol % BiI$_3$-doped CsBi$_4$Te$_6$.

Doping experiments involving BiI_3 gave results similar to those with SbI_3. Thus 0.3% BiI_3 - doped samples are highly conductive and p-type with a maximum thermopower at 210 K of ~+116 μV/K, see Figure 3. The power factor for such doped $CsBi_4Te_6$ is shown in Figure 4. The power factor has a maximum value of 33 μW/cm·K^2 at ~ 160 K. The maximum occurs at a rather low temperature and, although it is only half as large as the power factor of the SbI_3-doped $CsBi_4Te_6$ or the optimized Bi_2Te_3, it points to the possibility that this type of doping could lead to materials with high thermoelectric performance at cryogenic temperatures.

Solid Solutions

To meet the necessity for the design of a practical thermoelectric device, both p- and n-type materials are required and the best materials currently being used in thermoelectric applications are $Bi_{2-x}Sb_xTe_3$ and $Bi_2Te_{3-y}Se_y$ for p- and n-type materials, respectively. We are currently examining the same type of chemical manipulation on $CsBi_4Te_6$ to achieve both p- and n-type behavior. Additionally, a reduced thermal conductivity may be expected from the mass fluctuation caused by introducing Sb and Se, into $CsBi_4Te_6$. In this report we present preliminary results obtained from oriented ingot samples of $CsBi_{4-x}Sb_xTe_6$ and $CsBi_4Te_{6-y}Se_y$. It should be noted that the production of such ingots has not been optimized and they most likely contain imperfections and microcracks. This of course can affect critically the electrical conductivity of the sample, yet it does not significantly influence the thermopower.

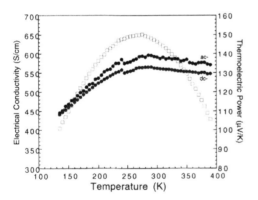

Figure 6. Variable temperature conductivity and thermopower data for $CsBi_{3.6}Sb_{0.4}Te_6$ and (B) $CsBi_4Te_{5.2}Se_{0.8}$.

All $CsBi_{4-x}Sb_xTe_6$ (x < 0.8) solid solutions showed p-type behavior with the maximum thermopower of ~150 μV/K at different temperatures. The thermopower data for all materials appear to be very dependent upon temperature and show a maximum between 250 and 350 K, see Figure 6. The $CsBi_{3.6}Sb_{0.4}Te_6$ ingot sample showed 600 S/cm and 150 μV/K at room temperature and the maximum power factor of 13.4 μW/cm·K^2 was obtained at 275 K. Based on the maximum values of S, the band gap of $CsBi_{4-x}Sb_xTe_6$ and $CsBi_4Te_{6-y}Se_y$ can be estimated from the formula $E_g \sim 2S_{max} \cdot T_{max}$ [5] to be 0.081 and 0.084 eV respectively.

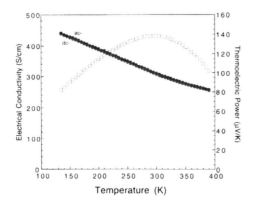

Figure 7. Variable temperature conductivity and thermopower data for $CsBi_4Te_{5.2}Se_{0.8}$.

The $CsBi_4Te_{6-y}Se_y$ sample showed a similar maximum in the thermopower at ~140 µV/K at 300 K. The electrical conductivity at room temperature was ~300 S/cm and showed a metal like dependence, Figure 7. Hall measurements are planned to obtain further insights into these samples such as carrier concentrations and mobilities.

In conclusion, the introduction of Sb and Se in the lattice of $CsBi_4Te_6$ has been demonstrated. The Sb and Se atoms are not uniformly distributed over all crystallographic sites. The introduction of various other dopants and further work on crystal processing is currently under way.

Acknowledgments. Financial support from the Office of Naval Research (N00014-98-1-0443) DARPA through ARO (DAAG55-97-1-0184) is gratefully acknowledged. The work made use of the SEM facilities of the Center for Electron Optics at Michigan State University. The work at Northwestern made use of the Central Facilities supported by NSF through the Materials Research Center (DMR-9632472).

REFERENCES

1. Duck-Young Chung, Tim Hogan, Paul Brazis, Melissa Rocci-Lane, Carl Kannewurf, Marina Bastea, Ctirad Uher and Mercouri G. Kanatzidis, *Science*, **287**, 1024 (2000).
2. L. Pauling, *"The Nature of The Chemical Bond"*, 3rd Ed. Cornell (1960).
3. P. Larson, B. Mahanti, D.-Y. Chung, M. G. Kanatzidis, See elsewhere in this volume.
4. Duck-Young Chung, Tim Hogan, Paul Brazis, Melissa Rocci-Lane, Carl Kannewurf, Marina Bastea, Ctirad Uher and Mercouri G. Kanatzidis, manuscript in preparation.
5. H. J. Goldsmid and J. W. Sharp, *J. Electron. Mater.* **28**, 869 (1999).

Quantum Wires and Dots

Carrier Pocket Engineering for the Design of Low Dimensional Thermoelectrics with High $Z_{3D}T$

Takaaki Koga[a], Stephen B. Cronin[b], and Mildred S. Dresselhaus[b,c]
[a]Division of Engineering and Applied Sciences, Harvard University, Cambridge, MA 02138
[b]Department of Physics and [c]Department of Electrical Engineering and Computer Science, Massachusetts Institute of Technology, Cambridge, MA 02139

ABSTRACT

The concept of carrier pocket engineering applied to Si/Ge superlattices is tested experimentally. A set of strain-symmetrized Si(20Å)/Ge(20Å) superlattice samples were grown by MBE and the Seebeck coefficient S, electrical conductivity σ, and Hall coefficient were measured in the temperature range between 4K and 400K for these samples. The experimental results are in good agreement with the carrier pocket engineering model for temperatures below 300K. The thermoelectric figure of merit for the entire superlattice, $Z_{3D}T$, is estimated from the measured S and σ, and using an estimated value for the thermal conductivity of the superlattice. Based on the measurements of these homogeneously doped samples and on model calculations, including the detailed scattering mechanisms of the samples, projections are made for δ-doped and modulation-doped samples [(001) oriented Si(20Å)/Ge(20Å) superlattices] to yield $Z_{3D}T \approx 0.49$ at 300K.

INTRODUCTION

Early work on low dimensional thermoelectricity used simple theoretical models, such as for an isolated 1D and 2D electron gas [1], and the results of these calculations predicted a significant enhancement in the thermoelectric figure of merit ZT within the quantum well. The earliest attempts to show this enhancement experimentally were performed on 2D superlattices grown by molecular beam epitaxy (MBE) and their transport properties were measured [2, 3]. In these early samples, the barrier layer was made much thicker than the quantum well thickness, ensuring good quantum confinement of carriers in the quantum well. In these samples it was shown that the thermoelectric power factor, $S^2\sigma$, of the quantum wells alone, neglecting the contribution from the barrier layers, was indeed enhanced. These experiments helped prove the principle, predicted theoretically, that thermoelectricity could be enhanced in low dimensional systems.

In terms of a practical device, thick barrier layers provide a parasitic thermal conduction path, and result in a much degraded ZT for the entire sample. In the present work, we concern ourselves only with increasing $Z_{3D}T$, which includes contributions from both the quantum wells and the barrier layers, that is the entire thermoelectric device. The systematic process by which low dimensional superlattices of given constituents are designed to optimize their 3D thermoelectric properties has been called "carrier pocket engineering"[4-7]. Within this framework, the large barrier widths that were used in previous proof-of-principle studies are greatly reduced to become comparable to the widths of the quantum wells, so that the electron wave functions are not strictly confined to the quantum wells. The design parameters that are optimized in this process include: (1) the thicknesses of the barrier layer and quantum well, denoted by d_B and d_W, respectively; (2) the growth direction of the superlattice [such as the (001) or (111) crystalline directions]; (3) the lattice strain of the quantum well and barrier layers as controlled by the composition and lattice constant of the substrate; and (4) the carrier density.

In this paper a brief description of the carrier pocket engineering calculations [6] are given, but the main focus will be on the recent experimental results for Si/Ge superlattice samples aimed at evaluating the carrier pocket engineering modeling concept for the design of high $Z_{3D}T$ superlattices.

CARRIER POCKET ENGINEERING CALCULATIONS IN THE Si/Ge SUPERLATTICE SYSTEM

In the carrier pocket engineering approach, the energy and the electronic wavefunctions of an electron in the periodic potential of the 2D superlattice (Si/Ge in this case) must be calculated. This calculation is carried out using the one-electron Kronig-Penney model, which assures that the wavefunction and its derivative are continuous at the quantum well-barrier interface. The resulting eigenvalues give the shift of the sub-band edges relative to the bulk band edge energies. Thus, by changing the parameters of the geometry of the superlattice, d_B and d_W, the sub-band edges can be *tuned* to achieve the optimum $Z_{3D}T$. As part of this calculation we must calculate the in-plane effective masses. We do this by taking the projection of the 3D constant energy surfaces on the plane of the quantum well. With these in-plane effective masses, the density of states is calculated [4, 6].

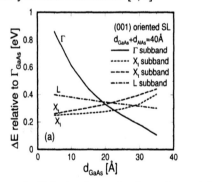
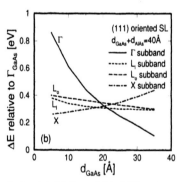

Figure 1. *Calculated energies for various sub-bands for (a) (001) and (b) (111) oriented GaAs/AlAs superlattices measured from the conduction band edge at the Γ-point in the Brillouin zone for bulk GaAs.*

The concept of carrier pocket engineering was first applied to the GaAs/AlAs system because the electronic band structure was well known and the fabrication of superlattices was well established, although ZT for bulk GaAs is very low ($ZT=0.0085$ at 300K). To illuminate the basic strategy of the carrier pocket engineering approach, we show in Fig. 1 the sub-band energies for the Γ, L, and X point sub-band edges as a function d_{GaAs}, the thickness of the GaAs layer, for (100) and (111) oriented GaAs/AlAs superlattices with a fixed thickness of the superlattice period ($d_B + d_W = 40$Å). For the Γ and L-point sub-bands, the GaAs layer is the quantum well, while for the X-point sub-bands, the AlAs layer is the quantum well, so that both layers of the superlattice are contributing to the transport and hence to $Z_{3D}T$. Optimization of ZT requires that the density of states at the Fermi level be maximized. We can expect that when the Γ, L and X-point band edges are at about the same energy, all the carrier pockets can contribute to $Z_{3D}T$. We note that when d_B and d_W of the superlattice both have a thickness of 20Å, all of the sub-bands have approximately the same energy. This results in a large density of states near the band edge, which has been shown to lead to enhanced thermoelectric performance [4]. Figure 1 demonstrates one of the main strategies of the carrier pocket engineering concept based on the GaAs/AlAs superlattice system.

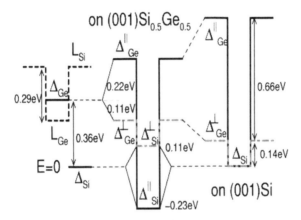

Figure 2. *Conduction band offset diagrams for (001) oriented Si/Ge superlattices. The left, middle, and right diagrams denote the band offsets for the unstrained layers, for a superlattice grown on a $Si_{0.5}Ge_{0.5}$ substrate, and for a superlattice grown on a Si substrate, respectively. The symbols $\Delta^{\parallel}$ and $\Delta^{\perp}$, respectively, denote carrier pockets parallel and perpendicular to the (001) growth direction.*

The transport coefficients, such as the Seebeck coefficient S, the electrical conductivity σ, and the electronic contribution to the thermal conductivity κ_e can be calculated for the whole superlattice using the Boltzmann transport equations for a 2D electron gas. In the generic presentation of the carrier pocket engineering concept, the simplest possible model of a constant relaxation time approximation and parabolic energy dispersion relations is used [4-7]. For application to specific materials systems, these approximations can be relaxed to obtain more quantitative models by including specific relaxation processes and non-parabolic dispersion relation effects, as appropriate.

The concept of Carrier Pocket Engineering was extended in its application to the Si/Ge superlattice system by considering the optimization of the additional parameter of lattice strain. Because of the excellent lattice matching between GaAs and AlAs, this effect is not important for the GaAs/AlAs superlattices. The use of lattice strain to control the conduction band offsets in the Si/Ge superlattice system allows one to adjust the band edges of the various carrier pockets to coincide with one another to maximize the overall density of states near the band edge. The use of the lattice strain effect to further augment the carrier pocket engineering approach is described from a theoretical standpoint in a previous publication [6,7].

Based on the calculations of Refs. [6,7], Si(20Å)/Ge(20Å) superlattice samples were chosen as a good system to demonstrate the carrier pocket engineering concept experimentally for an (001) oriented superlattice. Although (111) oriented samples were predicted to have an optimized $Z_{3D}T$ value four times larger than that for the (001) oriented superlattices, the (001) orientation was chosen for the proof-of-principle demonstration of the carrier pocket engineering concept on the basis of materials science considerations in the fabrication process.

EXPERIMENTAL DEMONSTRATION IN THE Si/Ge SUPERLATTICE SYSTEM

The samples used in this study were designed in accordance with the carrier pocket engineering concept and were grown by Dr. J. L. Liu at UCLA using MBE (molecular beam epitaxy). A schematic diagram of the Si/Ge superlattice samples used in this proof-of-principle study is shown in Figure 3. The samples were grown on Si-on-insulator (SOI) substrates to avoid a large contribution to the Seebeck coefficient from the Si substrate. The SOI substrate, which provides good electric isolation from the underlying Si substrate, is composed of an 1800Å layer of (001) oriented Si on top of a 3800Å layer of SiO_2, which, in turn, is grown on top of the (001) oriented thicker Si substrate. A $Si_{1-x}Ge_x$ (x: 0→0.5) graded buffer layer was grown on the SOI substrate using Sb (antimony) as a surfactant [8]. A homogeneous undoped $Si_{0.5}Ge_{0.5}$ buffer layer was then grown to relax the strain in the graded buffer layer to yield a (001) surface of $Si_{0.5}Ge_{0.5}$ that is fully relaxed through the lattice strain and would have a low misfit dislocation density. Finally, the structurally strain-symmetrized Si(20Å)/Ge(20Å) superlattices (with 100 periods, except for sample JL193, which has only 25 periods) were grown on top of the undoped (001) oriented $Si_{0.5}Ge_{0.5}$ buffer layer. The samples were doped n-type homogeneously throughout the Si(20Å)/Ge(20Å) superlattice part of the sample. The Sb dopant concentrations used for these samples are in the range between $1\times10^{18} cm^{-3}$ and $2\times10^{19} cm^{-3}$. The sample growth temperature of the Si(20Å)/Ge(20Å) layers was 350^0C to provide sharp superlattice interfaces by inhibiting interdiffusion problems.

Shown in Figures 4 and 5 are the Seebeck coefficient, S, electrical conductivity, σ, and Hall carrier concentration, n_{Hall}, for samples JL194, JL197 and JL199, measured as a function of temperature in the temperature range between 4.2K and 400K. Also plotted in these figures are the results of semi-classical models, based on the carrier pocket engineering model, assuming $n=5\times10^{18} cm^{-3}$, $7\times10^{18} cm^{-3}$ and $1.5\times10^{19} cm^{-3}$ (see Figure 5 for the measured Hall carrier densities) [9].

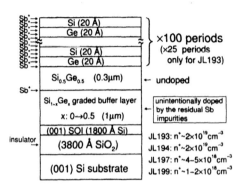

Figure 3. *A schematic diagram showing the sample structure for (001) Si(20Å)/Ge(20Å) superlattice samples grown on (001) oriented Si using a SOI substrate (see text).*

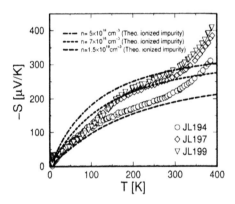

Figure 4. *Seebeck coefficient measured as a function of temperature for samples JL194, JL197 and JL199, together with the results of semi-classical models [9].*

The experimental measurements for $S(T)$ in Figure 4 show a basic linear T dependence at low T, followed by a saturation behavior until approximately 300K, followed by a further increase in $S(T)$ above 300K. The results show no evidence for a contribution to $S(T)$ from the silicon substrate. We find that the measured values for the Seebeck coefficient $S(T)$ are in fair agreement with the values predicted by theory up to 300K, using no fitting parameters. In the temperature range between 180 and 300K, we conclude that our theoretical models based on the carrier pocket engineering model successfully describe the temperature-dependent Seebeck coefficient for the (001) oriented Si(20Å)/Ge(20Å) superlattices used in the present study.

In the temperature range below 180K, the measured values of the Seebeck coefficient are somewhat smaller than those predicted theoretically. This discrepancy is probably due to the presence of additional scattering mechanisms that are not included in our semi-classical models. The importance of such additional scattering mechanisms is clearly evident in the measured electrical conductivity, shown in Figure 5, where the measured values for the conductivity for samples JL197 and JL199 are a factor of two to three lower than the theoretically predicted values [9]. If these additional scattering mechanisms have a $\tau \propto E^r$ dependence, with a value of r comparable to or smaller than 1, the values of the measured S for the superlattice samples should be reduced relative to the theoretically predicted values below 180K.

Above 300K the measured values for the Seebeck coefficient for samples JL194, JL197 and JL199 all diverge from the theoretical predictions. To investigate whether or not this divergence of the measured $S(T)$ from the theoretical model above 300K is due to a contribution from the substrate and/or buffer layers, samples were prepared and characterized with various number of periods. Three samples, which differ only in the number of superlattice periods, were fabricated and characterized to evaluate the effect of the substrate and the buffer layer on the measured transport properties [9]. It was found that for superlattices with 100 periods the contribution from the substrate and the buffer layers was negligible. Therefore, the exact physical origin of the observed discrepancy in $S(T)$ between theory and experiment above 300K is not known at this time. It is however suspected to be related to another scattering mechanism not included in the model or to a lower energy for the second sub-band edge in the actual superlattice samples than is predicted by the Kronig-Penney model.

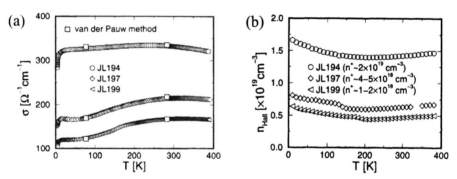

Figure 5. *(a) Electrical conductivity and (b) Hall carrier density measured as a function of temperature for samples JL194, JL197 and JL199. The nominal dopant concentrations for these n-doped superlattice samples are indicated in the legend for (b).*

From the experimentally measured Seebeck coefficient and electrical conductivity, we can estimate the $Z_{3D}T$ for these samples. The thermal conductivity of Si/Ge superlattices has been studied experimentally using a 3ω technique by Borca-Tasciuc, et al.[10] For homogeneously n-type doped Si(20Å)/Ge(20Å) superlattices, the cross-plane thermal conductivity, $\kappa_\perp$, was determined to be between approximately 1 and 1.5W/mK. Moreover, the temperature dependence of $\kappa_\perp$ was found to be very weak in the temperature range between 77 and 300K. Based on estimates of the measured $\kappa_\perp$ for such samples, we estimate that $\kappa_\parallel \approx$ 5W/mK, where $\kappa_\parallel$ denotes the thermal conductivity along the plane of the superlattice (in-plane thermal conductivity), that is constant with temperature for our superlattice samples JL194, JL197 and JL199.

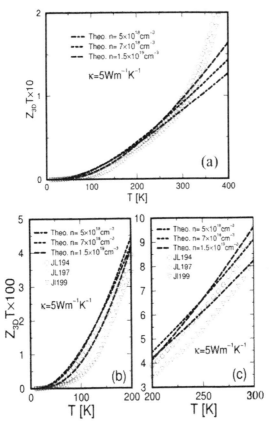

Figure 6. *The estimated $Z_{3D}T$ of samples JL194, JL197 and JL199 as a function of temperature, based on experimental measurements of S and σ and the estimated thermal conductivity $\kappa_\parallel$ of 5W/mK, independent of temperature. (b) and (c) show close-ups of the plot in figure (a), for temperature ranges below 300K. Also plotted are comparisons with calculations based on the carrier pocket engineering concept for the measured carrier concentrations of the three samples based on Fig. 5 [9].*

Gathering all these factors, we now estimate $Z_{3D}T$ for these samples as a function of temperature, based on the T independent estimate of 5W/mK for the thermal conductivity $\kappa_{||}$, and the results for $Z_{3D}T$ are plotted in Figure 6 as a function of temperature. We notice first that both the experimental and theoretical values for $Z_{3D}T$ are not sensitive to the doping level of the sample in the range of carrier concentration between $n=5\times10^{18}$cm^{-3} and 1.5×10^{19}cm^{-3} at 300K. This is because the carrier concentration for the homogeneously doped Si(20Å)/Ge(20Å) superlattice samples is already close to the optimum carrier concentration to achieve the highest $Z_{3D}T$ at 300K with the available sample quality. The rapid increase in $Z_{3D}T$ above 300K is identified with the corresponding behavior in $S(T)$ above 300K, discussed above (see Fig. 6(a)).

ALTERNATE DOPING PROFILES TO FURTHER INCREASE $Z_{3D}T$

The samples discussed in this study have all been homogeneously doped superlattices. However, by changing the doping profile, greater enhancement of the thermoelectric transport properties is predicted, based on a modification of the details of the scattering mechanisms, which can lead to a significantly increased thermoelectric figure of merit.

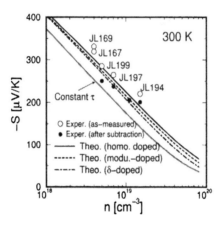

Figure 10. *The theoretically predicted values for the Seebeck coefficient for (001) oriented Si(20Å)/Ge(20Å) superlattices as a function of carrier concentration at 300K. The calculations were made using the semi-classical models based on the carrier pocket engineering concept, assuming: (1) homogeneous doping throughout the whole superlattice (solid curve), (2) modulation doping only in the Ge layers of the superlattice (dashed curve), and (3) δ–doping in the middle of each Ge layer in the superlattice (dash-dotted curve). Also shown in the figure are the as-measured values for the experimental S (open circles), the values for S obtained after subtracting the buffer layer and/or substrate contributions (filled circles) [9], and the values for S predicted by the constant relaxation time approximation model calculation (gray solid curve) for comparison.*

For the Si(20Å)/Ge(20Å) superlattice samples prepared in this study, screened ionized impurity scattering is the dominant scattering mechanism. This is because the dielectric constants of Si and Ge are relatively small ($\varepsilon \approx 11.7\varepsilon_0$ and $16.1\varepsilon_0$ for Si and Ge, respectively, where ε_o is the dielectric constant of free space) and because of the large carrier concentrations required by thermoelectricity. By doping the superlattice using a modulation doping scheme, where the dopants would be introduced uniformly only in the Ge layer, or using a δ-doping scheme where the dopants would be introduced only in a very localized thin layer that is placed in the middle of the Ge layer, we can reduce the contribution from impurity scattering, and this strategy turns out to be beneficial to thermoelectrics, due to the enhanced carrier mobilities.

Although the δ- and modulation-doping schemes are found to be beneficial for enhancing the values of the carrier mobility μ, these doping profiles are not necessarily advantageous for enhancing the values of S. Shown in Fig. 10 are the results of the theoretical calculation of the Seebeck coefficient for (001) oriented Si(20Å)/Ge(20Å) superlattices as a function of the carrier concentration n, assuming: (1) homogeneous doping throughout the Si and Ge superlattice layers (solid curve), (2) modulation doping only in the Ge layers (dashed curve), and (3) δ-doping only at the center of each Ge layer (dash-dotted curve). We indeed find a slight decrease in the theoretical values of S for δ- and modulation-doped superlattices relative to those for homogeneously doped superlattices for a given carrier concentration n. We also show, in Fig. 10, the experimental results for the homogeneously doped superlattice samples JL194, JL197 and JL199, and the calculated values of S as a function of n using the constant relaxation time approximation, for comparison.

We find the following features in Fig. 10. First of all, the predicted values of S that are obtained using the semi-classical model, including an energy-dependent $\tau(E)$, are greatly enhanced over the values predicted by the constant relaxation time approximation. This is because the momentum relaxation time $\tau(E)$ predicted using the semi-classical models increases with increasing energy. This property of $\tau(E)$ is mainly caused by the nature of ionized impurity scattering, where a $\tau(E) \propto E^{-3/2}$ dependence is predicted for pure ionized impurity scattering for 3D materials without including the screening effect due to the free carriers. The second feature that is observed in Fig. 10 is that, although the predicted values for S are rather insensitive to the details of the doping techniques that are considered here, the S values do show some differences among the various doping schemes. The main reason for the observed differences in the value of S among these three doping techniques comes from the relative contribution of ionized impurity scattering to the total scattering probability for the conduction electrons. It is predicted that the contribution of ionized impurity scattering to the total scattering probability is the largest in the homogeneously doped samples and is the smallest in the δ-doped samples among the three doping techniques considered here. Therefore the resultant $\tau(E)$ function for the δ-doped superlattice samples is expected to have the weakest energy dependence among those for the three doping schemes considered here.

Although the predicted values for S for the δ-doped and the modulation-doped (001) oriented Si(20Å)/Ge(20Å) superlattices are found to be smaller than those predicted for the homogeneously-doped samples, if the predicted values for the carrier mobility for the δ-doped and the modulation-doped samples are significantly larger than the corresponding values for the homogeneously doped superlattice samples, we can still achieve enhancements in the thermoelectric power factor $S^2\sigma$ and the resultant $Z_{3D}T$ relative to those predicted for the homogeneously doped samples, by introducing the δ- and/or modulation doping schemes, since the observed differences in the predicted values of S between the δ- and modulation-doped samples and the homogeneously doped samples are rather small.

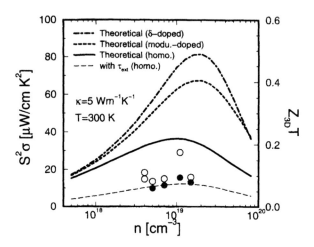

Figure 11. *Calculated $S^2\sigma$ (left scale) and $Z_{3D}T$ (right scale) vs carrier density n for alternate doping schemes that are explained in the text, shown together with: the as-measured values for the experimental $S^2\sigma$ (open circles) and the values for $S^2\sigma$ obtained after subtracting the buffer layer and/or substrate contributions (filled circles).*

Figure 11 shows the theoretically predicted values for the thermoelectric power factor $S^2\sigma$ (left scale) and $Z_{3D}T$ (right scale) for (001) oriented Si(20Å)/Ge(20Å) superlattice samples as a function of carrier concentration n at 300 K. These calculations were made using the semi-classical model based on the carrier pocket engineering concept, assuming: (1) homogeneous doping throughout the whole superlattice (solid curve), (2) modulation doping only in the Ge layers of the superlattice (dashed curve), and (3) δ-doping in the middle of each Ge layer in the superlattice (dash-dotted curve). Also shown in the figure are the as-measured values for the experimental $S^2\sigma$ (open circles), the values for $S^2\sigma$ obtained after subtracting the buffer layer and/or substrate contributions (filled circles) [9], and the calculated values for $S^2\sigma$ obtained using Matthiessen's rule to take into account the extrinsic scattering mechanisms that are present in the actual superlattice samples for the (001) oriented homogeneously doped

Si(20Å)/Ge(20Å) superlattices (thin dashed curve) [9] for comparison. The comparison between the filled circles and the thin dashed curve shows quite good agreement.

CONCLUSIONS

Utilizing the concept of carrier pocket engineering, parameters for the superlattice design were chosen to optimize $Z_{3D}T$. These samples were fabricated and characterized, and values for $Z_{3D}T$ were estimated in the temperature range between 4K and 400K. The carrier pocket engineering model explains well the values and temperature dependences of the transport properties of the samples below 300K, although more work needs to be done to explain the experimental results above 300K.

Further modeling indicates a potential for significant enhancement in the thermoelectric performance beyond those of the homogeneously doped samples of this study. Superlattice samples prepared utilizing the δ–doping or modulation-doping schemes are predicted to have a large increase (a factor of two to three, depending on the temperature and the carrier concentration) in the values of the calculated mobility, and hence in the power factor. The largest calculated value for $Z_{3D}T$ for the δ–doped sample at 300 K is 0.49 using a constant $\kappa_\parallel$=5W/mK for the value of thermal conductivity, which is a factor of more than two enhancement in $Z_{3D}T$ relative to the corresponding value for the homogeneously doped samples ($Z_{3D}T$ =0.22). Future investigations should be extended to an experimental proof-of-principle study of (111) oriented Si/Ge superlattices where the values of $Z_{3D}T$ are predicted to be enhanced by an additional factor of four relative to those for the (001) oriented Si/Ge superlattices (looking to a value of $Z_{3D}T \approx 2.0$ at 300K for δ- and/or modulation-doped (111) oriented Si/Ge superlattices), using a simple model based on the constant relaxation time approximation [6].

ACKNOWLEDGEMENTS

The authors would like to thank Dr. J.L. Liu and Prof. K.L. Wang at UCLA for providing high quality samples for this study, and also Prof. G. Chen, T. Borca-Tasciuc and Dr. G. Dresselhaus for valuable discussions. The authors gratefully acknowledge support from ONR under MURI subcontract #205-G-7A114-01 and the US Navy under contract N00167-98-K-0024.

REFERENCES

1. L.D. Hicks and M.S. Dresselhaus, *Phys. Rev. B* **47**, 12727 (1993).
2. T.C. Harman, D.L. Spears, and M.J. Manfra, *J. Electron. Mater.* **25**, 1121 (1996).
3. L.D. Hicks, T.C. Harman, X. Sun, and M.S. Dresselhaus, *Phys. Rev. B*, **53**, R10493 (1996).
4. T. Koga, X. Sun, S.B. Cronin, and M.S. Dresselhaus, *Appl. Phys. Lett.* **73**, 2950 (1998).
5. T. Koga, X. Sun, S.B. Cronin, and M.S. Dresselhaus, In *Thermoelectric Materials-The Next Generation Materials for Small-Scale Refrigeration and Power Generation Applications:*

MRS Symposium Proceedings, Boston, volume **545**, edited by T.M. Tritt, H.B. Lyon Jr., G.D. Mahan, and M.G. Kanatzidis, pages 375-380, Materials Research Society Press, Pittsburge, PA, 1999.
6. T. Koga, X. Sun, S.B. Cronin, and M.S. Dresselhaus, *Appl. Phys. Lett.* **75**, 2438 (1999).
7. T. Koga, X. Sun, S.B. Cronin, and M.S. Dresselhaus, In *The 18th International Conference on Thermoelectrics: ICT Symposium Proceedings, Baltimore*, Institute of Electrical and Electronics Engineer, Inc., Piscataway, NJ 09955-1331, 1999.
8. J.L. Liu, C.D. Moore, G.D. U'Ren, Y.H. Lou, Y. Lu, G. Jin, S.G. Thomas, M.S. Goorsky, and K.L. Wang, *Appl. Phys. Lett.* **75** 1586 (1999).
9. Takaaki Koga. *Concept and Application of Carrier Pocket Engineering to Design Useful Thermoelectric Materials Using Superlattice Structure.* PhD thesis, Harvard University, April 2000. Division of Engineering and Applied Sciences.
10. Theodorian Borca-Tasciuc, *et al*, *Thermal Conductivity of Symmetrically Strained Si/Ge Superlattices,* Nanostrucures and Superlattices, in press.

Half-Heusler Alloys and Quasicrystals

Effects of the Addition of Rhenium on the Thermoelectric Properties of the AlPdMn Quasicrystalline System

A. L. POPE[1], R. GAGNON[2], R. SCHNEIDMILLER[3], P. N. ALBONI[1], R. T. LITTLETON IV[4], B. ZAWILSKI[1], D. WINKLER[1], T. M. TRITT[1,4], J. STROM-OLSEN[2], J. KOLIS[3,4], S. LEGAULT[2]

1. Department of Physics and Astronomy, Clemson University, Clemson, SC 29634 USA
2. McGill University, Montreal, Canada
3. Department of Chemistry, Clemson University, Clemson, SC 29634 USA
4. Materials Science Department, Clemson University, Clemson, SC 29634 USA

ABSTRACT

Partially due to their lack of periodic structure, quasicrystals have inherently low thermal conductivity on the order of 1 - 3 W/m-K. AlPdMn quasicrystals exhibit favorable room temperature values of electrical conductivity, 500-800 $(\Omega\text{-cm})^{-1}$, and thermopower, 80 µV/K, with respect to thermoelectric applications. In an effort to further increase the thermopower and hopefully minimize the thermal conductivity via phonon scattering, quartenary $Al_{71}Pd_{21}Mn_{8-x}Re_x$ quasicrystals were grown. X-ray data confirms that the addition of a fourth element does not alter the quasiperiodicity of the sample. $Al_{71}Pd_{21}Mn_{8-x}Re_x$ quasicrystals of varying Re concentration were synthesized where x had values of 0, 0.08, 0.25, 0.4, 0.8, 2, 5, 6, and 8. Both thermal and electrical transport property measurements have been performed and are reported.

INTRODUCTION

Schlectmann and co-workers first discovered Quasicrystals in the early 1980's.[1] Recently they have been investigated for many applications due to their "forbidden" rotational symmetry.[2] Quasicrystals may be viewed as being structurally ordered but electronically they behave in a manner similar to that of amorphous materials. The structure of these materials has been extensively studied and only recently have large, stable quasicrystals been synthesized. With the synthesis of these materials, thermal and electrical transport of this unique class of materials has been studied. Quasicrystals exhibit electrical transport properties between those of a glass and a semi-metal. Thermal conductivity in these samples is similar to that of a glass (on the order of 1-3 W/m-K at room temperature), and is observed to remain low despite changes in composition or growth conditions. With inherently low thermal conductivity and favorable electrical transport, quasicrystals have the attributes of Glen Slack's "Phonon-Glass, Electron-Crystal" description of a good thermoelectric material.[3]

Quasicrystals exhibit a resistivity behavior that is typical of a semi-metal or in which semimetal-insulator transitions are prevalent ($\rho \approx 1\text{-}100$ Ω-cm; resistivity can be much higher for the AlPdRe system).[4] The electrical resistivity in these systems is observed to increase as the quasicrystalline perfection increases, contrary to Matthiessen's rule ($\rho = \rho_{PHONONS} + \rho_{IMPURITIES} + $

$\rho_{DEFECTS}$).[5] With increasing temperature the electrical resistivity decreases, and through the addition of impurities and defects a method for tuning the electrical resistivity is obtained.

The thermopower of quasicrystals can be explained by investigating the Fermi energy of the quasicrystal.[6, 7, 8] Diffusion thermopower can be defined as $S_D \sim T/eE_F$ where T is the temperature, e is the charge of the carrier and E_F is the Fermi energy. In a quasicrystal, E_F is much smaller than that of a typical metal resulting in a diffusion thermopower that is much larger than those observed in metals. These calculations are not relevant at low temperatures since electron-phonon scattering is not elastic at low and intermediate temperatures and phonon drag thermopower contributions can become predominant.

Recently Macià has theoretically evaluated the potential of quasicrystals for thermoelectric applications and has determined that due to the "spiky" features around the density of states near the Fermi level, a figure of merit of ZT ≈ 1.6 at room temperature is theoretically predicted for some quasicrystals.[9] The largest figure of merit to date in quasicrystals is ZT = 0.25 at T ≈ 550K (or ZT = 0.08 at T ≈ 300K, for the same sample) and agree with Macià's theory for the AlPdMn system.[10]

It is known that the transport properties of quasicrystals are observed to be very sensitive to composition. Small changes or variations in the quality or preparation of samples can greatly influence the thermal and electrical properties of these materials without significantly affecting their thermal conductivity. It has also been shown that small amounts of impurities in quasicrystals may substantially enhance thermopower while decreasing resistivity. With this in mind, a series of $Al_{71}Pd_{21}Mn_{8-x}Re_x$ quasicrystals were synthesized where x = 0, 0.08, 0.25, 0.4, 0.8, 2, 5, 6, and 8 in an effort to investigate the effect of substitutional doping in these materials. The AlPdRe system has been inferred to exhibit a semiconducting nature.[11,12] The hope is that this doping may yield a way in which to enhance thermopower in this system while maintaining reasonable electrical resistivity values. Thermopower in the AlPdMn system (S≤+85 µV/K) and the AlPdRe system (S≤+120 µV/K) are large and positive.[11, 12] Both icosahedral quasicrystalline materials form with comparable levels of Re or Mn, making a solid solution of the two quasicrystalline materials possible.[4]

EXPERIMENTAL PROCEDURE

$Al_{71}Pd_{21}Mn_{8-x}Re_x$ quasicrystals were synthesized both at Clemson University and McGill University. At McGill University, quasicrystals were synthesized using Aluminum (5N purity), Manganese (4N), Palladium (3N) and Rhenium (3N7). These starting materials were etched in acids to remove any oxide layer. Bulk materials were placed in a tri-arc system to homogenize samples a few times. The ingot was subsequently placed in a water-cooled Hukin crucible in a RF induction furnace. This method produces samples with few cracks. With the use of the tri-arc system samples can be turned and remelted several times without venting the chamber, which results in faster sample preparation, in addition to cleaner samples.

At Clemson, samples were synthesized in an argon atmosphere where stoichiometric amounts of elemental powders were measured and mixed in a vibrating mill. The homogenized mixture was pressed at 6000 lbs. The pellet was placed in an arc furnace and the chamber purged. An arc was first established to zirconium and subsequently extinguished to getter any moisture within the arc furnace. An arc was then established to the sample pellet and maintained until the sample was completely melted. The samples were placed into resistive furnaces and annealed. Rectangular samples were cut form each ingot with typical sizes of 6 x 2 x 2 mm^3.

Thermopower and resistivity were measured in a closed cycle refrigerator from 10 K to 300 K. Thermopower was measured using a standard differential technique while resistivity was simultaneously measured using a standard four-probe measurement. Thermal conductivity is measured using an absolute steady-state technique in a closed cycle refrigerator over a temperature range of 10 K to 300 K. Due to the small size of the AlPdRe quasicrystal, this sample was measured in the Parallel Thermal Conductance system.[13] All data was taken using high-speed computer data acquisition software and instrumentation.

Powder X-ray diffraction (XRD) was carried out on the series of $Al_{71}Pd_{21}Mn_{8-x}Re_x$ quasicrystals using Cu Kα radiation on a Scintag XDS-2000 Θ-Θ diffractometer. This diffactometer has a resolution of ~ 0.03°. High angle diffraction data were used to identify crystal structure of the samples as a function of annealing temperature and time. Zero-background quartz was used as the sample support for all X-ray work.

Electron probe microanalysis (EPMA) was used to determine the composition of all the samples. Sectioned samples were polished and adhered to aluminum substrates with double-sided conductive carbon tape. The microprobe data was collected on a Cameca SX-50 using a 15 keV accelerating voltage, a 60 nA beam current, and a minimum (1-2 μm) spot size. Backscattered electron images were obtained from each sample using EMPA.

RESULTS AND DISCUSSION

XRD data indicates that all of our samples of composition $Al_{71}Pd_{21}Mn_{8-x}Re_x$ were primarily quasicrystalline. Figure 1 shows a striking resemblance between Re substitution for x=0 and x=5. Peaks are shifted down in angle due to the addition of Re indicating a larger lattice due to the much larger size of Re. Additional peaks observed in the XRD for x=5 are of the decagonal quasicrystalline phase and were observed to form for compositions where x≥2.[14] The decagonal phase is much less prevalent in the bulk quasicrystal than the icosahedral phase.

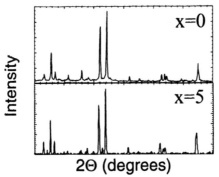

Figure 1: XRD diffraction data for $Al_{71}Pd_{21}Mn_{8-x}Re_x$. Small amounts of crystalline phases were found for x≥2.

Microprobe data on ternary AlPdMn and AlPdRe show samples with quasicrystalline composition and no secondary phases present (as determined from backscattered images). Ternary AlPdMn and AlPdRe are seen to have high thermopower values (Figure 2a). Thermopower in both systems has been measured over a much wider range of values; therefore, few conclusions should be drawn relative to the magnitude of the thermopower. Thermopower in the AlPdMn system increases almost monotonically with temperature, varying as $T^{0.4}$ at high temperatures. Thermopower in the AlPdRe system is seen to increase rapidly at low temperatures and then increases more slowly above 35 K, going to a $T^{0.4}$ dependence at higher temperatures (T>50 K). Below 40 K the conduction in AlPdRe is governed by variable-range hopping (S ~ $T^{1/2}$) which may explain the increase in magnitude of AlPdRe over AlPdMn. It is believed that the diffusion thermopower contribution to the thermopower is much larger in the AlPdRe quasicrystal that in AlPdMn. Phonon drag contributions to the thermopower in the samples are comparable.[15]

While thermopower values in these systems show marked similarities, values of the resistivity in the two systems are very different (Figure 1b). In the AlPdMn system, electrical resistivity increases with temperature until 70 K where the resistivity begins to sharply decrease as is seen in many quasicrystalline systems. Electrical transport in this system can be explained through weak localization and electron-electron interactions.[16] Resistivity in the AlPdRe system appears to conduct through a thermally activated process. Resistivity is initially high (24 mΩ-cm) and falls off rapidly to 7 mΩ-cm. The resistivity ratio, R = ρ (4.2K)/ρ (290K), was determined to be 3.5. Resistivity data does not fit that of a true semiconductor, but fits the theory of variable-range hopping. Variable-range hopping is expected to occur when localized states are present near the Fermi energy and the density of states at the Fermi energy is finite.[17] Plotting ln ρ vs. $1/T^{1/4}$ for AlPdRe, results in a linear fit for the entire temperature range (2K to 300K) indicating that variable-range hopping is indeed the method of conduction in these materials. The activation energy associated with variable range hopping in quasicrystals can be large since the density of states is small ($\Delta E \sim 1/N(E_F)$) resulting in the electrons hopping a short

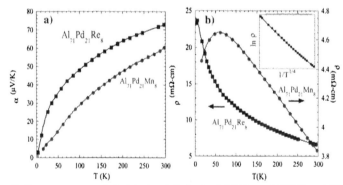

Figure 2: Thermopower and electrical resistivity for AlPdMn and AlPdRe. AlPdRe has a higher thermopower but also a higher electrical resistivity governed by variable-range hopping.

distance. It has been suggested that the AlPdRe system undergoes a metal-insulator transition.[18] The magnitude of the resistivity in AlPdRe is significantly higher than in the AlPdMn quasicrystalline system.

Electron microprobe analysis (EMPA) on the series of $Al_{71}Pd_{21}Mn_{8-X}Re_X$ quasicrystals showed varied results. For compositions where $X \leq 0.25$ the samples are quasicrystalline and no secondary phases are present. These samples have very low porosity. Once $X \geq 0.4$ the bulk (dark regions in Figure 3) of the sample is still of quasicrystalline composition, but secondary phases are present. The lighter regions are areas with higher average atomic number indicating that the light regions have more Re than Mn present. Through EMPA it has been determined that the secondary phase is primarily composed of Al, Pd, and Re. At X=0.4 small, rosette shaped secondary phases are present. It may be noted that some rosettes have fivefold symmetry and are compact and well formed, while other rosettes appear to be more stretched out. At $X = 0.8$, small, light-colored dendrites have formed from the secondary phases with several small remnant rosettes also of secondary phase still present. It appears that the rosettes observed for smaller Re concentrations have elongated from the nucleation site of the rosette and formed the dendritic structure observed for all higher doping of Re. At larger concentrations of Re, the samples become much more porous (porosity increasing from ~ 5% to ~ 50 - 70%) and the structure of the material is that of interpenetrating needles. For these larger Re concentrations the materials are once again primarily single phase and a mixture of Al, Pd, Mn and Re. There is, however, a small dendrite in the center of some of the bulk needles that still has a higher Re concentration than the rest of the material. Interestingly, the secondary phases do not appear as a decagonal phase in the XRD data until $X \geq 2$. These secondary phases for $X < 2$ may have been too small to be detected by XRD. Preferential formation of quasicrystalline AlPdRe over quasicrystalline AlPdMn may explain the observed secondary phases. Also, bulk AlPdRe is known to grow in platelets, which may explain the formation of dendritic structures of the secondary phase.

With the growth of secondary phases in the bulk quasicrystal, the resulting transport properties in the $Al_{71}Pd_{21}Mn_{8-X}Re_X$ system are affected. As the level of Re is increased in the $Al_{71}Pd_{21}Mn_{8-X}Re_X$ system, it is observed that the resistivity remains low, on the order of resistivity in the AlPdMn system (Table 1). It must be noted that for high X, the porosity of these materials increases, leading to an error in the absolute value of resistivity. In order to evaluate the change in resistivity due to the addition of Re, $R = \rho(150K)/\rho(300K)$ was evaluated. It was seen that R~1.15 for $X \leq 0.8$ indicating a resistivity slowly decreasing with increasing temperature. For $2 \leq X \leq 6$, R is practically linear but in the AlPdRe samples (X=8) the slope increases dramatically(R=1.55). The change in R-values indicates the presence of three different regions in which differing conduction occurs.

Due the naturally high resistivity values in AlPdRe, it could be that the conduction takes place preferentially through the less resistive material. Even at large doping of Re, resistivity is seen to remain low. For $X \leq 5$, resistivity increases with temperature, peaks (T < 7 0K) and then continues to decrease with increasing temperature. For $X = 6$, resistivity decreases with temperature over the entire temperature range but does not display the activated behavior characteristic of the AlPdRe quasicrystal or the high resistivity by this family of

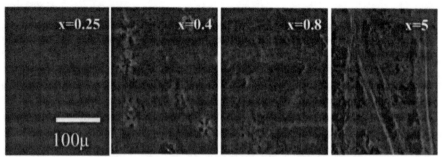

Figure 3: Backscattered images of $Al_{71}Pd_{21}Mn_{8-x}Re_x$ from EMPA show that for higher concentrations of Re, a secondary phase is observed. As X increases, small three or five-fold rosettes develop which appear to elongate into dendrites for higher Re concentrations. For X = 5, the bulk of the material forms a matrix of needles with Re rich dendrites present in the center.

quasicrystals. The variability in the resistivity values is also due to the fact that sample compositions differ in the amounts of Al and Pd present (see Table 1 for EMPA composition).

As the Re level is changed in this quasicrystalline family, changes are observed in the thermopower. AlPdRe and AlPdMn both have reasonably high thermopower values (70 and 64 µV/K respectively). With the addition of X = 0.08 Re, thermopower is seen to reduce to 27 µV/K. With larger Re doping ($0.25 \leq X \leq 0.8$) thermopower is seen to drop even further (~10 µV/K). However, it is noted that with additional Re ($2 \leq X \leq 6$) the thermopower begins to increase again though it never exceeds the thermopower of the parent AlPdRe system (Table 1). Parent materials, AlPdMn and AlPdRe, increase rapidly with temperature and then begin to saturate ($S \sim T^{0.4}$). Low levels of Re doping cause the thermopower to have extremely low thermopower and then begin to sharply increase with increasing temperature ($S \sim T^{1.5}$). Quasicrystals with large amounts of Re ($2 \leq X \leq 6$) exhibit a thermopower behavior that increases almost linearly with temperature ($S \sim T$ indicating diffusion thermopower). The different characteristics of thermopower correspond with the change in slope of the resistivity. The value of the thermopower is affected by small amounts of Re in the AlPdMn system.

Thermal conductivity in quasicrystals is known to be inherently low. Thermal conductivity in AlPdRe ($\lambda \sim 1.3$) is significantly lower than thermal conductivity in AlPdMn ($\lambda \sim 1.9$). When Re is doped into the AlPdMn system, thermal conductivity is seen to have values lower than ternary AlPdMn. With the introduction of Re into the AlPdMn system, secondary phases are formed. It may be that the interface between the different phases induces interface scattering (much like grain boundary scattering) and retards the heat conduction. Given the structure of the interfaces of the secondary phase, there is considerable area covered by the interface to yield an effect. Some part of the reduction in thermal conductivity could also be due to enhanced alloy scattering. Decagonal quasicrystals are periodic in two dimensions and periodic in one dimension and this may be affecting the thermal transport for $X \geq 2$, causing the thermal conduction to increase. [19]

Doping	Microprobe Composition	S (μV/K)	ρ (mΩ-cm)	λ (W/mK)	ρ(150)/ρ(300)
x=0	$Al_{69.2}Pd_{22.6}Mn_{8.2}$	64	2.5	2.2	1.13
x=0.08	$Al_{71.04}Pd_{21.19}Mn_{7.63}Re_{0.14}$	27	3.2	1.95	1.22
x=0.25	$Al_{70.21}Pd_{22.04}Mn_{7.51}Re_{0.23}$	11	3.8	1.8	1.12
x=0.4	$Al_{71.14}Pd_{21.34}Mn_{7.28}Re_{0.24}$	9.5	3.2	1.7	1.19
x=0.8	$Al_{72.02}Pd_{21.33}Mn_{6.14}Re_{0.51}$	12	2	1.85	1.18
x=2	$Al_{69.68}Pd_{20.16}Mn_{7.67}Re_{2.48}$	50	3.26	1.85	1.02
x=5	$Al_{70.09}Pd_{19.49}Mn_{4.52}Re_{5.9}$	60	1.18	2	1.07
x=6	$Al_{71.14}Pd_{21.03}Mn_{1.86}Re_{5.97}$	34	2.15	2.8	1.06
x=8	$Al_{70.1}Pd_{21}Re_{8.9}$	70	7.1	1.3	1.55

Table 1: Thermal and electrical transport properties for $Al_{71}Pd_{21}Mn_{8-x}Re_x$ where x is the nominal doping level of Re at T=300K.

SUMMARY

The addition or substitution of Re to the AlPdMn system, $Al_{71}Pd_{21}Mn_{8-x}Re_x$, drastically changes the electrical and thermal transport properties of these quasicrystalline materials. The results of this study seem to indicate that the addition of Re causes the growth of secondary quasicrystalline phases in the material. It is observed that thermoelectric properties are substantially changed with impurity levels of Re. Thermal conductivity is seen to decrease when small amounts of Re are added to the AlPdMn system, probably due to interface scattering and possibly alloy scattering.

Resistance in the AlPdRe system is much higher than in the AlPdMn system. With the addition of Re to the AlPdMn system, electrical resistivity remains on the order of AlPdMn. The difference in conduction might be attributed to secondary phases. It is believed that due to the high resistance of the AlPdRe, conduction occurs primarily through the AlPdMn matrix.

The mechanisms governing the thermopower in this system are difficult to understand. Thermopower is observed to be very sensitive to doping with small amounts of Re causing thermopower to decrease. Thermopower begins to increase again with enhanced levels of Re. The variable range hopping mechanism observed to govern the thermopower in the AlPdRe quasicrystal is not observed to be the mechanism for conduction in any of the Re doped samples.

There are still many unanswered questions in this system. More investigations are required to observe the changes of the system more closely. Higher doping of Re, $6 \leq X \leq 8$, are necessary to see if further enhancement of the thermopower is observed while thermal conductivity and resistivity remain low. Further studies need to be performed in order to understand the evolution of the secondary phases, which should give insight into how the transport properties are changing. Calculations of the density of states may also give insight into theoretical thermopower calculations. Heat capacity measurements are in progress to help determine these values. Also, the effects of composition on the ternary quasicrystal on the electrical and thermal properties have not been fully investigated. Discerning the various effects of composition, doping, porosity, microstructure, and even quasicrystalline quality, on the transport properties still have to be investigated and unraveled. As such, these materials still remain a challenge, and obviously much work lies ahead.

ACKNOWLEDGMENTS

We acknowledge support for this work from the ONR, the ARO, DARPA and NSF/EPSCoR: (ONR/DARPA #N00014-98-0444 and ARO/DARPA #DAAG55-97-0-0267 and NSF/ETS/96-30167). We would like to thank Jim McGee at the University of South Carolina for his assistance with EMPA. One of us (ALP) would like to acknowledge assistance from the Deans Scholarship Program at Clemson University.

REFERENCES:

1 D. Shechtman, I. Blech, D. Gratias and J.W. Cahn. "Metallic Phase with Long-Range Orientational Order and No Translational Symmetry" Phys. Rev. Lett. **53** (1984) 1951.
2. P.A. Thiel and J. M. Dubois. Materials Today, **2**, 3 (1999) and M. Brown, Technical Insights' *Futuretech*, No **253**, Wiley Press (April 5, 1999)
3. G. A. Slack, in *CRC Handbook on Thermoelectrics*, Rowe ed (1995), ref 2, 407
4. Q. Guo and S. J. Poon. "Metal-insulator transition and localization in quasicrystalline $Al70.5Pd21Re8.5-xMnx$ alloys" Phys. Rev. B. **54** (1996) 12793.
5. S. J. Poon "Electronic Properties of quasicrystals" Adv. Phys. 41 (1992) 303.
6. S. E. Burkov, A. A. Varlamov, and D. V. Livanov. "Electronic transport in quasicrystals" Phys. Rev. B **53** (1996) 11504.
7. J. L. Wagner, B. D. Biggs, and S. J. Poon. "Band-structure effects on the electronic properties of icosahedral alloys" Phys. Rev. Lett. **65** (1990) 203.
8. W. Hansch. "Temperature dependence of the diffusion thermopower in metals" Phys. Rev. B **31** (1985) 3504.
9. Enrique Macià. "May quasicrystals be good thermoelectric materials?" submitted to Physical Review Letters
10. A.L. Pope, T. M. Tritt, M.A. Chernikov, M. Feuerbacher. "Thermal and Electrical transport Properties of the single-phase quasicrystalline material: $Al_{70.8}Pd_{20.9}Mn_{8.3}$" Applied Physics Letter 75 (1999) 1854.
11. F. Morales and R. Escudero. Bulletin of the American Physical Society. 44 No 2 (1999).
12. S.J. Poon, F. S. Pierce, Q. Guo, and P. Volkov. "Transport and Electronic Prperties of Insulating Quasicrystalline Alloys and an Approximant" <u>Proceedings of the 6th International Conference on Quasicrystals (ICQ6)</u> Eds. S. Takeuchi and T. Fujiwara. Singapore, World Scientific. in press.
13. Bartosz M. Zawilski, Roy T. Littleton IV, and Terry M. Tritt, *Measurement of the thermal conductivity of small samples using a parallel thermal conductance technique*, to be submitted to Rev. of Sci. Instrum.
14. K. Hiraga et. al. Jap. Jpur. of App. Phys. **30** (1991) 2028.
15. W. Hansch. Phys. Rev. B **31** (1985) 3504. and A. P. Tsai, et al, Philosophical Magazine Letters, Vol 61, No 1, 9-14,1990.
16. H. Akiyama, et. al., Journal of the Physics Society of Japan. 62 No 2, 639-646 (1993).
17. N. F. Mott. "Conduction in Non-Crystalline Materials" Oxford Science Pub (1987).
18. F. S. Pierce, S. J. Poon, Q. Guo. Science **261** (1993).
19. C. Janot. Quasicrystals: A Primer. Clarendon Press (1997).

Effect of substitutional doping on the thermal conductivity of Ti-based Half-Heusler compounds

S.Bhattacharya[1], V. Ponnambalam[2], A.L.Pope[1], Y.Xia[2], S.J.Poon[2], R.T.Littleton IV[1], T.M.Tritt[1]
1. Department of Physics and Astronomy, Clemson University, Clemson, SC, USA
2. Department of Physics, University of Virginia, Charlottesville, VA, USA

ABSTRACT

Half-Heusler alloys with the general formula $TiNiSn_{1-x}Sb_x$ are currently being investigated for their potential as thermoelectric (TE) materials.[1, 2, 3, 4] These materials exhibit high thermopower (40-250μV/K) and low electrical resistivity values (0.1 - 8mΩ-cm) which yields a relatively large power factor ($\alpha^2\sigma T$) of (0.2 - 1.0) W/m•K at room temperature. The challenge is to reduce the relatively high thermal conductivity ($\approx$ 10 W/m•K) that is evident in these materials. The focus of this research is to investigate the effect of Sb-doping on the Sn site and Zr doping on the Ti site on the thermal conductivity of TiNiSn. Highly doped half-Heusler alloys have shown marked reduction in thermal conductivity to values on the order of 3.5 - 4.5 W/m•K. Systematic determination of thermal conductivity in a variety of these doped materials as well as Sb and Zr doped TiNiSn are presented and discussed.

INTRODUCTION

There has been a renewed interest in thermoelectricity with the possibility of optimizing the electronic and transport properties of both new and existing novel materials. New thermoelectric (TE) materials for applications such as refrigeration and power generation are heavily being investigated. Applications for power generation are of interest to the automotive industry for waste heat recovery for power conversion to enhance fuel efficiency utilizing an environmentally friendly energy source. Refrigeration is used in beverage coolers, cooling of electronics and opto-electronics, as well as temperature stabilization of biological samples. The TE materials and devices are of interest not only because of their reliability and durability, but also because of the technology being environmentally friendly. The efficiency of a TE material is given by the dimensionless parameter ZT, the figure of merit, of the material and is given by[5]:

$$ZT = \frac{\alpha^2 \sigma T}{\lambda}$$

where, α is the Seebeck coefficient or thermopower, σ is the electrical conductivity which is the reciprocal of electrical resistivity (ρ), λ is the thermal conductivity (comprised of the lattice (λ_L) and electrical contributions (λ_E). It is very desirable to have a high figure of merit (ZT $\approx$ 3 - 4). However current state-of-the-art materials have ZT $\approx$ 1 (Bi_2Te_3 has ZT ~ 1 at 400 K)[6]. A "good TE" material should have a large Seebeck coefficient, high electrical conductivity and a low thermal conductivity in order to have a high ZT. Generally, semiconductors or semimetals are chosen for research since they tend to fulfill the above requirements of having high thermopower and a favorable electrical conductivity.

Apart from the current state-of-the-art materials Bi_2Te_3 and SiGe, several new materials like skutterudites and clathrates[7], quasicrystals[8] and pentatellurides[9] have been investigated with the hope of obtaining high ZT values over either new or existing temperature regimes. One new class of materials under investigation for TE applications is the half-Heusler alloys. These materials are intermetallic alloys with the general formula MNiSn (M=Zr, Hf, Ti). These materials show some promise for high temperature TE power generation applications. Half-

Heusler alloys have a MgAgAs[10] type crystal structure, forming four interpenetrating fcc sublattices with Ni sublattice vacant. Heusler alloys (e.g. MNi$_2$Sn) differ from half-Heusler alloys in having the Ni sublattice fully occupied and these materials are also metallic. Half Heusler alloys, on the other hand, are small band gap semiconductors with a gap of (0.1-0.2ev)[11, 12, 13] in the density of states[1]. These intermetallics are known to be primarily of a paramagnetic nature, which make the system even more fascinating. Half-Heusler alloys have very high thermopower (40 - 250μV/K) and yet exhibit reasonably low electrical resistivity (0.1 - 8 mΩ-cm). The challenge in these materials is reducing the high thermal conductivity (≈ 10 W/m•K) that they exhibit. Others have reported similar values. Uher et. al.[1] found the thermal conductivity of ZrNiSn and HfNiSn to be as high as 20W/m•K at room temperature. But the thermal conductivity of isoelectronic Zr$_{0.5}$Hf$_{0.5}$NiSn was found to be as low as 6W/m•K and the thermal transport properties were significantly dependent on annealing conditions. Hohl et. al.[14] found the thermal conductivity of ZrNiSn to be about 9W/m•K and HfNiSn to be about 7W/m•K at room temperature. But the isoelectronic Zr$_{0.5}$Hf$_{0.5}$NiSn had a lower thermal conductivity of about 4.5W/m•K.

The thermal conductivity in this system consists mainly of lattice contribution with a small electronic contribution, in contrast to the more ideal situation where $\lambda_E \sim \lambda_L$ in order to obtain a "good thermoelectric". The focus of this research is to investigate the effect of introducing scattering centers via chemical substitution to reduce the lattice thermal conductivity. Extensive doping with Sb on the Sn site in TiNiSn system has been performed, where phonon scattering mechanisms by lattice disorder or grain boundaries is employed in hopes of reducing the lattice thermal conductivity[4]. A system of TiNiSn$_{1-x}$Sb$_x$ has previously been investigated and is found to have a high room temperature power factor ($\alpha^2 \sigma T$) of (0.2 - 1.0 W/m•K) with power factor of 1.0W/m•K for X = 0.05[15]. Taking this optimal Sb-doping compound TiNiSn$_{0.95}$Sb$_{0.05}$ as our base system, we have measured the electronic and thermal transport properties of a system Ti$_{1-Y}$Zr$_Y$NiSn$_{0.95}$Sb$_{0.05}$ as a function of temperature. In this paper the thermal conductivity of Ti$_{1-Y}$Zr$_Y$NiSn$_{0.95}$Sb$_{0.05}$ and other highly doped systems are reported.

EXPERIMENTAL PROCEDURE

Alloys of different compositions were prepared by arc melting appropriate quantities of elements together. Titanium (4N purity), Zirconium (3N purity), Antimony (5N purity), Tin (5N purity) and Nickel (5N purity) metal powders are mixed together and pressed into a pellet. This pellet was arc melted on a water-cooled copper hearth under argon atmosphere. The resulting button was remelted two or three times after turning upside down to ensure homogeneity. Then the button was wrapped in a Ta foil and sealed in an evacuated quartz tube for annealing. A short term annealing at 900C for 14 hours and a long term annealing at 750C for 1 week were carried out. A small rectangular piece (~2mm X 2mm X 8mm) was cut from each ingot of different compositions to measure resistivity and thermopower.

Resistivity and thermopower are measured simultaneously in a closed cycle helium cryostat from 10 K to 300 K. Samples are mounted on custom designed mounting pucks, which can be directly plugged into the system[16]. Resistivity is measured by a standard four-probe method where current is injected through one pair of leads and voltage is measured using the other pair. The direction of current is then reversed to subtract off the thermals. Thermopower is the negative gradient of voltage over gradient of temperature. The temperature gradient is measured by a 3mil Au-Fe (0.7 at% Fe) vs. Chromel differential thermocouple, which is embedded in two copper blocks soldered to the two ends of the sample. Voltage difference is measured by a different set of voltage leads attached to these Cu blocks when the current is set to zero. Aluminum flux and Ostalloy 244 (In 52%, Sn48%) are used to make contacts with low contact resistance (< 2 Ω). Also proper attention is given to see that the sample is well thermally sunk to the base. High temperature resistivity and thermopower are measured in a similar method using a different probe[17].

Thermal conductivity is measured from 10 K to 300 K using another custom designed sample puck that plugs into the cold finger of a closed cycle refrigerator. Thermal conductivity is measured using a steady state technique.[18] The sample is mounted on a stable temperature copper base. Two #38 Cu-wires are attached to the sample using stycast epoxy on which a 1mil Cn-Cromega-Cn thermocouple is attached. A 120Ω strain gauge is attached with a thin layer of epoxy to the top of the sample to provide power ($P = I^2R$) and the temperature gradient (ΔT) is measured using the thermocouple. The base temperature is stabilized and controlled using two heaters, one coarse control and one fine control heater. After the temperature is stable to within +/- 50 mK, then a power vs. ΔT sweep is performed and then the base temperature is changed to a new value. The data is measured to within an uncertainty of +/- 10% as determined from measuring several standards. Much of this is due to determination of the effective sample length (between thermocouple leads) and cross sectional area. Evacuating the sample chamber to a vacuum of 10^{-6} Torr minimizes heat conduction loss through convection or via gas transfer. The radiation loss is minimized with a series of guards and heat shields as described in reference 18.

RESULTS AND CONCLUSIONS

Thermal conductivity measurements and high temperature resistivity and thermopower measurements have been performed on a series of TiNiSn$_{1-X}$Sb$_X$ samples. Figure 1(a) shows a graph of lattice contribution to thermal conductivity vs. temperature of TiNiSn$_{1-X}$Sb$_X$ for low concentrations of Sb; X = 0.0, 0.005, 0.02, 0.03, 0.05.

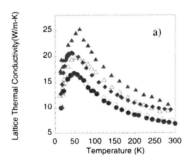
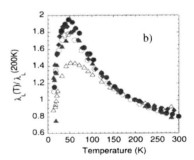

Figure: 1(a) Lattice contribution to Thermal Conductivity (b) Normalized Lattice Thermal Conductivity vs. Temperature of TiNiSn$_{1-X}$Sb$_X$ where X= 0.0 (○), 0.005 (◆), 0.02 (●), 0.03 (▲) and 0.05 (△).

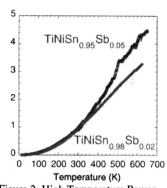
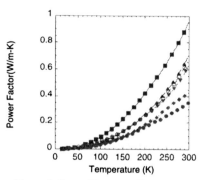

Figure 2: High Temperature Power Factor ($\alpha^2\sigma T$) vs. Temperature of $TiNiSn_{0.98}Sb_{0.02}$ and $TiNiSn_{0.95}Sb_{0.05}$.

Figure 3: Power Factor vs. Temperature for $Ti_{1-Y}Zr_YNiSn_{0.95}Sb_{0.05}$ with Y=0.0(■), 0.25(▲), 0.5(▼), 0.6(♦), 0.75(●),1.0(○).

Pure TiNiSn exhibits a thermal conductivity of about 8 W/m•K at room temperature. Thermal conductivity of TiNiSn measured by Bell Lab group is about 9W/m•K at room temperature, which is in good agreement with our results. For these low concentrations of Sb doping on Sn site of $TiNiSn_{1-X}Sb_X$ with X = 0.005, 0.02, 0.03 and 0.05 the room temperature values of thermal conductivity range from approximately 6 - 15 W/m•K. The results shown in Figure 1(a) are somewhat inconclusive, with the lattice contribution to thermal conductivity actually increasing for small amounts of Sb substitution. The samples have peaks in the thermal conductivity at about 50 - 60 K. Using the Wiedemann Franz law ($\lambda_E = L_0\sigma T$, where L_O = 2.45 x 10^{-8} V^2/K^2 is the Lorenz number), the electronic contributions to thermal conductivity have been calculated. The lattice contributions are calculated by subtracting the electronic part from the total thermal conductivity. Figure 1(b) shows a plot of the lattice thermal conductivity normalized to the values at 200K (where any radiation effects will be low). As seen in this figure, all the samples show very similar temperature dependence. The effect of disorder and scattering is somewhat evident by observing the thermal conductivity at the peak, requiring X ≈ 0.05 before much reduction is evident with magnitudes of thermal conductivity increasing with increased amount of doping. Thus doping TiNiSn with small concentrations of Sb did not significantly reduce the lattice thermal conductivity but recall that the band structure of these materials is very sensitive to small amounts of Sb, with the resistivity reducing an order of magnitude with small Sb concentrations[15].

High temperature resistivity and thermopower measurements are also performed on some of the promising samples in this series. $TiNiSn_{0.98}Sb_{0.02}$ and $TiNiSn_{0.95}Sb_{0.05}$ have power factors ($\alpha^2\sigma T$) of about 3.5W/mK and 4.5W/mK at about 600K as shown in Figure 2. The power factor tends to increase substantially with increasing temperature.

Investigations of the effect of substitutional doping of Zr on the Ti was next performed, but with large concentrations of Zr for Ti. Keeping the optimal concentration of Sb to be 0.05 unchanged, we measured the electronic and thermal transport properties on a new series, $Ti_{1-Y}Zr_YNiSn_{0.95}Sb_{0.05}$ with Y = 0.0, 0.25, 0.5, 0.6, 0.75 and 1.0, i.e., we alloyed the two metals Zr and Ti by varying the amount of Zr at Ti site. A plot of power factor ($\alpha^2\sigma T$) is shown in the Figure 3. The samples with no doping at the Ti or Zr sites have actually the highest power factors, with decreasing power factors with increased Zr doping at Ti site. Thermopowers in this series of samples are negative, ranging from (45-70) µV/K with the highest thermopower being for $TiNiSn_{0.95}Sb_{0.05}$. The resistivity exhibits a semimetallic nature, which increases with increasing temperature. Resistivity values lie between (0.15-0.2) mΩ-cm at room temperature with the highest value for Y = 0.75. Though the half-Heusler alloys possess very high thermopower and favorable electrical resistivity, the main hurdle in these materials is decreasing their high lattice contribution to thermal conductivity. A plot of lattice contribution to thermal conductivity vs. temperature is shown in the Figure 4(a) for a series, $Ti_{1-Y}Zr_YNiSn_{0.95}Sb_{0.05}$ with Y = 0.0, 0.25, 0.5, 0.6, 0.75 and 1.0. As seen in Figure 4(a), the end elements of the series, $TiNiSn_{0.95}Sb_{0.05}$ and $ZrNiSn_{0.95}Sb_{0.05}$ have the highest values of lattice thermal conductivity. For concentrations in between these end members the lattice thermal conductivity is lowered by approximately 50%, most likely due to mass fluctuation or alloy scattering. Figure 4(b) shows a plot of the normalized lattice thermal conductivity, normalized to the values at 200K as before (where any radiation effects will be low). As seen in this figure, the samples appear to show very similar temperature dependence. The overall effect of the substitution is to disorder the lattice, as evident by the large reduction in the thermal conductivity peak, which tells something of the crystal quality, and also reducing the overall thermal conductivity. It could be that even at concentrations of 0.25, 0.5, 0.6 and 0.75, some form of ordering is occurring.

At Y = 1.0 or $ZrNiSn_{0.95}Sb_{0.05}$, room temperature thermal conductivity is observed to be about 16 W/m•K. The temperature dependence exhibited by the curve and the room temperature values are in excellent agreement with the thermal conductivity of ZrNiSn as measured by several groups. The minimal amount of Sb doping in our sample should have little effect.

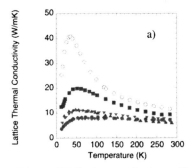 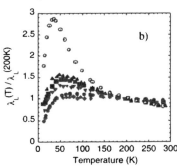

Figure 4(a). Lattice contribution to Thermal Conductivity (b) Normalized Lattice Thermal Conductivity vs. Temperature of $Ti_{1-Y}Zr_YNiSn_{0.95}Sb_{0.05}$ with Y=0.0 (■), 0.25 (▲), 0.5 (▼), 0.6 (♦), 0.75 (●) and 1.0 (☉).

SUMMARY

Substitutional doping of the TiNiSn$_{1-X}$Sb$_X$ and Ti$_{1-Y}$Zr$_Y$NiSn$_{0.95}$Sb$_{0.05}$ systems of the Half-Heusler alloys has been performed. The thermal conductivity of these doping levels has been systematically measured over a wide temperature range (10K<T<300K). The addition of Sb, substituting on the Sn site, greatly affects the electrical transport and leads to a slight increase in thermal conductivity. However, large concentrations of Zr substitution for Ti reduce the overall thermal conductivity by at least 50%. The effect of disorder is evident in the reduction of the low-temperature peak in the thermal conductivity (due to depletion of Umklapp scattering of the phonons) which is reduced by as much as a factor of four for the mixed Zr-Ti compounds. Half-Heusler alloys are currently being investigated for their potential as thermoelectric materials. These materials exhibit very high power factor, however the challenge remains to reduce the relatively high thermal conductivity ($\approx$ 10 W/m•K) observed in these materials.

ACKNOWLEDGMENTS

We acknowledge support at Clemson for this work from the Office of Naval Research, the Army Research Office, DARPA and the National Science Foundation, NSF/EPSCoR: (ONR #N00014-98-0271 and ONR/DARPA #N00014-98-0444 and ARO/DARPA #DAAG55-97-0-0267 and NSF/ETS/96-30167). Research at the University of Virginia was supported by NSF#DMR97-00584.

REFERENCES

1. Uher, C. et at, Phys Rev. B **59**, No.13 (1999) pp.8615-8621
2. Hohl,H.et al , J.Phys.:Condens, Matter Vol.11 No. 7 (1999), pp.1276-1277
3. Browning, V.M. et al , 1998, MRS Symposium Proc., Vol. **545**, pp.403-412
4. Uher, C. et al 1998, MRS Symposium Proc., Vol. **545**, pp.247-258
5. H.J. Goldsmid, Thermoelectric Refrigeration, Prenum Press, 1964
6. Tritt, T.M., "Holey and Unholey Semiconductors", Science Vol. **283**, No. 5403 (1999) pp. 804-805
7. G.S.Nolas, G.A.Slack, D.T.Morelli, T.M.Tritt, A.C.Ehrlich, J.appl.Phys.79 (8),4002(1996)
8. A.L.Pope, T.M.Tritt, M.A.Chernikov, M.Feuerbacher, Appl.Phys.L.,**75**, v.13, pp1854-1856(1999)
9. R.T.Littleton Iv et.al., Appl.Phys.L.,**72**, 2056-8 (1998)
10. Jeitschko W. 1970 Metall. Trans. A1 3159
11. Aliev F.G., et.al.. 1989 Z.Phys. B **75** 167
12. Aliev F.G., Kozyrkov V.V., Moschalkov V.V., Scolozdra R.V. and Durczewski K., 1990 Z. Phys. B **80** 353
13. Cook B.A., Harringa J.L., Tan Z.S. and Jesser W.A. 1996 Proc. ICT'96: 15th International Conference on Thermoelectrics ed. Caillat T et al IEEE Catalog No. 96th 8169 p122
14. Hohl H. et.al., Proc. 1997, MRS Symposium Proc., Vol. **478**, pp.109-114.
15. S.Bhattacharya,V.Ponnambalam, A.L.Pope,Y.Xia,S.J.Poon,T.M.Tritt,"Effect of Sb-Doping on the thermoelectric properties of Ti-based Half-Heusler compounds, TiNiSn$_{1-x}$Sb$_x$", manuscript in progress
16. A.L.Pope, R.T.Littleton,T.M.Tritt, "Apparatus for Rapid Measurements of Resistivity and Thermopower from 10K to 300K, to be submitted to RSI
17. R.T.Littleton IV, J.Jeffries, M. Kaeser, T.M.Tritt, Rev. Sci. Instrum. (in progress)
18. A.L.Pope, B.Zawilski,T.M.Tritt, "Thermal Conductivity Measurements on Removable Sample Mounts", to be submitted to RSI

High Temperature Thermal Conductivity Measurements of
Quasicrystalline $Al_{70.8}Pd_{20.9}Mn_{8.3}$

Philip S. Davis and Peter A. Barnes,
Auburn University, Auburn AL 36849
Cronin B. Vining,
ZT Service Inc., 2203 Johns Circle, Auburn, AL 36830
Amy L. Pope, Bob Schneidmiller, Terry M. Tritt and Joe Kolis,
Clemson University, Clemson, SC 29634

ABSTRACT

We report measurements of the thermal conductivity on a potential high temperature thermoelectric material, the quasicrystal $Al_{70.8}Pd_{20.9}Mn_{8.3}$. Thermal conductivity is determined over a temperature range from 30 K to 600 K, using both the steady state gradient method and the 3ω method. Measurements of high temperature thermal conductivity are extremely difficult using standard heat conduction techniques. These difficulties arise from the fact that heat is lost due to radiative effects. The radiative effects are proportional to the temperature of the sample to the fourth power and therefore can lead to large errors in the measured thermal conductivity of the sample, becoming more serious as the temperature increases. For thermoelectric applications in the high temperature regime, the thermal conductivity is an extremely important parameter to determine. The 3ω technique minimizes radiative heat loss terms, which will allow for more accurate determination of the thermal conductivity of $Al_{70.8}Pd_{20.9}Mn_{8.3}$ at high temperatures. The results obtained using the 3ω method are compared to results from a standard bulk-thermal-conductivity-technique on the same samples over the temperature range, 30 K to 300 K.

INTRODUCTION

Thermoelectric devices are typically used in two distinct ways, either as a refrigerator or as an electric generator. For example, thermoelectric refrigerators can be used to cool electronics at room temperature, while thermoelectric generators are used to generate electricity at high temperatures, ~ 700 – 800 K, on deep space probes. These demands for thermoelectric devices require materials that are "thermoelectrically efficient" at the temperatures of use. Currently, Bi_2Te_3 and $Si_{1-x}Ge_x$ are the "thermoelectrically efficient" materials of choice in these respective applications.

A "thermoelectrically efficient" material is one in which the dimensionless figure of merit, ZT, is a maximum, where

$$ZT = \frac{S^2 \sigma}{\kappa} T \qquad (1)$$

Here S is the Seebeck coefficient, σ is the electrical conductivity and κ is the thermal conductivity. In order to increase the figure of merit, the numerator, which is also called the power factor, $S^2\sigma$, of Equation 1 should be made as large as possible and the denominator should be made as small as possible. We will mainly be concerned with measuring and minimizing the denominator or the thermal conductivity.

QUASICRYSTALS (POSSIBLE THERMOELECTRIC MATERIALS?)

Quasiperiodic structures, or quasicrystals, are non-crystalline materials with perfect long-range order, but with no three-dimensional periodicity ingredient, not even the underlying lattice of the incommensurate structures.[1] Theoretically, the results of quasiperiodicity could lead to interesting new transport properties within these structures. Since quasicrystals do not possess a typical lattice constant, and thus a very large number of atoms in a unit cell, they tend to have very low thermal conductivities. These thermal conductivities are on the order of an amorphous glass and exhibit a glass-like temperature dependence of the thermal conductivity.

The recent discovery[2] of a stable icosahedral phase in the Al-Mn-Pd system, $Al_{70.8}Pd_{20.9}Mn_{8.3}$, has lead to new opportunities in the study of the transport properties of these quasicrystals. Initially most of the studies performed on this material were related to its structual characteristics. Recently there has been some work done on the transport properties of these quasicrystals.[3] These results, suggest that the quasicrystal $Al_{70.8}Pd_{20.9}Mn_{8.3}$ may be a good high temperature thermoelectric.

We are not aware of any data on the thermal conductivity at higher temperature in the range of 300 K to 600 K, for these thermodynamically stable icosashedral AlPdMn quasicrystals. The thermal conductivity results from two different techniques, a standard steady-state temperature gradient method from 30 K to 300 K and a transient AC method "the 3-Omega method" from 30 K to 600 K, will be presented and discussed below.

EXPERIMENT

Quasicrystals were synthesized at Clemson University. Stoichiometric amounts of elemental Al, Pd, and Mn powders were weighed out in an argon atmosphere and subsequently mixed in a vibrating mill. The homogenized mixture was loaded into 1 cm pellet die and pressed to 6000 lbs. The pellet and zirconium were placed separately into an arc furnace, which was evacuated and refilled with argon. An arc was first established to the zirconium to getter any water within the arc furnace to reduce the chance of oxidation of the quasicrystal. The arc was then established to the sample pellet and maintained until the sample was completely melted. The sample was placed into an alumina crucible that was placed inside a quartz tubing vessel. The quartz vessel was placed into a resistive heating furnace and annealed to eliminate secondary phases. The sample was sectioned with a wire saw using boron carbide as a cutting agent. Left over sample sections were ball milled for powder X-ray diffraction measurements.

THERMAL CONDUCTIVITY MEASUREMENTS.

We employed a transient method (3ω-method) to determine the thermal conductivity of the quasicrystal over the temperature range, 30 K to 600 K, then compared this data with the data obtained from a standard temperature gradient method on the bulk crystal. The 3ω-method is a technique that involves reading an AC voltage at a frequency three times the driving frequency of the circuit and 60-80 dB lower than the driving signal. Our lock-in amplifier, with a built-in signal generator, provides the driving signal to the circuit and reads the 3-omega signal. The 3-omega signal is obtained by placing the sample into one arm of a Wheatstone bridge, then reading the difference of the voltage across the sample and the matching variable resistor (Figure 1). The DC output signal from the lock-in, which is a function of frequency, yields data leading to the thermal conductivity. One advantage that this technique has over a standard temperature gradient method in measuring thermal conductivity, is the insensitivity of the 3-omega method to radiation loss effects at high temperature.[4]

The sample was prepared as follows. The surface of the sample was polished to a mirror finish, and then a thin film of polyimide, several microns thick, was applied to the prepared surface of the quasicrystal. This film electrically isolates the heater/thermometer line from the substrate, which allows us to assume that all the power is input in the heater/thermometer line and not the sample. On top of the polyimide the heater/thermometer line is deposited through a shadow mask.

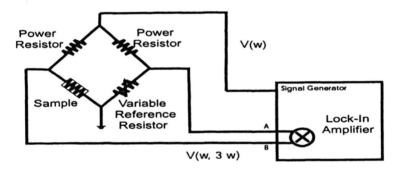

Fig.1 Schematic diagram of the circuit used in the 3-Omega method.

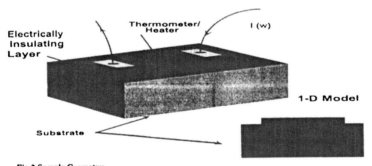

Fig.2 Sample Geometry

The material used in the line is nickel, because it has a reasonably large coefficient of resistance. This property of nickel provides for a more sensitive measure of the surface temperature oscillations that arise during the experiment (Figure 2).

The data analysis for our sample in principle requires the solution of Fourier's heat conduction Law for both layers with appropriate boundary conditions. Since our polyimide insulating layer is very thin, several microns, it is thermally bridged at low frequencies. In other words, since the thermal waves penetrate ~80-100 times this depth at low frequencies, it is assumed that we have a one layer problem. The derivation[5] of the functional dependence of the 3ω voltage on frequency for this one-dimensional problem is summarized below. The third harmonic AC voltage measured on the sample is

$$V_{3\omega,rms} = \frac{p V_{sample,rms} dR/dT}{4 R_{sample} \sqrt{2 \lambda \rho C_p}} \omega^{-1/2} \qquad (2)$$

Here ρ, C_p, and λ are the density, heat capacity and thermal conductivity of the material respectfully. p and dR/dT are the input power density and coefficient of resistance of the heater line. Thus, to obtain the thermal conductivity of the sample we plot the 3ω voltage versus the inverse square root of frequency and then take the slope of this line.

In Figure 4, the 3ω voltage, $V_{3\omega}$ is plotted as a function of the inverse square root of frequency, $\omega^{-1/2}$ at constant temperature, T. From this graph it is clear that there two distinct slopes, each of which contain information on the sample. In the high frequency limit the slope contains the physical information of the polyimide layer and in the low frequency limit the slope contains the information on the $Al_{70.8}Pd_{20.9}Mn_{8.3}$ quasi-crystal. With the data for heat capacity and density it is straightforward to obtain the thermal conductivity of the sample at each temperature.

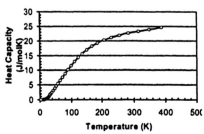

Fig. 3 Quasicrystal Heat Capacity. Above 400K the value is assumed to saturate at 3R

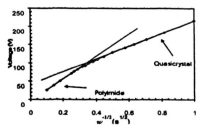

Fig.4 3-Omega voltage. Asymptotic lines provide the slope and therefore the thermal conductivity of the materials. Low frequency limit gives the thermal conductivity information of the quasicrystal. High frequency limit gives the information on the polyimide insulating layer.

The heat capacity data is shown in Figure 3. The density was taken as 0.069 atoms/A^3, at room temperature and an assumed linear coefficient of expansion of 1.2 x 10^{-5} (1/K). There are many possible errors involved in making thermal conductivity measurement with these two methods. Potential errors in the 3ω are: an error associated with the one-dimensional modeling of the thermal waves and a small error occurs in assuming that the electrically insulating layer is thermally bridged. The errors associated with these assumptions are thought to be relatively small, ~ 5-10%. The remaining error in the value of the thermal conductivity calculated from the 3ω method comes from the uncertainty in the various parameters in Equation 2.

The errors involved with temperature gradient methods are fairly well known, but we will enumerate them here again. Radiation loss is probably the best known cause of error in the temperature gradient method. This error is large at high temperatures but negligible at low temperatures. Care must be taken to minimize thermal conduction of heat through the various wires connecting the sample to the measurement apparatus. Finally there is an error in the measurement of the dimensions of the sample. All these errors combined contributed an uncertainty in the measurement in the low temperature regime of about 10%. In Figure 5 the thermal conductivity results are shown for the $Al_{70.8}Pd_{20.9}Mn_{8.3}$ quasicrystal over the temperature range, 10K-300K. In Figure 5, thermal conductivity results obtained from the 3ω-method and from the standard temperature gradient technique are shown for the temperature range, 10K-300K. The 3ω-data is ~10-15% higher than the temperature gradient method data over this range. In Figure 6, the thermal conductivity results are shown for the temperature range, 300-600 K. The results for the 3ω method generally agree with the Wiedermann-Franz approximation, but with a slightly higher slope and the same ~10-15% offset. The Wiedemann-Franz approximation is done assuming that the electronic contribution increases as $\sigma = \sigma_o + \alpha T$ and that the lattice portion is constant. The reasons for the discrepancy in the data between the Wiedemann-Franz approximation and 3ω over the low temperature range can be attributed to the uncertainties in the methods employed. Janot has predicted that thermal conductivity should increase as $T^{1.5}$, above room temperature.[6] Our data does not agree with this theory.

Another possible explanation for the difference in the data originates in the structure of the material itself. The technique used to synthesize the quasicrystal leaves many small voids in the sample. All of these voids take part in the conduction of heat in the standard temperature gradient method. Since the 3ω technique only samples a thin section at the surface of the sample, which has been polished to a mirror-like finish, it is thought that a smaller percentage, by volume, of these voids are being sampled, thereby increasing the thermal conductivity at least at low temperatures. Since, at low temperatures the data from the two methods matches in temperature dependence, it is thought that the effect of these voids is temperature independent. At high temperatures the effect of these voids is not clear.

Further investigation of high temperature thermal conductivity measurements in quasicrystalline systems is necessary. It is necessary to measure the high temperature electrical conductivity so that the electronic contribution can be determined. With this data, the assumptions of a constant lattice contribution and a linearly increasing conductivity can be determined. We have seen, however, that thermal conductivity increases at low temperature, peaks, experiences a flat plateau before it begins rising at higher temperatures. It is also

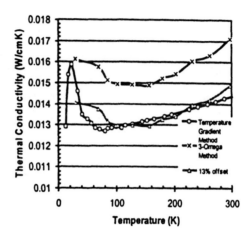

Fig.5 Low temperature thermal conductivity. The two methods coincide with a 13% offset.

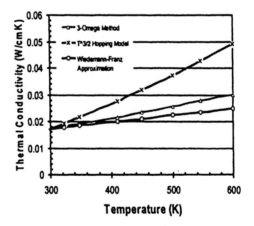

Fig.6 High Temperature thermal conductivity. The 3-omega data follows the Wiedemann-Franz approximation with a slightly stronger temperature dependence. The hopping model thermal conductivity with a $T^{3/2}$ dependence is not seen.

encouraging to observe very good agreement of the two techniques within the overlap region of temperature, with each technique also yielding the same temperature dependence and similar absolute values.

ACKNOWLEDGEMENTS

Auburn would like to thank the Army Research Office for their grant (#DAAG 55-97-10010), which is being monitored by Dr. Jack Rowe. Clemson would like to acknowledge support for this work from the Office of Naval Research, the Army Research Office, DARPA and NSF/EPSCoR: (ONR #N00014-98-0271 and (ONR/DARPA #N00014-98-0444) and (ARO/DARPA #DAAG55-97-0-0267 and NSF/ETS/96-30167). One of us (ALP) acknowledges support form a Clemson University Dean's Scholars Award.

REFERENCES

[1] C. Janot, "Quasicrystals: A Primer", 2nd ed. (Oxford University Press, Cambridge, 1994), p.1
[2] C. Beeli, H. U. Nissen, and J. Robadey, Philos. Mag. Lett. **63**, 87 (1991)
[3] A.L. Pope, R.T.Littleton, J. Jeffries, T.M. Tritt, M. Feuerbacher, R. Gagnon, S. Legault, J. Strom-Olsen, "18th International Conference on Thermoelectrics", eds. A.Ehrlich, p. 417-420, (1999)
[4] D.G. Cahill and R.O. Pohl, Phys Rev. B **35**, 4067 (1987)
[5] J. Foley, Masters Thesis, Auburn University, (1999)
[6] C. Janot, Phys. Rev. B **53**, 181 (1996)

TE Theory

Theoretical Evaluation of the Thermal Conductivity in Framework (Clathrate) Semiconductors

Jianjun Dong[1], Otto F. Sankey[1], Charles W. Myles[2], Ganesh K. Ramachandran[3], Paul F. McMillan[3], and Jan Gryko[4]
[1]Department of Physics and Astronomy, Arizona State University, Tempe, AZ 85287,
[2]Department of Physics, Texas Tech University, Lubbock, TX 79409,
[3]Department of Chemistry, Arizona State University, Tempe, AZ 85287,
[4]Department of Physical and Earth Sciences, Jacksonville State University, Jacksonville, AL 36265.

ABSTRACT

We have calculated the room temperature thermal conductivity in semiconductor germanium clathrates using statistical linear-response theory and an equilibrium molecular dynamics (MD) approach. A key step in our study is to compute a realistic heat-current $J(t)$ and a corresponding auto-correlation function $<J(t)J(0)>$. To ensure convergence of our results and to minimize statistical fluctuations in our calculations, we have constructed large super-cell models (2944 atoms) and have performed several independent long time simulations (>1,500 ps in each simulation). Our results show an unexpected "oscillator" character in the heat-current correlation function of the guest-free Ge clathrate frameworks. This is absent in the denser diamond phase and other with simple structural frameworks. We seek to interpret these results using lattice dynamics information. A study of the effects of the so-called "rattling" guest atoms in the open-framework clathrate materials is in progress.

INTRODUCTION

Recent experiments reveal that certain *crystalline* forms of Ge-based clathrates (e.g. $Sr_8Ga_{16}Ge_{30}$) have an abnormally low, *glass-like* thermal conductivity (κ) [1]. This makes these materials very promising for the design of high ZT thermoelectric materials. However, the mechanism of this reduction of κ is still not fully understood. Slack and co-workers have suggested [2] that the small guest atoms (e.g. Sr) located inside the open framework (e.g. clathrate) behave like loosely bound "rattlers", and that their low-frequency rattling phonon modes scatter the heat-carrying acoustic phonons of the framework. No quantitative theoretical calculation has yet been reported to support this conjecture. We have recently performed a detailed first-principles theoretical study of the lattice dynamics of alloyed Ge clathrates, which showed that interaction between the framework acoustic phonon branches and the Sr-related rattling modes is possible because of avoided crossing [3]. These results provide the lattice dynamics basis to interpret the Slack-proposed rattler scattering mechanism.

It is, however, possible that the large reduction in κ is caused by multiple mechanisms. The novel open-framework of the clathrates may also be relevant. In a broader context, the conventional transport formalism in condensed matter physics has been developed based on our understanding of simple crystals, such as diamond. This formalism has not been extensively tested with complex materials systems, such as the framework semiconductors.

The unique character of clathrate framework phonon spectra has been discussed in our previous study [4]. Here, we repeat the lattice dynamics study with a simple classical Tersoff potential [5]. The phonon spectra calculated using this approach shown in Fig. 1. We find semi-quantitative agreement between the Tersoff potential results and those of our previous first-principles study [4]. This indicates that the simple Tersoff potential captures the fundamental physics we are interested in here. Since both (guest-free) Ge clathrates and diamond phase Ge contain 100% four-fold coordinated Ge atoms, the phonon properties of the Ge_{46} clathrate are comparable with those of diamond phase Ge. The spectrum of the diamond phase (2-atoms per primitive cell) (Fig. 1a) is well known: three acoustic branches in the low and medium frequency region and three optic branches in the high frequency region. If a 64-atom super-cell is used instead, the spectra can be easily interpreted in terms of "Brillouin zone folding" (Fig. 1b). However, the framework clathrate clearly exhibits some new features (Fig. 1c), besides those similar to zone folding effects. The optical branches in Ge_{46} (Fig. 1c) are "flatter" than the folded acoustic branches in diamond phase Ge (Fig. 1b), which means that they are less effective for heat transport. Some optic modes in Ge_{46} have relatively larger dispersion, and can be approximately traced as folded acoustic modes. These "acoustic-like" phonon modes are "resonant" with other optic modes. Because of the avoided crossing effect, these branches bend flat at the "crossing points" and some small energy gaps are created. The details about whether and how these unique features in the phonon spectra alter the transport properties are not clear.

In this study, we wish to achieve a better understanding of the heat conduction mechanism in framework clathrate materials and to try to gain insights into the possible "tuning" of the thermal conductivity of such materials in order to optimize their thermoelectric properties. As a first step, we limited our study to Ge diamond and the pure (guest-free) Ge_{46} clathrate. We find an unexpected "oscillator" feature in the time-correlation function of the heat current in the Ge_{46} clathrate framework and we interpret this result in the context of lattice dynamics. The study of effect of guest atoms on thermal conductivity is in progress, and will be reported elsewhere [6].

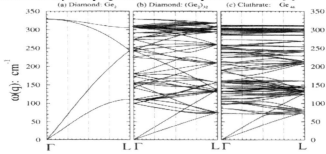

Fig.1 The Tersoff potential calculated phonon dispersion $\omega(\vec{q})$ of (a) Ge diamond in a 2-atom primitive FCC unit cell, (b) Ge diamond in a 64-atom simple cubic super-cell, and (c) Ge_{46} clathrate (type-I simple cubic framework). Γ and L are the labels of (0,0,0) point and $\pi/2L$ (1,1,1) point respectively, where L is the lattice size of the cubic unit cell.

COMPUTATIONAL METHODOLOGIES

In our current study, the tensor thermal conductivity (κ) is calculated based on the Kubo formula of statistical linear-response theory [7],

$$\kappa_{\alpha\beta} = \frac{V}{K_B T^2} \int_0^\infty <J_\alpha(t) J_\beta(0)> dt, \quad \ldots (1)$$

where J_α (or J_β) is the heat current (thermal energy flux) along the α (or β) Cartesian axis at the equilibrium temperature T, and $<J_\alpha(t) J_\beta(0)>$ is the corresponding time correlation function. V is the volume of the system, and K_B is the Boltzmann constant. Since both the diamond and clathrate structures are cubic, κ in our study is a scalar and the calculated results are averaged over three Cartesian directions.

We use equilibrium molecular dynamics (MD) simulations to evaluate the heat current function $\vec{J}(t)$, which is [8],

$$\vec{J} = \frac{1}{V}(\sum_i \vec{v}_i E_i + \sum_{ij}(\vec{r}_j - \vec{r}_i)(\vec{v}_j \frac{\partial E}{\partial \vec{r}_j})), \quad \ldots (2)$$

where E_i is the total energy of the ith atom, and $\vec{r}_i$ and $\vec{v}_i$ are the position and velocity vectors of the ith atom respectively.

The advantage of this MD approach is that anharmonic phonon scattering is naturally built into the interaction potentials, and no further assumption is required. Therefore, it is simple and straightforward. A disadvantage of the method is that it is classical and is expected to be less accurate at low T. Similar implementations of this approach have been adopted to study the thermal conductivity of crystals with point defects [9], amorphous materials [10], and hydrate clathrates [11]. There are several factors that significantly influence the quality of such calculations. (1) The time correlation functions of heat current decay slowly for most crystalline materials, and the integration in Eq. (1) shows poor convergence. We have found that to obtain a good ensemble average, the number of simulation steps must be of the order of a million. (In our studies, a time step of 1fs is used.) This rules out the possibility of using very accurate yet computationally intensive, first-principles MD method. In this study, we adopt a simple Tersoff potential [5], and believe it is a reasonable approximation for addressing the fundamental physics of the transport properties. (2) A large super-cell model has to be constructed in the MD simulation to have a realistic sampling of the k-point phonon modes. A small size model (e.g. a 2x2x2 super-cell model) may introduce finite-size artifacts.

THERMAL CONDUCTIVITIES: DIAMOND vs. FRAMEWORK CLATHRATE

We first studied the thermal conductivity of the diamond structure Ge crystal at room temperature ($T=300K$). It can be used to test our numerical implementation by comparison with experiments and to compare with those of the framework semiconductors.

We start with a 512-atom super-cell model of Ge diamond at zero temperature. Then, we raise the temperature to 300K, and equilibrate the system, first under constant temperature, then under constant energy conditions for about 80,000 time steps at each stage. After this equilibration period, we allow it to evolve (total energy conserved) using

Newton dynamics for 1,500,000 time steps (1,500 ps). We calculate the heat current vectors at each simulation time step using Eq. (2). The time averaged heat current correlation functions (Eq. (2)) are calculated over a 300 ps range using the 1,500,000 sampled MD simulation results. We find that in each of calculations described above, statistical fluctuations in the correlation function are not negligible for long times ($t > 100$ ps) and that multiple calculations starting from different configurations must be averaged to minimize statistical errors. In the study of Ge diamond, nine independent simulations were performed. The averaged normalized correlation function (divided by $<J(0)J(0)>$) is plotted in Fig. 2a.

As expected, the heat current correlation function in Ge diamond decays for long time. It is not totally surprising that this decay in Ge diamond does not strictly follow an exponential law. However, an approximate fitting to a $\beta^*\exp(-t/\tau)$ function gives an effective lifetime parameter $\tau \approx 50 - 60$ ps (depending on the choice of fitting range). We also notice that the normalized correlation function rapidly drops from unity to about 0.4 within less than 0.1 ps. This rapid lose of correlation is not fully understood. However, it is seen in other studies such as in crystalline SiC [9].

Using Eq. (1), we find that thermal conductivity, $\kappa_{diamond}$, of Ge diamond at 300K is about 114.5 $Wm^{-1}K^{-1}$ at 300K, which is about twice the experimental measured value of 62 $Wm^{-1}K^{-1}$ [12]. Nevertheless, the agreement is reasonable since we have not explicitly fit any anharmonicity parameters. Also our calculations assume a perfect single crystal with no defects or grain boundaries. The calculated results reflect an upper-bound for the thermal conductivity.

The thermal conductivity of Ge_{46} clathrate is calculated using the same numerical techniques. The averaged heat current time correlation function is shown in Fig. 2b. In the present study, a large 2944-atom model (a 4x4x4 super-cell, with a unit cell of 46 atoms) is used in the MD simulations, and the final results are averaged over 3 independent 1,500,000 step simulations.

We obtain the unexpected results that the correlation function for the Ge clathrate does not follow a simple decay function; rather it decays like a damped oscillator. Even though the envelope of the correlation function in the Ge clathrate is remarkably similar to that of the Ge diamond phase, the cancellation due to these oscillations in the time integration in Eq. (1) significantly reduces the thermal conductivity.

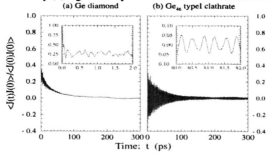

Fig.2 The normalized time auto-correlation functions of (a) Ge diamond, and (b) Ge_{46} clathrate. The results are calculated using equilibrium MD simulations over a time range of 1,500 ps.

Therefore, we find a much smaller thermal conductivity in the pure (guest-free) Ge clathrate than in diamond phase Ge. Our theoretical $\kappa_{clathrate}$ at 300K is 12.8 Wm^{-1}K^{-1}, almost 1/9 of the calculated $\kappa_{diamond}$ at the same temperature. This is the most significant result of our study.

No pure Ge$_{46}$ clathrate has yet been synthesized. The most relevant experimental data is the thermal conductivity of metal-doped, alloyed Ge clathrates, such as Sr$_8$Ga$_{16}$Ge$_{30}$, Eu$_8$Ga$_{16}$Ge$_{30}$, and Sr$_4$Eu$_4$Ga$_{16}$Ge$_{30}$, for which κ ranges from 0.6 to 1.0 Wm^{-1}K^{-1} [1]. Experiments find that κ for compound Ge clathrates containing guests is reduced by a factor of ≈ 60 compared to diamond Ge. Measurements do not address what proportion of this decrease is due to the guest "rattlers" and what proportion is due to the framework itself. Our calculations suggest that the framework itself (without guest atoms and alloy disorder) reduces the thermal conductivity by a factor of 9. From this we conclude that the framework and the guest contribute comparable amounts to the reduction of thermal conductivity in the experiments.

In order to better understand the origin of the significant differences in the time correlation functions of the Ge diamond and clathrate structures, we have calculated the (frequency domain) power spectra (Fig.3). We should point out that the "wiggles" at short time in the Ge diamond correlation function (Fig. 2a) are not random noise since they barely change when the function is averaged over 9 independent calculations. These "wiggles" show their signatures in the non-zero frequency part of the power spectra (Fig. 3a). These frequencies are within the phonon frequency region. However, the weight of non-zero frequency components is much smaller than the dominant weight of the zero frequency components.

The most important changes in the power spectra (Fig. 3b) of the Ge$_{46}$ correlation function are the significant decrease of in the zero frequency peak height and that several relatively high peaks in the finite-frequency region (again, within the phonon frequency range) appear. The large peaks explain the oscillator features in the time correlation function. However, the physical origin of the peaks is not clear. The two dominant peaks are located near 96 cm^{-1} and 137 cm^{-1} respectively. On examining the phonon spectrum in detail (Fig. 1c) we have found four different phonon eigen-modes near these two frequencies, which satisfy the energy and momentum conservation conditions for U-process phonon scattering. The existence of such four-phonon scattering is plausible. However, we cannot explain why this phonon scattering gives such a large weight in the power spectrum; presumably many other phonon scattering processes are also possible.

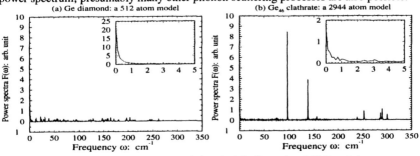

Fig.3 Power spectra of time auto-correlation of (a) Ge diamond, and (b) Ge$_{46}$ clathrate.

CONCLUSIONS

The room temperature thermal conductivity of Ge diamond and Ge_{46} clathrate materials has been calculated using the Kubo formula and equilibrium MD simulation techniques. Theory shows that the intrinsic thermal conductivity κ in the framework clathrate semiconductor is only about 1/9 of $\kappa_{diamond}$. This reduction in the intrinsic thermal conductivity $\kappa_{clathrate}$ can be understood in terms of the rapid oscillations found in the time heat current correlation function. A frequency analysis shows that the oscillation frequencies of the correlation function lie within phonon frequency range. This suggests that heat current in the clathrate strongly couples with several specific phonon modes. The nature of this phenomenon needs further analysis. The effect of guest rattlers has been neglected in this report, and will be reported elsewhere.

ACKNOWLEDGEMENTS

This work is supported by the NSF-ASU MRSEC (Grant No. DMR-96-32635) and the NSF (Grant No. DMR-99-86706). One of the authors (JD) would like to thank Dr. J.L. Feldman (Naval Research Lab) and Dr. J.S. Tse (NRC, Canada) for insightful discussions. CWM thanks the Department of Physics and Astronomy at Arizona State University for their hospitality while a portion of this work was done.

REFERENCES

1. G.S. Nolas, J.L. Cohn, G.A. Slack, and S.B. Schujman, *App. Phys. Lett.* **73**, 178180 (1998); J.L. Cohn, G.S. Nolas, V. Fessatidis, T.H. Metcalf, and G.A. Slack, *Phys. Rev. Lett.* **82**, 779-782 (1999).
2. G.A. Slack, in CRC Handbook of Thermoelectrics, ed. M. Rowe (CRC Press, Boca Raton, FL, 1995), pp. 407.
3. J. Dong, O.F. Sankey, G.K. Ramachandran, and P.F. McMillan, *J. Appl. Phys*, in press (2000).
4. J. Dong and O.F. Sankey, *J. Phys.: Cond. Matt.* **11**, 6129-6145 (1999).
5. J. Tersoff, *Phys. Rev. B* **39**, 5566-5568 (1989).
6. J. Dong and O.F. Sankey, unpublished.
7. R. Kubo, *Rep. Prog. Phys.* **29**, 255 (1966).
8. R.J. Hardy, *Phys. Rev.* **132**, 168-177 (1963).
9. J. Li, L. Porter, and S. Yip, *J. Nuc. Mater.* **255**, 139-152 (1998).
10. Y.H. Lee, R. Biswas, C.M. Soukoulis, C.Z. Wang, C.T. Chan, and K.M. Ho, *Phys. Rev. B* **43**, 6573-6580 (1991); M.D. Kluge, J.L. Feldman, and J.Q. Broughton, Molecular dynamics simulations of thermal conductivity in insulating glasses, Phonon Scattering in Condensed Matter VII, ed M. Meissner and R.O. Pohl (Springer-Verlag, 1993) pp.225-226.
11. R. Inoue, H. Tanaka, and K. Nakanishi, *J. Chem. Phys.* **104**, 9569-9577 (1996).
12. G.A. Slack and C.J. Glassbrenner, *Phys. Rev.* **120**, 782 (1960).

ELECTRONIC STRUCTURE OF $CsBi_4Te_6$

P. Larson*[†], S.D. Mahanti*[†], D-Y Chung**[†], and M.G. Kanatzidis**[†]
*Department of Physics and Astronomy, **Department of Chemistry, and
[†]Center for Fundamental Materials Research,
Michigan State University, East Lansing, MI 48824

ABSTRACT

Recently, $CsBi_4Te_6$ has been reported as a high-performance thermoelectric material for low temperature applications with a higher thermoelectric figure of merit (ZT $\sim$ 0.8 at 225 Kelvin) than conventional $Bi_{2-x}Sb_xTe_{3-y}Se_y$ alloys at the same temperature. First-principle electronic structure calculations within density functional theory performed on this material give an indirect narrow-gap semiconductor. Dispersions of energy bands along different directions in k-space display large anisotropy and multiple conduction band minima close in energy, characteristics of a good thermoelectric material.

INTRODUCTION

In recent years many new materials have been synthesized and their physical properties investigated in search of enhanced thermoelectric (TE) properties.[1] Recently, $CsBi_4Te_6$ has been reported to have a larger figure of merit (ZT $\sim$ 0.8 at 225 Kelvin) than conventional $Bi_{2-x}Sb_xTe_{3-y}Se_y$ alloys near this temperature[2]. $CsBi_4Te_6$ and another ternary compound $BaBiTe_3$ belong to a class of materials whose crystal structures are formed from Bi_2Te_3 blocks connected by either Bi-Bi or Te-Te bonds with planes of Cs or Ba ions separating the two-dimensional Bi-Te networks.(Figure 1) Electronic structure calculations performed on Bi_2Te_3 and $BaBiTe_3$[3, 4] have led to a detailed understanding of the gap structure, bonding, and transport properties of these materials. Therefore, a similar calculation for $CsBi_4Te_6$ promises to provide insights into this new material as well.

CRYSTAL STRUCTURE AND METHOD

The crystal structure $CsBi_4Te_6$ (Figure 1) contains one-dimensional (1D) $[Bi_4Te_6]$ lath-like ribbons running parallel to the growth axis (normal to the plane of the page) with width and height of 43 a.u. and 23 a.u. respectively[2]. These $[Bi_4Te_6]$ blocks containing 40 atoms in the unit are connected by Bi-Bi bonds, bonds not usually seen in chalcogenides[2]. (The Bi atoms are represented as small filled circles, the Te atoms as small open circles, and the Cs atoms as large filled circles.) The conventional unit cell of this material would be monoclinic (Space group: $C\,2/m$ (#12)) with 44 inequivalent atoms/unit cell (88 total atoms/unit cell) (Figure 1). [A = 97.425 a.u., B = 8.264 a.u., C = 27.424 a.u., $\alpha = 90°$, $\beta = 101.438°$, $\gamma = 90°$]. Since the computation time in electronic structure calculations varies approximately as the cube of the number of electrons of the inequivalent atoms in the unit cell, an alternate triclinic cell (Space group: P-1) with 22 inequivalent atoms/unit cell (44 total atoms/unit cell) was used instead in our electronic structure calculations. (Figure 2a) [A = 49.233 a.u., B = 49.233 a.u., C = 27.424 a.u., $\alpha = 101.438°$, $\beta = 101.438°$, $\gamma = 9.693°$]. Figure 2b gives the Brilloun Zone for the triclinic cell where the ΓX direction is parallel to the growth axis, ΓV is parallel to Bi_2Te_3 network connected by the Bi-Bi bonds, and ΓZ is nearly perpendicular to the Cs layers.

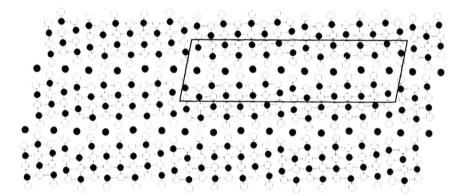

Figure 1. Crystal structure of $CsBi_4Te_6$. The monoclinic unit cell is indicated by the box.

Electronic structure calculations were performed using the Full-Potential Linearized Augmented Planewave (FLAPW) method[5] within density-functional theory (DFT)[6] using the generalized gradient approximation (GGA) of Perdew, Burke, and Ernzerhof[7] for the exchange and correlation potentials. The calculations were performed using the WIEN97 package.[8] The atomic radii were kept constant at 2.8 a.u. to minimize the regions between the spheres. Convergence of the self-consistent iterations was performed with 13 k-points in the reduced Brillouin zone to within 0.0001 Ry with a cutoff between valence and core states of -2.0 Ry. Scalar relativistic corrections were included in the calculation along with spin-orbit interactions which were included in a second variational procedure.[9]

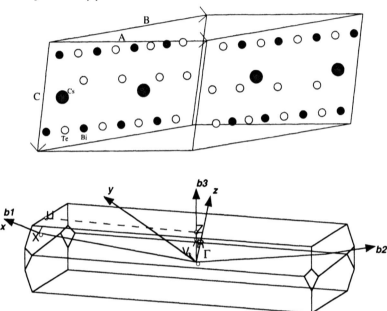

Figure 2. a) Reduced triclinic unit cell of CsBi$_4$Te$_6$ (the bonds are not shown) and b) its Brillouin zone

RESULTS

The band structure obtained from this calculation shows an indirect gap semiconductor with a band gap of approximately 0.37 eV. The direct band gap is at the Γ point and is slightly larger, about 0.4 eV. The band gap in this material has not been measured directly, but has been estimated as 0.11 eV.[2] While spin-orbit interaction effects played a significant role in moving the bands in Bi$_2$Te$_3$ and in the final gap structure, they appear to be less important in the two ternary systems, less so in CsBi$_4$Te$_6$. In BaBiTe$_3$, a direct band gap of about 0.4 eV changed to an indirect band gap of 0.26 eV.[3] Why spin-orbit interaction plays a less important role in CsBi$_4$Te$_6$ needs further study.

The band structure for CsBi$_4$Te$_6$ along different symmetry directions is given in Figure 3. The top of the valence band occurs at the Γ point while the bottom of the conduction band is at a general point in the Brillouin zone (denoted as C*). We also find several subsidary minima in the lowest conduction band whose energies are slightly above that of C*. They appear along the ΓZ, RV, and ZR directions. While the symmetry of this crystal is low, which means the degeneracy of each off-axis minimum is low, there exist several such minima in the lowest conduction band which will contribute to the transport. By a simple argument[3], one can show that S could increase for multiple hole or electron pockets which then increase the dimensionless figures of merit, $ZT = \sigma S^2 T/(\kappa_L + \kappa_e)$. This may suggest a high ZT value for the electron-doped $CsBi_4Te_6$. However, the nondegenerate valence band maximum at the Γ point cannot explain the observed high ZT values in hole-doped samples, which may be explained by the large anisotropy of the charge transport (discussed below). There exists a peak in the valence band between UZ approximately 0.2 eV below the top of the valence band which may also contribute to the transport in the hole doped samples at higher temperatures.

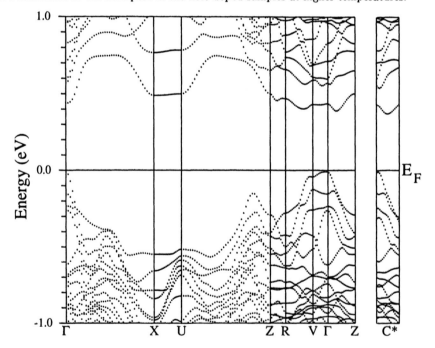

Figure 3. Band structure of $CsBi_4Te_6$

Dispersion along ΓX (along the growth direction of the crystal) and ΓZ (through the Cs$^+$ layer) appears to be much larger than the dispersion along ΓV (through the Bi$_2$Te$_3$ network connected by B-Bi bonds). Further analysis is necessary to understand the importance of these dispersions on the electron- and hole-doped transport in CsBi$_4$Te$_6$ and determine what roles these play in enhancing the thermopower.

An orbital analysis of the electronic states shows that the top of the valence band at the Γ point consists mainly of Te 5p orbitals, very similar to the results found for Bi$_2$Te$_3$ and BaBiTe$_3$.[3] The bottom of the conduction band consists mainly of the Bi 6p orbitals, primarily coming from the atoms comprising the Bi-Bi bond which connects the 1D [Bi$_4$Te$_6$] laths together. This is similar to BaBiTe$_3$ where the bottom of the conduction band is formed (along with Bi 6p orbitals) from the Te 5p orbitals associated with atoms forming a Te-Te bond between [Bi$_4$Te$_6$] laths[3]. In both these cases, the formation of bonding-antibonding states, either for Bi-Bi or Te-Te, seems important in the formation of the band gap in these semiconductors.

CONCLUSIONS

The electronic structure calculations give CsBi$_4$Te$_6$ as a narrow-gap semiconductor with $E_g \sim$ 0.37 eV, the calculated band gap larger than the estimated value (from experiment) of approximately 0.1 eV.[2] It is possible that because of the doping, one is not measuring the intrinsic gap of this compound. The calculated band structure shows strong anisotropic dispersions consistent with its highly anisotropic crystal structure. The large thermopower observed in the electron doped samples may be understood by the many low lying minima along several directions.

One cannot, however, explain the observed large thermopower in the hole-doped systems [2] using the above argument for electrons, but it may be understood through the large anisotropy of the band structure.[10] We find that the band dispersion along the Bi$_2$Te$_3$ network normal to the growth axis appears much weaker than the dispersion through the Cs$^+$ layers or along the growth axis. The unusual Bi-Bi bonds found in this material seem important for the formation of the band gap as the Bi 6p orbitals of this Bi-Bi bond contribute dominantly to the states near the bottom of the conduction band. The flat top valence band in the ΓV direction, along the Bi$_2$Te$_3$ network, may indicate a large effective mass of the carriers which may further enhance the thermoelectric properties.[10, 11] Further analysis of these results, including calculation of effective masses and performing transport calculations of this material which are underway, is necessary to understand the relationship between the anisotropic dispersion seen in this material and its measured enhanced thermoelectric properties.

ACKNOWLEDGEMENT

This work was partially supported by DARPA Grant No. DAAG55-97-1-0184.

References

[1] *Thermoelectric Materials-The Next Generation Materials for Small-Scale Refrigeration and Power Generation Applications.* edited by T.M. Tritt, M.G. Kanatzidis, G.D. Mahan, and H.B. Lyon, Jr., MRS Symposia Proceedings No. **545** (Materials Research Society, Pittsburgh, 1999); F.J. DiSalvo, Science **285**, 703 (1999); T.M. Tritt, Science **283**, 804 (1999).

[2] D-Y Chung, T. Hogan, P. Brazis, M. Rocci-Lane, C. Kannewurf, M. Bastea, C. Uher, M.G. Kanatzidis, Science 287, 1024 (2000); P. Brazis, M. Rocci, D-Y Chung, M.G. Kanatzidis, C.R. Kannewurf. *Thermoelectric Materials-The Next Generation Materials for Small-Scale Refrigeration and Power Generation Applications.* edited by T.M. Tritt, M.G. Kanatzidis, G.D. Mahan, and H.B. Lyon, Jr., MRS Symposia Proceedings No. **545**, 75 (Materials Research Society, Pittsburgh, 1999); M. G. Kanatzidis, D-Y Chung, L. Iordanidis, K-S Choi, P. Brazis, M. Rocci, T. Hogan, C.R. Kannewurf, *ibid.*, 233 (Materials Research Society, Pittsburgh, 1999).

[3] P. Larson, S.D. Mahanti, M.G. Kanatzidis. Phys. Rev. B **61**, 8162 (2000)

[4] S.K. Mishra, S. Satpathy, O. Jepsen, J. Phys.: Condens. Matter **91**, 461 (1997).

[5] D. Singh, *Planewaves, Pseudopotentials, and the LAPW Method* (Kluwer Academic, Boston, 1994).

[6] P. Hohenberg and W. Kohn, Phys. Rev. **136**, B864 (1964); W. Kohn and L. Sham, *ibid.* **140**, A1133 (1965).

[7] J.P. Perdew, K. Burke, and M. Ernzerhof, Phys. Rev. Lett. **77**, 3865 (1996).

[8] P. Blaha, K. Schwarz, and J. Luitz, WIEN97 (Vienna University of Technology, Vienna, 1997).

[9] D.D. Koelling and B. Harmon, J. Phys. C **10**, 3107 (1977); P. Novak (unpublished).

[10] L.D. Hicks, T.C. Harman, and M.S. Dresselhaus, Appl. Phys. Lett. **63**, 3230 (1993); L.D. Hicks and M.S. Dresselhaus, Phys. Rev. B **47**, 16631 (1993); L.D. Hicks and M.S. Dresselhaus, *ibid.*, 12727 (1993).

[11] P. Larson, S.D. Mahanti, S. Sportouch, and M.G. Kanatzidis, Phys. Rev. B **59**, 15660 (1999).

WHERE SHOULD WE LOOK FOR HIGH ZT MATERIALS: SUGGESTIONS FROM THEORY.

M. Fornari [1,2], D. J. Singh [1], I. I. Mazin [1] and J. L. Feldman [1]
[1] Naval Research Laboratory, Washington DC [2] George Mason University, Fairfax VA.

ABSTRACT

The key challenges in discovering new high ZT thermoelectrics are understanding how the nearly contradictory requirements of high electrical conductivity, high thermopower and low thermal conductivity can be achieved in a single material and based on this identifying suitable compounds. First principles calculations provide a material specific microscopic window into the relevant properties and their origins. We illustrate the utility of the approach by presenting specific examples of compounds belonging to the class of skutterudites that are or are not good thermoelectrics along with the microscopic reasons. Based on our computational exploration we make a suggestion for achieving higher values of ZT at room temperature in bulk materials, namely n-type $La(Ru,Rh)_4Sb_{12}$ with high La-filling.

INTRODUCTION

The quest for a material with high thermoelectric efficiency consists of optimizing both thermal and electronic properties in order to balance the lattice thermal conductivity κ_L with the electronic conductivity σ, the Seebeck coefficient S and the electronic thermal conductivity κ_e. The goal is to obtain a figure of merit $ZT = \sigma S^2 T / (\kappa_L + \kappa_e)$ as high as possible at the operating temperature of the device, *i.e.* 300 K or below for coolers.

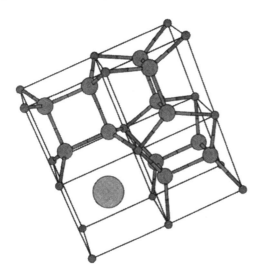

Figure 1

Half of the BCC conventional cell of the skutterudite structure. Three octants are occupied by pnictogen "square" rings. The remaining octant is filled with a rare earth. The metal atoms occupy a cubic sublattice.

The quantities involved in Z are strongly coupled to each other. A good thermoelectric material should be a high effective mass semiconductor (high S) with considerable carrier density after doping (high σ) and with complex unit cell in order to make low κ_L possible. Filled skutterudites are among the promising materials. They satisfy the above criteria and indeed $La(Fe,Co)_4Sb_{12}$ and $CeFe_4Sb_{12}$ have shown p-type high figures of merit [1,2], though unfortunately not at temperatures applicable to cooling devices. On the other hand, the skutterudite structure is very stable upon chemical substitutions and alloying and new interesting compounds with low thermal conductivity and optimal electronic properties at lower temperature could be found. Using first principles calculations we performed a systematic computation of the electronic properties of La-filled skutterudites, LaM_4Pn_{12} (M=Fe, Ru, Os and Pn=P, As, Sb). The comparison between band structures of different compounds emphasized trends and pointed out a particularly promising composition $La(Ru,Rh)_4Sb_{12}$ for which a ZT possibly exceeding unity at room temperature and below may be found [3].

LATTICE STRUCTURE AND COMPUTATIONAL APPROACH

The conventional cubic cell (Fig. 1) of the binary skutterudites, MPn_3, contains 16 atoms (space group Im3). The metal atoms occupy a cubic sublattice; six octants are filled with a Pn_4 ring oriented according to the cubic symmetry and the remaining two are empty. In the filled skutterudites, RM_4Pn_{12}, the empty octants contain a filling atom, often a rare earth *e.g.* R=La, Ce.

We have used linearized augmented plane wave (LAPW) density functional *ab initio* calculations (DFT) in the framework of the local density approximation (see Ref. [3,7] for computational details) to calculate the band structure and density of states (DOS). In order to evaluate the effects of alloying, the virtual crystal approximation (VCA) has been used: the virtual crystal for the alloy has been realized adding one electron in valence and modifying

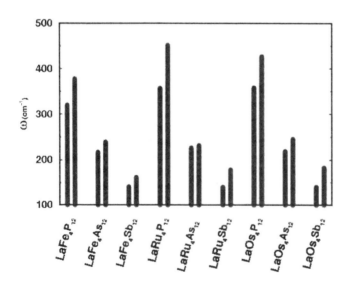

Figure 2

Frequencies of the two A_g modes associated with the full-symmetry pnictogen vibrations. Note the lower frequencies in antimonides.

accordingly the atomic number of the transition metal (Z→Z+0.25). The validity of VCA is verified *a posteriori* considering the rigid band behavior. Transport properties were studied using kinetic transport theory.

The size and shape of the pnictogen ring is determined by two symmetry independent parameters (x,y). Our calculations are performed using, if available, the experimental structure, otherwise we (A_g modes) that correspond to Raman active modes. Pnictogen ring related

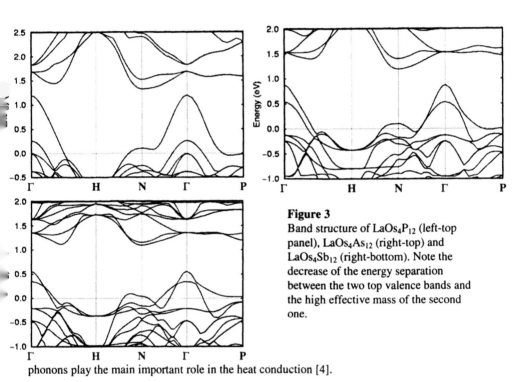

Figure 3
Band structure of $LaOs_4P_{12}$ (left-top panel), $LaOs_4As_{12}$ (right-top) and $LaOs_4Sb_{12}$ (right-bottom). Note the decrease of the energy separation between the two top valence bands and the high effective mass of the second one.

phonons play the main important role in the heat conduction [4].

THE SEARCH

The experimental result that filling the cages in $CoSb_3$ with La (and alloying with Fe) leads to a drop of about an order of magnitude [1, 2] in the thermal conductivity is our starting point. A clear explanation is still lacking but the comparison between the A_g modes frequencies (Fig. 2) strongly suggests that the lower frequencies of A_g modes in antimonide compounds may favor a coupling with La vibrations (50-100 cm^{-1}) that does not occur as strongly in the phosphide and arsenide because of the frequency mismatch. In fact, even if La is more tightly bound and has higher frequencies in the P and As compounds because of the cage's smaller volume, the strong coupling with the A_g modes does not seem to occur in the arsenide and phosphide. The overall effect is detrimental in the thermal conductivity, which is substatially higher [5].

In addition La filling has important contributions both on qualitative and quantitative features of the band structure. These effects [6] rest on the interaction between La related f-resonances that appear in the conduction band (about 3 eV above the Fermi level) and that push down, via hybridization, the highest p-Pn band. Depending on the particular starting unfilled material band structure the result can be very different: moving from CoP_3 to $LaFe_4P_{12}$ a gap is opened [7], from $CoSb_3$ to $LaFe_4Sb_{12}$ the gap increases and the two highest valence bands become very close in energy (the energy separation at Γ in LaFe4Sb12 is about 6 meV, [6]). After p-doping both bands participate in transport. The key point is that the effective mass of the second band is high, favoring high thermoelectric power (S) at reasonable doping levels and hence high values of Z.

Even though $LaOs_4Pn_4$ compounds have little practical relevance, because of the difficulty in using Os, it is interesting to analyze the effects of pnictogen substitution in the Os pnictides. The phosphide (Fig. 3a) is a semiconductor with a small indirect gap (E_G=0.15 eV) that becomes larger in the arsenide (E_G=0.33 eV, Fig. 3b) and antimonide (E_G=0.54 eV, Fig. 3c). The two top valence bands become energetically closer (0.93 -> 0.34 -> 0.21 eV), and in $LaOs_4Sb_{12}$ the second one shows high effective mass. The material might promise good p-type TE properties. Anyway, in $LaOs_4Sb_{12}$ the energy separation between the two top valence bands is still larger than in $LaFe_4Sb_{12}$ making this last skutterudite more useful. Also, the conduction band (CB) is subject to considerable changes upon filling and the trend in Os pnictides resembles the main features of all the La filled skutterudites we have investigated. In the unfilled materials the bottom of the CB occurs at Γ point (even if in some cases, e.g. $CoSb_3$, this should be interpreted carefully because of the semimetallic or zero gap character of the band structure). In addition to the f-resonance effect, La filling shifts the minimum of the CB from Γ to the zone boundary (generally N). The result is a multivalley multidegenerate CB which is very useful for obtaining high thermopower with high carrier concentrations after n-doping. Although only from a theoretical point of view, $LaOs_4P_{12}$ shows the key features of a promising TE material: (i) expected low thermal conductivity, (ii) multivalley bands, (iii) high effective masses.

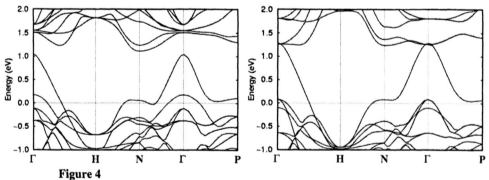

Figure 4
Band structures of $LaRu_4P_{12}$ (left) and $LaRu_4As_{12}$ (right). In the phosphide the La-f interaction is not enough strong to open a gap.

In Fig. 4 and Fig. 5 we show the effect of pnictogen substitution for $LaRu_4Pn_{12}$. The f-resonance interaction is not enough strong to open a gap in the phosphide, and also when the gap is opened in the arsenide the two highest valence bands are too far apart in energy to be used in transport at a reasonable p-doping. The conduction band seems more interesting because the two minima (occurring at N and in Γ-H direction) are multidegenerate and in LaRu$_4$Sb$_{12}$ are very close in energy resulting in a very promising CB structure for TE n-type applications. Also, since this is an antimonide, κ_L is expected to be quite low and particularly lower than in phosphide and arsenide.

A RESULT: La(Ru,Rh)$_4$Sb$_{12}$

LaRu$_4$Sb$_{12}$ is an indirect gap semiconductor (E_G=0.16 eV) with two "degenerate" minima (ΔE=10 meV) in the conduction band occurring at the N (6-fold degenerate) point and just apart from the Γ-H symmetry line (24-fold degenerate). The effective masses in the cubic directions are m$^*_x$=1.60, m$^*_y$=0.36, m$^*_z$=2.00 for the N point and m$^*_x$=12.12, m$^*_y$=0.24, m$^*_z$=3.90 for the other minimum, revealing strong anisotropy. Due to the high effective masses (the average value is <m*>=3.3) the maximum Seebeck coefficient will be high and because of the multivalley character also the carrier density and, hence σ is expected to be favorable.

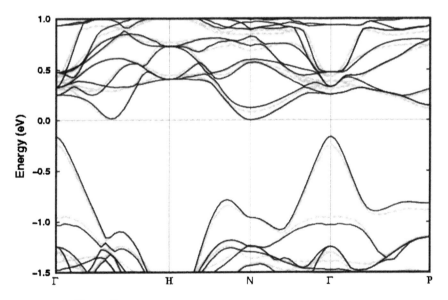

Figure 5
Band structure of LaRu$_4$Sb$_{12}$ (black) and LaRhRu$_3$Sb$_{12}$ (gray). The reference energies are set at the bottom of the respective conduction bands.

To be more quantitative we used standard kinetic theory in the constant scattering time approximation to compute the temperature dependence of the thermopower at different doping levels for La(Ru,Rh)$_4$Sb$_{12}$. The integration over the Brillouin zone has been performed using a very fine mesh. We computed the DOS, N(ε) and Fermi velocities v(ε) using a smooth Fourier interpolation of the bands.

The results are in Fig. 6. It is clearly shown that the Seebeck coefficient has a quite stable value in excess of 150 µV/K in the important temperature window between 100-300 K for a reasonably high carrier concentration of 4.6 x 10^{19} cm^{-3}. For comparison, La(Co,Fe)$_4$Sb$_{12}$ reaches the same high S but at T=600 K, [2]. To evaluate the effect on the mobility of the Rh-Ru alloying needed to control the carrier concentration we compare in Fig. 5 the band structures of LaRu$_4$Sb$_{12}$ and La(Ru$_{0.75}$Rh$_{0.25}$)$_4$Sb$_{12}$. They are almost identical for the relevant bands confirming *a posteriori* the validity of VCA and suggesting very low disorder scattering by Ru-Rh alloying and thus high mobility.

As we have shown before, La filling is a key point in the optimization of ZT and in our calculations we have considered a perfect crystal with two La atoms in each conventional cubic cell. In reality, possible La vacancies can strongly affect both the lattice and the electronic properties. To investigate the sensitivity of the mobility to La filling, we have calculated the band structure for the related unfilled material: RhSb$_3$ (Fig. 7). The band structure is inverted producing a zero gap material and the global features are drastically different from the

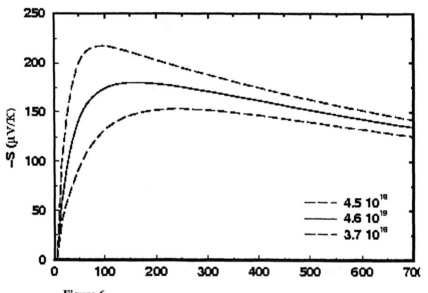

Figure 6
Seebeck coefficient for n-type LaRhRu$_3$Sb$_{12}$ at different carrier densities. The maximum value occurs below room temperature.

completely filled LaRu$_4$Sb$_{12}$ meaning that high La filling is required in order to maintain a reasonable mobility. To reach high La filling in LaCo$_4$Sb$_{12}$ new growth techniques have been developed and we are confident that they could also apply to ruthenates. [8]

We like to underline that the drastic difference between the band structures of RhSb3 (Fig. 7) and LaRu4Sb12 is not related to the f-resonance induced by La. The main role in this effect is

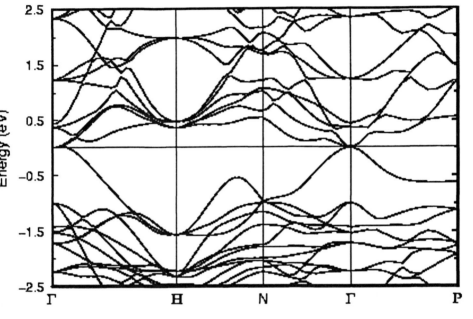

Figure 7
Band structure of RhSb3. From comparison with Fig. 5 it appears clear that La filling must be very high in order to avoid detrimental effects on the TE properties.

played by the different ionicity of the bonds involved in the material which causes a charge rearrangement when electrons are transferred from the rare earth to the metal. Thus, a partial substitution of La with other trivalent ions (*e.g.* Yb) should be as good in this case, and this fact could potentially be used to optimize the thermal conductivity.

SUMMARY AND CONCLUSIONS

Comparing the band structures of different skutterudite compounds we have elucidate the effect of La filling and of pnictogen substitutions. These calculations of the band structures of chemically different skutterudites, in addition to a better understanding of trends, point to a new promising specific composition LaRu$_3$RhSb$_{12}$ with an expected ZT about unity below room temperature and at n-type carrier concentration in the 10^{19} cm^{-3} range.

This expectation is based on (1) experimental evidence of low thermal conductivity in similar compounds, (2) the multivalley conduction band structure with high effective masses, and (3) computed thermopowers S > 150 μV/K for T below room temperature. In addition our first principles calculations predict low Ru-Rh scattering and high La-vacancy scattering. The detrimental effect due to incomplete filling needs to be limited using modern growth techniques.

ACKNOWLEDGMENTS

We are grateful for helpful discussion with T. Caillat, F.J. Di Salvo and, J. P. Fleurial. We thank ONR and DARPA for supporting this work.

REFERENCES

[1] J.-P. Fleurial, A. Borshchevsky, T. Caillat, D. Morelli, and G. P. Meisner in Proc. 15th Int. Conf. On Thermoelectrics, p. 91 (IEEE Press, Piscataway, 1996).
[2] B. C. Sales, D. Mandrus, and R. K. Williams, Science **272**, 1325 (1996).
[3] M. Fornari, and D. J. Singh, Appl. Phys. Lett. **74**, 3666 (1999)
[4] J. L. Feldman, and D. J. Singh, Phys. Reb. **B61**, 9209 (2000)
[5] A. Watchrapasorn, R. C. De Mattei, R. S. Feigelson, T. Caillat, A. Borshchevsky, G. J. Snyder, and J.-P. Fleurial, J. Appl. Phys. **86**, 6213 (1999)
[6] D. J. Singh, and I. I. Mazin, Phys. Rev. **B56**, R1650 (1997).
[7] M. Fornari, and D. J. Singh, Phys. Rev. **B59**, 9722 (1999).
[8] H. Takizawa, K. Miura, M. Ito, T. Suzuki, and T. Endo, J. Alloys Compd. **282**, 79 (1999)

Enhancement of Power Factor in a Thermoelectric Composite with a Periodic Microstructure

Leonid G. Fel, Yakov M. Strelniker, and David J. Bergman

School of Physics and Astronomy, Raymond and Beverly Sackler Faculty
of Exact Sciences, Tel Aviv University,
Tel Aviv 69978 Israel

ABSTRACT

The thermoelectric power factor has been calculated for a two-constituent composite medium, where one constituent is a "high quality thermoelectric" while the other constituent is a "benign metal", with large electrical and thermal conductivities but poor thermoelectric properties. It was recently discovered that, in such a mixture, the power factor could be greatly enhanced by an appropriate choice of microstructure. Here we report on a study of three periodic microstructures with cubic point symmetry under rotations: *simple cubic* (SC), *body centered cubic* (BCC), and *face centered cubic* (FCC) arrays of identical spheres of the benign metal embedded in the high quality thermoelectric host. We show detailed results for these microstructures in the case where the benign metal constituent is Copper, while the high quality thermoelectric constituent is the thermoelectric alloy $(Bi_2Te_3)_{0.2}(Sb_2Te_3)_{0.8}$.

1. Recently, it was shown that the thermoelectric power factor of a high quality thermoelectric material can be enhanced by mixing it together with a "benign metal" (i.e., a metal with large electrical and thermal conductivities but a small Seebeck coefficient, and hence with poor thermoelectric properties) in the form of a two-constituent composite medium [1]. This was demonstrated in some very simple microstructures (see Fig. 1). One of those was a *parallel slabs* sandwich, which is a very anisotropic medium. The other was an isotropic *coated spheres assemblage*. The last one is a very artificial kind of microstructure, where the system is made up entirely of coated spheres, with a spherical core of the benign metal constituent and a concentric spherical shell of the high quality thermoelectric constituent, and with each sphere having the *same core-to-shell volume ratio*, even though those spheres must come in many different sizes in order to fill up the entire volume. A realistic microstructure would have to have some disorder in the positions, sizes, and shapes of the constituent grains and possible coatings. As a first step towards a consideration of such realistic microstructures, we have performed detailed calculations on some periodic microstructures. Those are much easier to carry out, using algorithms previously developed for evaluating the bulk effective electrical conductivity of periodic composites [2]. Here we report on studies of three microstructures that exhibit a cubic point symmetry under rotations: simple cubic (SC), body centered cubic (BCC), and face centered cubic (FCC) arrays of identical spheres of the benign metal embedded in the high quality thermoelectric host.

2. The thermoelectric power factor W is a material parameter defined by

$$W \equiv \sigma \alpha^2,$$

where σ is the electrical conductivity at zero temperature gradient while α is the thermoelectric coefficient or Seebeck coefficient or absolute thermopower. The value of W is of crucial importance in the effective utilization of waste heat for electric power generation

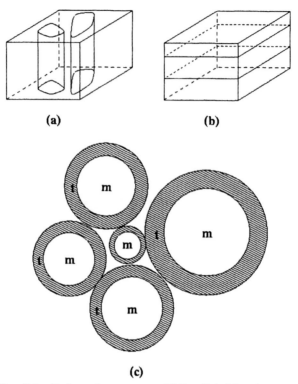

Figure 1: *(a) Parallel cylinders microstructure. (b) Parallel slabs microstructure. (c) Coated spheres assemblage or Hashin-Shtrikman microstructure [3], made entirely of coated spheres with a spherical core of the benign metal constituent* **m**, *and a concentric spherical shell of the high quality thermoelectric constituent* **t**. *The coated spheres must come in an infinite hierarchy of descending sizes in order to fill up the entire volume, but they all have the same core-to-shell volume ratio, equal to the overall volume ratio. All three microstructures are exactly solvable.*

using thermoelectric energy conversion devices: The maximum electric power is dissipated in a load when the load resistance equals the internal resistance of the thermoelectric $R = L/(A\sigma)$, and that power is given by

$$P_{max} = \frac{AW(\delta T)^2}{4L}.$$

Here $\delta T \equiv T_h - T_l$ is the temperature difference between the two heat reservoirs, while L and A are the thickness and cross section of the thermoelectric block or cylinder which connects those two reservoirs. Under "waste heat" conditions, the maximum cross section that is available for such a block is fixed, as is the temperature difference δT, while the minimum thickness of the block is determined by requirements of mechanical strength or stability. One way to increase P_{max} is by increasing the power factor W. Another way to increase P_{max}, up to ∞ in principle, is by using a sandwich microstructure, like the one shown in Fig. 1(b), where most of the thickness is occupied by slabs of a fictitious material with infinite electrical and thermal conductivities, while the high quality thermoelectric occupies only an infinitesimal thickness: The entire temperature drop δT and voltage drop would then appear across the infinitesimally thin thermoelectric layer, i.e., $L = 0$. In reality, we are of course limited to finite conductivities. Nevertheless, that extreme fictitious situation indicates that it might be advantageous to use such a sandwich microstructure with a thin layer of high quality thermoelectric and a thick layer of benign metal, with very high electrical and thermal conductivities. However, a nontrivial calculation is needed in order to find the optimum volume fractions, as well as in order to prove that this particular microstructure is better than any other [1]. The enhancement of P_{max} obtained in this or in any other microstructure can be conveniently characterized by keeping the total thickness L fixed at its nominal value, and ascribing the enhancement entirely to a *bulk effective value* of the power factor W_e for the composite microstructure.

In Ref. [1] it was shown that, in composite mixtures of a "high quality thermoelectric" and a "benign metal", it is often possible to achieve values for W_e which are considerably greater than that of the pure high quality thermoelectric constituent. When that is possible, then the greatest enhancement of W_e is always achieved in the parallel slabs microstructure. At the same time the greatest decrease of W_e is always achieved in the parallel cylinders microstructure. The thickness ratio of the two types of slabs in the sandwich, at maximum enhancement of W_e, is usually such that the high quality thermoelectric is present as a very thin film in series with a much thicker slab of the benign metal constituent [1].

In contrast with the enhancement of W_e, the bulk effective thermoelectric figure of merit Z_e or Z_eT can never be enhanced by making a composite mixture [4,5]. However, it is interesting to note that even when a very large enhancement of W_e is achieved, the penalty paid, in the form of a reduction of Z_e compared to its value in the high quality thermoelectric, is quite modest. For example, in a sandwich where the benign metal constituent (denoted by the subscript m) is Cu, and the high quality thermoelectric constituent (denoted by the subscript t) is the alloy $(Bi_2 Te_3)_{0.2}(Sb_2Te_3)_{0.8}$, the relevant properties of the pure constituents at $T=300$ K are: $\sigma_m=5.8\times10^7$ (Ohm m)$^{-1}$, $\gamma_m=394$ W/(m K) [6], $\alpha_m=1.84\times10^{-6}$ V/K [7], $W_m=1.96 \times 10^{-4}$ W/(m K^2), $Z_mT = 1.5 \times 10^{-4}$; $\sigma_t=3.70\times10^4$ (Ohm m)$^{-1}$, $\gamma_m=1.49$ W/(m K), $\alpha_t=2.5\times10^{-4}$ V/K [8], $W_t=2.31\times 10^{-3}$ W/(m K^2), $Z_tT = 0.87$, where γ is the bulk thermal conductivity at zero electric field. The bulk effective values W_e and Z_eT for the parallel slabs sandwich at maximum enhancement of W_e are: $W_{e\,max} = 0.180$ W/(m K^2) $\cong 78\times W_t$, $Z_{e\,max}T \cong 0.36$, and the volume fraction of the high quality thermoelectric alloy at that composition is only $p_{t\,max}=0.0034$. We note that at maximum enhancement, the power factor of the Cu-$(Bi_2 Te_3)_{0.2}(Sb_2Te_3)_{0.8}$ mixture is ten times greater than the value quoted in Ref. [9] for the so-called "best thermoelectric" YbAl$_3$. A similar enhancement of W_e is also achievable in the coated spheres assemblage: In that case it was found that $W_{e\,max} = 0.1785$ W/(m K^2), and the volume fraction of the high quality thermoelectric alloy is $p_{t\,max} = 0.01$ [10].

3. Composite media are usually either disordered on the microscale, as in the case of granular or polycrystalline media, or partially ordered, as in the case of fiber reinforced materials. It is only quite recently that highly ordered periodic dielectric composites have been fabricated in the context of a search for materials where a photonic band gap might be observed [11]. In the expectation that thermoelectric composites will also soon be available, we have considered a number of periodic microstructures. We have studied three basic microstructures that have cubic point symmetry under rotations: simple cubic (SC), body centered cubic (BCC) and face centered cubic (FCC) arrays of identical spheres of the benign metal embedded in the high quality thermoelectric host.

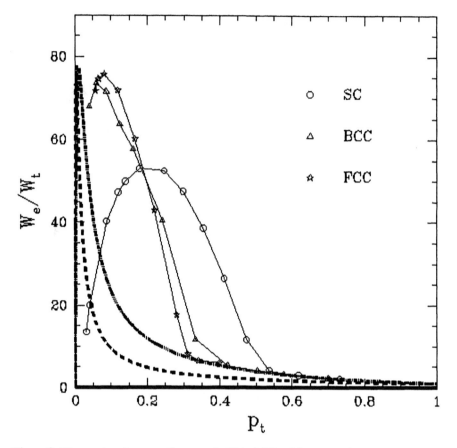

Figure 2: Thermoelectric power factor ratio $W_e(p_t)/W_t$ of the composite Cu-$(Bi_2Te_3)_{0.2}(Sb_2Te_3)_{0.8}$: W_t=2.31× 10^{-3} W/(m K^2), W_m=1.96 × 10^{-4} W/(m K^2). For the parallel slabs microstructure we find $p_{t\,max}$=0.0034, $W_{e\,max}$ =0.180 W/(m K^2). Results for that microstructure are shown By the <u>thick dashed line</u>. The <u>other thick line</u> shows results for the coated spheres assemblage microstructure. Results for the periodic microstructures are shown as <u>open circles</u> (SC), <u>open triangles</u> (BCC), and <u>open stars</u> (FCC), with thin lines drawn through the points to serve as a visual guide.

In order to do this we first apply a decoupling transformation, which transforms the coupled-fields thermoelectric transport problem into uncoupled conductivity-like problems in two different composites with the same microstructure. Those two simple-conductivity-like problems are then treated using an algorithm developed earlier for periodic composites [2]. That algorithm, which is based on a Fourier-space representation of an integral equation for the potential, can tackle any kind of periodic microstructure, i.e., it is not limited to arrays of non-overlapping spherical inclusions.

In Fig. 2 we present the results of numerical calculations of the bulk effective thermoelectric power factor $W_e(p_t)$ for the three cubic microstructures mentioned above and for two exactly solvable microstructures (parallel slabs and coated spheres), where p_t is the

Table: The maximal power factor $W_{e\,max}$ and some related quantities in the case of the composite Cu-$(Bi_2Te_3)_{0.2}(Sb_2Te_3)_{0.8}$ for the different microstructures.

Structure	p_{touch}	$p_{t\,max}$	$W_{e\,max}$, W/(m K^2)	$W_{e\,max}/W_t$	$p_{touch}/p_{t\,max}$	$TZ_e(p_{t\,max})$
Slabs	-	0.0034	0.180	77.9	-	0.365
Coated Spheres	-	0.01	0.179	77.3	-	0.362
SC	0.476	0.181	0.123	53.2	2.63	0.208
BCC	0.319	0.06	0.171	74	53.16	0.263
FCC	0.259	0.082	0.175	75.7	31.58	0.290

volume fraction of the thermoelectric host. All curves for the cubic microstructures achieve their maxima at a value of p_t where neighboring metallic spheres overlap. The degree of overlap is characterized by the ratio $p_{touch}/p_{t\,max}$, where p_{touch} is the value of p_t at which neighboring metallic spheres just touch each other. The values of $W_{e\,max}$, p_{touch}, $p_{t\,max}$, and $p_{touch}/p_{t\,max}$, which is always greater than 1, are given in the Table. We note that the values of $W_{e\,max}$ for the BCC and FCC microstructures are not much less than the greatest possible value of $W_{e\,max}$, which is achieved in the parallel slabs microstructure (see Fig. 2).

In Fig. 3 we present the results of numerical calculations of the bulk effective thermoelectric figure of merit $\Delta_e(p_t)$ of the composite Cu-$(Bi_2Te_3)_{0.2}(Sb_2Te_3)_{0.8}$

$$\Delta_e \equiv \frac{TZ_e}{1+TZ_e} = \frac{T\sigma_e \alpha_e^2}{\gamma_e}.$$

The values of $TZ_e(p_{t\,max})$ for the various microstructures are shown in the Table.

4. Evidently, details of the microstructure have an important effect on the final results for both $W_e(p_t)$ and $\Delta_e(p_t)$ or $TZ_e(p_t)$. The results for the coated spheres assemblage are very close to those for the parallel slabs microstructure, where the maximum enhancement of W_e can be achieved. The results for SC microstructure are considerably worse, although even there a large enhancement of W_e is attainable. The results for BCC and FCC microstructures lie in-between those for coated spheres and SC. This reflects the fact that the microgeometry of BCC and FCC is closer to that of the coated spheres assemblage, especially when p_t is very small. We note that values of TZ_e that are greater than $TZ_e(p_{t\,max})$ are easily attainable by using a value of p_t that is slightly greater than $p_{t\,max}$. The penalty for that is a somewhat reduced value of W_e.

ACKNOWLEDGMENTS

Partial support for LGF and YMS came from the Bureau for Absorption in Science of the State of Israel.

REFERENCES

1. D. J. Bergman and L. G. Fel, J. Appl. Phys. **85**, 8205 (1999).
2. D. J. Bergman and K. J. Dunn, Phys. Rev. B **45**, 13262 (1992).

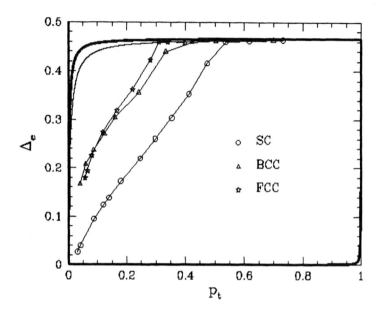

Figure 3: *Figure of merit* $\Delta_e(p_t)$ *of the composite* Cu-$(Bi_2Te_3)_{0.2}(Sb_2Te_3)_{0.8}$: $\Delta_t = 0.466$, $\Delta_m = 1.5 \times 10^{-4}$. *The* <u>upper thick solid line</u> *shows results for the parallel slabs sandwich microstructure. The* <u>lower thick solid line</u> *shows results for the parallel cylinders microstructure. The* <u>thin solid line</u> *shows results for the coated spheres assemblage microstructure. Results for the periodic microstructures are shown as* <u>open circles</u> *(SC),* <u>open triangles</u> *(BCC), and* <u>open stars</u> *(FCC), with thin lines drawn through the points to serve as a visual guide.*

3. Z. Hashin and S. Shtrikman, J. Appl. Phys. **33**, 3125 (1962).
4. D. J. Bergman and O. Levy, J. Appl. Phys. **70**, 6821 (1991).
5. D. J. Bergman, in Proc. 16th Int'l Conf. on Thermoelectrics, Dresden, Germany, 1997, publ: IEEE, Piscataway, NJ, p. 401.
6. H. Fabian, in Ullmann's Encyclopedia of Industrial Chemistry, 5th Edition, Vol. **A7**: Chlorophenols to Copper Compounds, eds. W. Gerhartz, Y. S. Yamamoto, F. T. Campbell, R. Pfefferkorn, and J. F. Rounsaville, VCH Verlagsgesellschaft mbH, Weinheim, Germany, pp. 471-523 (Copper).
7. V. Tegeder and Ch. Tegeder, in Ullmann's Encyclopedia of Industrial Chemistry, 5th Edition, Vol. **A26**: Surgical Materials to Thiourea, eds. B. Elvers, S. Hawkins, and W. Russey, VCH Verlagsgesellschaft mbH, Weinheim, Germany, pp.621-631 (Thermoelectricity).
8. T. Ohta, A. Yamamoto, and T. Tanaka, in Proc. 13th Int'l Conf. on Thermoelectrics, USA, 1994, publ: IEEE, Piscataway, NJ, p. 267.
9. G. D. Mahan and J. O. Sofo, Proc. Nat. Acad. Sci. USA, Appl. Phys. Sci. **93**, 7436 (1996); G. D. Mahan, in Proc. 16th Int'l Conf. on Thermoelectrics, Dresden, Germany, 1997, publ: IEEE, Piscataway, NJ, p. 21.
10. D. J. Bergman and L. G. Fel, in Proc. 18th Int'l Conf. on Thermoelectrics (ICT99), Baltimore, MD, 1999.
11. E. Yablonovitch, T. J. Gmitter and K. M. Leung, Phys. Rev. Lett. **67**, 2295 (1991).

New Materials II

Connections between Crystallographic Data and New Thermoelectric Compounds

B. C. Sales, B. C. Chakoumakos, and D. Mandrus
Solid State Division, Oak Ridge National Laboratory
Oak Ridge, TN 37831-6056, U. S. A.

ABSTRACT

New bulk thermoelectric compounds are normally discovered with the aid of simple qualitative structure-property relationships. Most good thermoelectric materials are narrow gap semiconductors composed of heavy elements with similar electronegativities. The crystal structures are usually of high symmetry (cubic, hexagonal, and possibly tetragonal), and often contain a large number of atoms per unit cell. In the present work a new structure-property relationship is discussed which links atomic displacement parameters (ADPs) and the lattice thermal conductivity of clathrate-like compounds. For many clathrate-like compounds, in which one of the atom-types is weakly bound and "rattles" within its atomic cage, room temperature ADP information can be used to *estimate* the room temperature lattice thermal conductivity, the vibration frequency of the "rattler", and the temperature dependence of the heat capacity. ADPs are reported as part of the crystal structure description, and hence APDs represent some of the first information that is known about a new compound. For most ternary and quaternary compounds, all that is known is its crystal structure. ADP information thus provides a useful screening tool for the large and growing crystallographic databases. Examples of the use and limitations of this analysis are presented for several promising classes of thermoelectric materials.

INTRODUCTION

Atomic displacement parameters (ADPs) measure the mean-square displacement amplitude of an atom about its equilibrium position in a crystal. In the description of a new crystalline compound, crystallographers normally tabulate the room temperature ADP values for each distinct atomic site in the structure[1-4]. The various ADP values thus comprise some of the first information that is known about a new crystalline compound. The value of the mean square atomic displacement can be due to the vibration of the atom or to static disorder. The effects that this parameter can have on various physical properties, however, have not been widely recognized. In particular, ADPs are not normally used by solid state physicists or chemists as a guide in the search for new compounds with specific properties. ADPs are still regarded by many scientists as unreliable, since in many of the earliest structure determinations, ADPs often became repositories for much of the error in the structure refinement. In addition, crystallographers have not always reported ADP information using a consistent definition,[1] adding further confusion as to the usefulness of ADPs. The purpose of this article is to illustrate that when properly determined, ADPs can be used as a guide in the search for crystalline materials with unusually low lattice thermal conductivities. These materials are of particular interest in the design of thermoelectric compounds with improved efficiencies.[5-8] The analysis discussed below is restricted to clathrate-like compounds in which one of the atom types "rattles" about its equilibrium position in the crystal substantially more than the other atoms in the structure. Clathrate-like compounds include the partially filled or filled skutterudites (e.g. $CeFe_4Sb_{12}$, see Fig. 1); semiconducting compounds with the ice clathrate structure (e.g. $Sr_8Ga_{16}Ge_{30}$), some ternary tellurides (Tl_2SnTe_5), and LaB_6. LaB_6 is a good metal and is of no interest for thermoelectric applications but does provide an interesting example of the type of information that can be extracted from room temperature ADP values.[9]

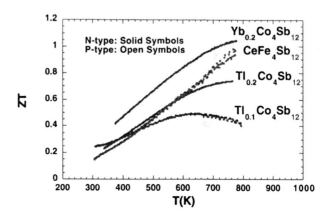

Fig. 1 ZT versus temperature for some partially filled and filled skutterudite compounds. The absolute ZT values are believed to be accurate to within 20 %.

ELEMENTARY THEORY OF ATOMIC DISPLACEMENT PARAMETERS

Atomic displacement parameters (ADPs) measure the mean-square displacement amplitudes of an atom about its equilibrium position in a crystal. In general there is no reason to assume that the displacements are the same in all directions, or that they bear any particular relation to the crystallographic axes. For this reason crystallographers typically report ADP information as a 3x3 matrix, U_{ij}, that allows for anisotropic displacements. In the description of a new crystalline compound, crystallographers normally tabulate the room temperature ADP matrix for each distinct atomic site in the structure.[1-4] The various ADP values thus comprise some of the first information that is known about a new crystalline compound. Often, an isotropic ADP value, U_{iso}, is given for each site. U_{iso} corresponds to the mean square displacement averaged over all directions and is given by one third of the trace of the diagonalized U_{ij} matrix. U_{iso} is a scalar which makes it easy to qualitatively compare the relative displacements of different atom-types in the structure. Sometimes U_{iso} is the only ADP information given if the full U_{ij} matrix cannot be extracted from the x-ray or neutron diffraction data set or if there are no significant anisotropic displacements. The U_{ij} data are often expressed in crystal structure figures by drawing ellipsoids around each atom. The surface of each ellipsoid corresponds to surfaces of constant probability. Normally the 50 percent ellipsoid is drawn corresponding to a 50 per cent probability of finding the atom inside the ellipsoid. The value of the mean square atomic displacement can be due to the vibration of the atom and/or to static disorder.

In a neutron or x-ray diffraction experiment, thermal vibrations of the atoms reduce the intensity of the Bragg reflections but do not effect the width. The scattered intensity, I, of a typical Bragg peak is qualitatively given by:

$$I = I_0 \exp[-1/3 <u^2> (\Delta k)^2] \qquad (1)$$

where I_0 is the scattered intensity from a rigid lattice (no vibrating atoms), $<u^2>$ is the mean square displacement of an atom about its equilibrium position, and Δk is the magnitude of the scattering vector (which increases as the sine of the scattering angle)[3]. In the physics literature the exponential factor is often referred to as the Debye-Waller factor. Atoms in a crystal vibrate more at higher temperatures which implies that $<u^2>$ increases monotonically with temperature.

The intensity of X-rays (or neutrons) scattered by a crystal is the sum of the Bragg scattering and the thermal diffuse scattering (TDS). TDS corresponds to scattering in which one or more phonons are excited. As the temperature is raised, the overall intensity from Bragg scattering decreases with a corresponding increase in TDS.

ADP values can be reliably determined using powder neutron diffraction and single crystal x-ray or neutron diffraction. The analysis of neutron data is usually easier for two reasons. First, the neutron wavelength used is typically the order of 1-2 Å which is much larger than the interaction distance between the neutron and the atomic nucleus. The nuclear scattering cross section is, therefore, a scalar with no angular dependence. X-rays scatter from the electron clouds around the nucleus and the resulting atomic form factor does have an angular dependence. This means that the intensity I_0 (Eq. 1) depends on angle. Since the ADP information is in the angular dependence of the scattered intensity ($<u^2> (\Delta k)^2$) (see Eq. 1), it is sometimes difficult to separate ADP information from atomic-form-factor effects. Second, for most compounds, the absorption correction for neutrons is small, but for x-rays absorption must be carefully determined to obtain good ADP values, particularly for compounds with heavy elements.

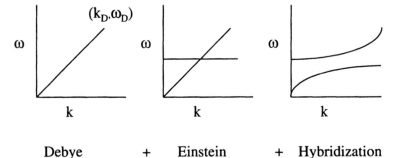

Fig. 2. *Schematic illustration of the dispersion behavior of a Debye solid and a Debye solid with a localized Einstein mode. The qualitative effects on the dispersion curves of the interaction between the Einstein mode and the acoustic phonons of the solid are also sketched.*

INTERPRETING ADP DATA

Interpreting the meaning of the ADP information requires a model. A reasonable model for clathrate-like compounds treats the "rattler" as an Einstein oscillator (quantum harmonic oscillator) and the other framework atoms as a Debye solid.[9,10] This approximation is illustrated schematically in Fig 2b.

To keep the analysis simple the ADP data for each atom type is converted to a scalar U_{iso} which measures the mean square atomic displacement averaged over all directions. If the value of U_{iso} is mostly due to atomic motion (rather than static disorder), and the Debye and Einstein temperatures are less than about 600 K, *room temperature* ADP data can be used to estimate the Einstein and Debye temperature of the clathrate-like compound using Eqs. 2 and 3 given below,

$$\Theta_D(K) = 208 / (U_{iso}^{av} (\text{Å}^2)/0.01 \cdot m_{av}/100)^{1/2} \qquad (2)$$

and

$$\Theta_E(K) = 120 / (U_{iso}^{rattler}/0.01 \cdot m_{rattler}/100)^{1/2} \qquad (3)$$

where, Θ_D (K) is the Debye temperature in Kelvin, U_{iso}^{av} (Å^2) is the weighted average of room temperature values of U_{iso} for each framework atom-type in the compound given in units of Å^2, and m_{av} is the average mass of a framework atom in the compound given in amu. Similarly, the Einstein temperature of the rattler, Θ_E (K) is given by Eq. 3 with $U_{iso}^{rattler}$ given in Å^2 and the rattler mass given in amu.

As an example, consider the filled skutterudite $LaFe_4Sb_{12}$. The room temperature U_{iso} data reported by Braun and Jeitschko[11] for this compound are: 0.0165 Å^2 for La, 0.003 Å^2 for Fe and 0.004 Å^2 for Sb. The weighted average mass of the framework atoms (Fe and Sb) is 105 amu/atom and the average U_{iso} is 0.00375 Å^2. Using these values in Eq. 2 results in a Debye temperature of 331 K. From Eq. 3 the Einstein temperature for the La "rattler" is 79 K.

Within the Debye model the velocity of sound of all of the phonon modes are equal (slope of line in Fig 2a). If the Debye temperature is known, the average velocity of sound, v, is also known and vice-versa since,

$$v = \omega_D / K_D = \Theta_D 2\pi k_B / [h (6\pi^2 n)^{1/3}] \qquad (4)$$

where n is the number of atoms per unit volume

With these same approximations the temperature dependence of the total molar heat capacity of a clathrate-like compound should approximately be given by:

$$C_v^{Clathrate}(T) = f\, C_{Debye}(T) + (1-f)\, C_{Einstein}(T) \qquad (5)$$

where, f is the fraction of framework atoms and (1-f) is the fraction of rattling atoms, and

$$C_{Debye}(T) = 9N_A k_B \left(\frac{T}{\Theta_D}\right)^3 \int_0^{\Theta_D/T} dx \frac{x^4 e^x}{\left(e^x - 1\right)^2} \tag{6}$$

and

$$C_{Einstein}(T) = 3N_A k_B \left(\frac{\Theta_E}{T}\right)^2 \frac{e^{\Theta_E/T}}{\left[e^{\Theta_E/T} - 1\right]^2} \tag{7}$$

ESTIMATION OF LATTICE THERMAL CONDUCTIVITY FROM ADP DATA

The simplest expression for the lattice thermal conductivity of a solid is given by an expression adapted from the kinetic theory of gases,[3]

$$\kappa_{Lattice} = 1/3\, C_v v_s\, d \tag{8}$$

where C_v is the heat capacity per unit volume, v_s is the velocity of sound and d is the mean free path of the heat carrying phonons. In a more realistic treatment of lattice thermal conductivity, which is discussed below, the mean free path (or relaxation time) and heat capacity depend on frequency and temperature, but for the present analysis C_v depends only on temperature, while v_s and d are treated as scalars.

Room temperature ADP data can be used to estimate the Debye temperature and the Debye velocity of sound. This analysis, which can be applied to all compounds, provides an estimate of the room temperature values of C_v and v_s that appear in Eq. 8. To estimate the lattice thermal conductivity using Eq. 8, however, requires a value for d. In general there is no easy way to estimate the value of d at room temperature using just crystallography data. Hence, for most compounds there is no obvious way to estimate the lattice thermal conductivity to better than about a factor of 5.[9,12]

For clathrate-like compounds, it has been experimentally observed by several groups[13-15] that as relatively small concentrations of rattlers (La, Ce, or Tl) are added to the skutterudite structure (Co_4Sb_{12}) there is an extremely rapid decrease in the lattice thermal conductivity. The mean free path, d, of the heat carrying phonons in these compounds is determined by the various scattering mechanisms in the crystal such as acoustic phonons, grain boundaries, electron-phonon scattering, static defects, voids and "rattlers". Resonant scattering by quasi-localized "rattlers" appears to be the dominant scattering mechanism responsible for the rapid decrease in the thermal conductivity as small amounts of Tl, La or Ce are placed in the voids. This mechanism is believed to be similar to the resonant scattering described by Pohl[16] for insulating crystals and by Zakrzewski and White[17] in insulating organic clathrates. It has been demonstrated that mass fluctuation scattering is much too weak to explain the rapid decrease in thermal conductivity.[13] The thermal resistivity of the lattice (1/thermal conductivity) at room temperature is shown in Fig. 3 as Tl is added to the voids in Co_4Sb_{12}. Thermal resistivity is shown rather than thermal conductivity because as a first approximation, the scattering rates for different scattering processes should add (Mathiesson's rule). There is a rapid initial increase in the thermal resistance, followed by a gradual saturation of the thermal resistance as higher concentrations of Tl are added to the voids. Within experimental error, there is no clear maximum in the thermal resistance data as a function of Tl concentration, and the maximum thermal resistance occurs near complete filling for both the Fe and Sn compensated compounds. The maximum attributed to mass fluctuation scattering by Meisner et al.[14] as a function of Ce filling is not observed in the present experiments.

If the Tl atoms are treated as localized Einstein oscillators as suggested by Keppens et al.[18], then the heat carrying phonon mean free path, d, should be a function of the distance between the Tl atoms in the crystal. The simplest estimate of the phonon mean free path is therefore the average distance between the Tl atoms. This implies that the phonon scattering from the Tl is so strong that d attains a minimum distance given by the average Tl–Tl separation. The scattering of acoustic phonons by the Tl should be a maximum when the acoustic phonon and rattling frequency are equal[16]; however, even at resonance it seems physically unlikely that d could be less than the Tl-Tl separation distance. This simple argument suggests that if the role of other scattering mechanisms is minimal, that the thermal resistivity should vary as $x^{1/3}$, where x is the Tl concentration (the average spacing between Tl atoms varies as $x^{-1/3}$). The additional thermal resistance generated as Tl is added to Co_4Sb_{12} reasonably follows a $x^{1/3}$ behavior (Fig 3), even though part of the thermal resistance is due to electron-phonon scattering and other scattering mechanisms.[15]

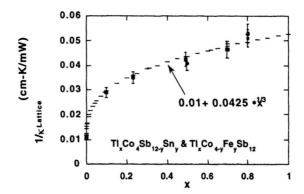

Fig. 3. Variation of the room temperature lattice thermal resistivity vs. the fraction of the voids filled with the rattler Tl. The average separation distance between Tl rattlers varies as $x^{-1/3}$. The square (circles) refer to charge compensation with Sn (Fe). For all but the two lowest Tl concentrations, $x \approx y$.

A plausible approximation for d in clathrate-like compounds is therefore the average separation distance between the rattlers, which *is* known from the crystallography data. At room temperature, this argument works well for the Tl-filled skutterudites (Fig 3). It also works for the filled skutterudites such as $LaFe_4Sb_{12}$. Using the measured thermal conductivity, heat capacity and an average value for the velocity of sound yields a mean free path of d = 9 Å (Sales et al. 1997). The nearest-neighbor distance of the La atoms in $LaFe_4Sb_{12}$ is 7.9 Å. The real test of this hypothesis is whether this analysis gives good estimates of the room temperature lattice thermal conductivity for a variety of clathrate-like compounds. For many clathrate-like systems replacing d in Eq. 8 by the average distance between the rattlers predicts a room temperature lattice thermal conductivty in suprisingly good agreement with experiment (Table 1). This means that for clathrate-like compounds *room temperature crystallography data can be used to provide a reasonable estimate of the lattice thermal conductivity*. This is significant since a low lattice thermal conductivity is a requirement for a good thermoelectric material.

RATTLER SEPARATION DISTANCE AND MEAN FREE PATH

A more realistic model of lattice heat conduction of a solid, within the Debye approximation, has been described by Callaway[20] and Klemens[21]. The lattice thermal conductivity, $\kappa_{Lattice}$, is given by:

$$\kappa_{Lattice} = 1/3 \int_0^{\omega_D} v^2 \tau(\omega,T) \frac{dC}{d\omega} d\omega \qquad (9)$$

with

$$\tau^{-1}(\omega,T) = \sum_i \tau_i^{-1}(\omega,T) \qquad (10)$$

where, ω_D is the Debye frequency, v is the Debye velocity of sound, τ_i is the relaxation time for the ith phonon scattering mechanism, T is the temperature in Kelvin, and $dC/d\omega$ is the specific heat per angular frequency. Within the Debye model, the specific heat per angular frequency is obtained from Eq. 6 by a change of variables, replacing x by $\hbar\omega/2\pi k_B T$. Notice that since v is a constant in the Debye model, the phonon mean free path is just:

$$d(\omega,T) = v\, \tau(\omega,T) \qquad (11)$$

In a crystal various processes can scatter phonons. For the present purposes we will follow the approach described by Pohl[16], and Walker and Pohl[22] and consider the minimum number of scattering mechanisms that can account for the experimental data. For a solid with no resonant scattering (no rattlers) the normal scattering mechanisms are grain boundary scattering, τ_B^{-1}, isotope or mass fluctuation scattering, τ_{iso}^{-1} (Klemens[21]) and phonon-phonon scattering (umklapp and normal), $\tau^{-1}_{U,N}$ (Walker and Pohl[22]). These various scattering terms are given by:

$$\tau_B^{-1} = v/L \qquad (12)$$

$$\tau_{iso}^{-1} = V_0 \Gamma \omega^4 / 4\pi = C\,\omega^4 \qquad (13)$$

with
$$\Gamma = \sum_i f_i \left(1 - \frac{m_i}{m}\right)^2 \qquad (14)$$

and
$$\tau^{-1}_{U,N} = B\omega^2 T\, e^{-b/T} \qquad (15)$$

with V_o is the atomic volume, L is the average grain size, m_i is the mass of the ith atom, m is the average atomic mass, and f_i is the relative concentration of the ith species. Equation 15 is a phenomenological expression that accounts for both umklapp and normal phonon-phonon scattering.[22] These three scattering processes (Eqs. 12-15) can account for the temperature dependence of the thermal conductivity of compounds with no rattlers, such as Co_4Sb_{12}. The resonant scattering by the rattlers can be phenomenologically described by a function that is proportional to the concentration of rattlers, and is peaked at the Einstein frequency of the rattler:

$$\tau^{-1}_{resonant}(\omega) = A_0 f(\omega - \omega_E) = A_0 T^2 \frac{\omega^2}{(\omega^2 - \omega_E^2)^2} \qquad (16)$$

where A_0 is proportional to the rattler concentration and the particular form of the function f is taken from Walker and Pohl.[22]

With regard to the clathrate-like compounds, simple calculations using only Eqs. 12-16 *do not* reproduce the behavior shown in Fig 3. The calculated thermal resistance increases linearly with the rattler concentration A_0 and does not saturate as is indicated in Fig 3.

One of the key ideas to understanding the data displayed in Fig 3 is the concept of a minimum thermal conductivity or a maximum thermal resistance first proposed by Slack in 1979.[23] In any solid, Slack proposed that it does not make sense to consider a mean free path for the heat carrying phonons that is less than an interatomic spacing ($d \approx 3$ Å). This hypothesis, which is born out by experiment,[24] suggests that for a crystalline compound the minimum thermal conductivity corresponds to a glass with the same composition for which $d \approx 3$ Å $= 3 \times 10^{-8}$ cm. This means that there is a cutoff for the maximum scattering rate given by $\tau^{-1}_{max} = v/3 \times 10^{-8}$ s^{-1}, where the velocity of sound is given in cm/s. For materials with thermal conductivities that approach within an order of magnitude or so of the minimum value, Eq 10 is replaced by:

$$d(\omega, T) = [\sum_i v\tau_i^{-1}(\omega, T)]^{-1} + d_{min} \qquad (17)$$

where $d_{min} \approx 3$Å.

To see if this approach produces reasonable results, the temperature dependence of the thermal conductivity of the unfilled skutterudite Co_4Sb_{12} (no rattlers) was fit using Eqs 9 and 17. The measured Debye temperature (307 K), velocity of sound (2.93×10^5 cm/s) and grain size (10^{-3} cm) were used as input parameters. The three constants (C, B and b) in Eqs. 13 and 15 were adjusted to give a fit to the experimental data (Fig 4). The values used were $C = 1.58 \times 10^{-42}$ sec^3,

B= 3.87x10^{-18} sec/K, and b= 150 K (These values are in the same range as found by Walker and Pohl,[22])

The resonant phonon scattering due to the addition of Tl rattlers was then modeled using the same parameters used to fit the Co_4Sb_{12} data plus a resonant term (Eq. 16) with $A_0 = A$ 7.42 x 10^{32} sec^{-3}K^{-2} where A is dimensionless and is proportional to the concentration of Tl rattlers. An Einstein frequency was taken from the experimental data[15] and corresponded to an Einstein temperature of 55 K. The effect on the thermal conductivity of increasing the strength of the resonant scattering is shown in Fig. 5. In this simple model there is expected to be a dip in the thermal conductivity at a temperature between 10 and 20 K for an Einstein temperature of

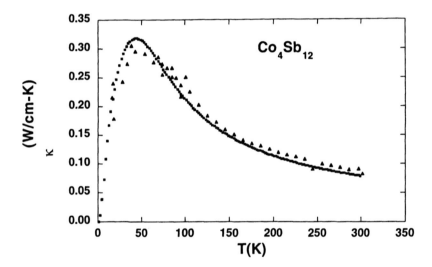

Fig. 4. Fit to the lattice thermal conductivity of a polycrystalline $CoSb_3$ sample with and average grain size of 0.001 cm using Eqs. 17, and 9-15. See text for details.

55 K. Careful thermal conductivity measurements in this temperature range have not been made for the Tl filled skutterudites. A dip has been recently seen, however, in a related compound, $Sr_8Ga_{16}Ge_{30}$.[25] The motivation, however, for calculating the lattice thermal conductivity as a function of Tl concentration was to see if the behavior in Fig 3 could be understood. The room temperature thermal resistance from the calculated data shown in Fig 5 is plotted in Fig 6 vs. the amplitude of the resonant scattering (the parameter A). A least squares fit of a power law to the data yields an exponent of 0.35. This exponent is relatively constant for large variations in A and is surprisingly close to the value of 1/3 obtained by simply using the average distance between rattlers for d. The calculated and measured thermal resistivity data can be compared directly if the proportionality constant is determined between A and the Tl concentration x. The room temperature thermal resistance of the $Tl_{0.8}Co_4SnSb_{11}$ sample is about 50 cm-K/W. This

value is obtained in the model calculation with a value of A≈20. A comparison between the calculated and measured thermal resistance agrees to within the experimental error.[9]

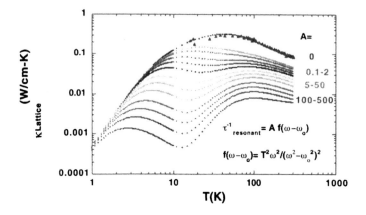

Fig. 5. Log of lattice thermal conductivity vs. log T calculated using Eqs.17, and 12-16. The strength of the resonant scattering and the concentration of rattlers are proportional to the parameter A (Eq. 16). The other parameters were the same as used to fit the Co_4Sb_{12} data (The triangles on the A=0 curve are the same Co_4Sb_{12} data shown in Fig 4.). See text for details

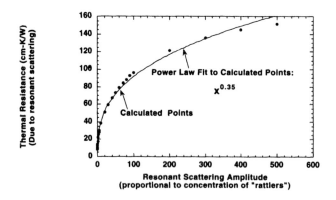

Fig. 6. Calculated thermal resistance at room temperature vs. the resonant scattering amplitude, A. The room temperature thermal resistance is from the calculated data shown in Fig 5. A is proportional to the concentration of rattlers, x.

Table 1 Comparison between thermodynamic quantities estimated using room temperature x-ray crystallography data (ADP columns) and the same quantities measured either directly or inferred from another technique.

	LaFe$_4$Sb$_{12}$		Tl$_2$SnTe$_5$		Ba$_8$Ga$_{16}$Ge$_{30}$	
	ADP[11]	Other	ADP[26]	Other	ADP[27]	Other
Θ_D (K)	299	305	159	160	275	300
Θ_E (K)	79	80	38	30	51	50
v_s (m/s)	2886	3007	1488	1500	2123	2300
d (Å)	7.9	9	6.45	6.5	5.38	12.3
κ_L (W/cm-K)	0.014	0.017	0.0039	0.004	0.007	0.016

ACKNOWLEDGEMENTS

It is a pleasure to acknowledge useful conversations with V. Keppens, T. Caillat,, J. W. Sharp, G. P. Meisner, G. Nolas and N. R. Dilley regarding this work and the technical assistance of J. A. Kolopus. This research was sponsored in part by a Cooperative Research and Development Agreement with Marlow Industries and in part by the Division of Materials Sciences, U. S. Department of Energy Contract No. DE-AC05-96OR22464

REFERENCES

1. W. F. Kuhs, *Acta Cryst.* **A48**, 80 (1992).
2. J.D. Dunitz, V. Schomaker, and K. N. Trueblood, *J. Phys. Chem.* **92**, 856 (1988).
3. C. Kittel, *Introduction to Solid State Physics, Third Edition* (John Wiley and Sons, New York 1968), pp. 69-70, p.186.
4. B. T. M. Willis and A. W. Pryor, *Thermal Vibrations in Crystallography* (Cambridge University Press, London 1975).
5. G. D. Mahan, B. C. Sales, and J. W. Sharp, *Physics Today*, **50**, No. 3, 42 (1997).
6. G. A. Slack in *CRC Handbook of Thermoelectrics*, edited by D. M. Rowe (Chemical Rubber, Boca Raton, FL, 1997) pp. 407-440.
7. H. J. Goldsmid, *Electronic Refrigeration* (Pion Limited, London 1986).
8. B. C. Sales, *MRS Bulletin* **23**, 15 (1998).

9. B. C. Sales, B. C. Chakoumakos and D. Mandrus, Chapter 3 in *Semiconductors and Semimetals: Thermoelectric Materials Research,* Edited by Terry M. Tritt (Academic Press, San Diego, 2000).
10. B. C. Sales, B. C. Chakoumakos, D. Mandrus and J. W. Sharp, *J. Solid State Chem.* **146**, 528 (1999).
11. D. J. Braun and W. Jeitschko, *J. Less-Common Metals*, **72**, 147 (1988).
12. H. J. Goldsmid, *Electronic Refrigeration* (Pion Limited, London, 1986).
13. G. S. Nolas, J. L. Cohn, and G. A. Slack, *Phys. Rev. B.* **58**, 164 (1998).
14. G. P. Meisner, D. T. Morelli, S. Hu, J. Yang and C. Uher, *Phys. Rev. Lett.* **80**, 3551 (1998).
15. B. C. Sales, B. C. Chakoumakos and D. Mandrus *Phys. Rev. B.* **61**, 2475 (2000).
16. R. O. Pohl, *Phys. Rev. Lett.* **8**, 481 (1962).
17. M. Zakrzewski and M. A. White, *Phys. Rev. B.* **45**,2809 (1992).
18. V. Keppens, D. Mandrus, B. C. Sales, B. C. Chakoumakos, P. Dai, R. Coldea, M. B. Maple, D. A. Gajewski, E. J. Freeman and S. Bennington, *Nature* **395**, 876 (1998).
19. B. C. Sales, D. Mandrus, B. C. Chakoumakos, V. Keppens and J. R. Thompson, *Phys. Rev. B.* **56**, 15081 (1997).
20. J. Callaway, *Phys. Rev.* **113**, 1046 (1959).
21. P. G. Klemens *in Solid State Physics* **7**, 1,(1958), Edited by F. Seitz and D. Turnbull (Academic Press, New York).
22. C. T. Walker and R. O. Pohl, *Phys. Rev.* **131**, 1433 (1963).
23. G. A. Slack, *in Solid State Physics* **34**,1 (1979) Edited by H. Ehrenreich, F. Seitz, and D. Turnbull (Academic Press, New York).
24 D. G. Cahill, S. K. Watson, and R. O. Pohl, *Phys. Rev. B.* **46**, 6131 (1992)
25. J. L. Cohn, G. S. Nolas, V. Fessatidis, T. H. Metcalf, and G. A. Slack, *Phys. Rev. Lett.* **82**, 779 (1999).
26. V. B. Agafonov, B. Legendre, N. Rodier, J. M. Cense, E. Dichi, and G. Kra, *Acta Cryst.* **C47**, 850 (1991).
27. B. Eisenmann, H. Schafer, and R. Zagler, *J. Less-Common Metals* **118**, 43 (1986).

Investigation of the thermal conductivity of the pentatellurides ($Hf_{1-x}Zr_xTe_5$) using the parallel thermal conductance technique.

B. M. Zawilski,[1] R. T. Littleton IV,[2] Terry M. Tritt,[1,2] D. R. Ketchum[3] and J. W. Kolis[3]
1 Department of Physics and Astronomy
2 Materials Science and Engineering Department
3 Department of Chemistry
Clemson University, Clemson, SC 29634 USA

ABSTRACT

The pentatelluride materials ($Hf_{1-x}Zr_xTe_5$) have recently garnered much interest as a potential low temperature thermoelectric material. Their power factor exceeds that of the current Bi_2Te_3 materials over the temperature range 150 K < T < 350 K. A formidable challenge has been the capability of measuring the thermal conductivity of small needle-like samples (2.0 x 0.05 x 0.1 mm^3) such as pentatellurides ($Hf_xZr_{1-x}Te_5$) due to heat loss and radiation effects. However in order to fully evaluate any material for potential thermoelectric use, the determination of the thermal conductivity of the material is necessary. We have recently developed a new technique called the parallel thermal conductance (PTC) technique to measure the thermal conductivity of such small samples. In this paper we describe the PTC method and measurements of the thermal conductivity of the pentatelluride materials will be presented for the first time. The potential of these materials for low temperature thermoelectric applications will be further evaluated given these results as well as future work and directions will be discussed.

INTRODUCTION

Recently there have been extensive efforts to develop new thermoelectric (TE) materials especially below room temperature.[1,2] Higher efficiency thermoelectric materials at lower temperatures could eventually lead to TE modules that would be extremely beneficial in localized cooling of many electronic devices.[3] The efficiency of a thermoelectric material is proportional to the dimensionless figure of merit:

$$ZT = \frac{\alpha^2 \sigma T}{\lambda} = \frac{\alpha^2 T}{\rho \lambda}, \quad (1)$$

where; α is the Seebeck coefficient (or thermoelectric power), σ is the electrical conductivity, ρ (1/σ) is the electrical resistivity, T is the temperature, and λ is the thermal conductivity. The thermopower needs to be sufficiently large so a large Peltier effect is observed. Electrical resistivity should be minimized to reduce the Joule heating (I^2R) contribution. The thermal conductivity in a good thermoelectric material should also be minimized so that a temperature difference, ΔT, may be established and maintained across the material. The numerator or power factor, $\alpha^2 \sigma T$, can typically be tuned through chemical doping and substitution. The total thermal conductivity exhibits contributions from both the lattice and an electronic part ($\lambda_{TOT} = \lambda_L + \lambda_E$, respectively). The electronic thermal conductivity is related to σ through the Wiedemann-Franz law [($\lambda_E = L_o\sigma T$) where L_o is the Lorentz number ($L_O = 2.45 \times 10^{-8}$ V^2/K^2)].

Obtaining a large Seebeck coefficient at low temperatures is one of the more important issues for discovering or optimizing thermoelectric materials for low temperature applications. Heavy fermion materials, Kondo systems, as well as quasi-one-dimensional materials are a few of a number of systems that are candidates for low temperature thermoelectric materials. We have recently been investigating a class of materials known as pentatellurides (HfTe$_5$ and ZrTe$_5$).[4,5,6] Both parent materials, HfTe$_5$ and ZrTe$_5$, exhibit a resistive peak, $T_p \approx 80$ K for HfTe$_5$ and $T_p \approx 145$ K for ZrTe$_5$. In addition, each material displays a large positive (p-type) thermopower ($\alpha \geq +125$ μV/K) around room temperature, which undergoes a change to a large negative (n-type) thermopower ($\alpha \leq -125$ μV/K) below the peak temperature. The temperature at which the thermopower crosses zero, T_o, correlates well with the peak temperature of the resistivity, T_P. Doping with Se on the Te site (HfTe$_{5-x}$Se$_x$) has yielded a power factor that exceeds that of the Bi$_2$Te$_3$ materials over the temperature range 150 K < T < 350 K. Results on the figure of merit of these pentatelluride materials are presented in another paper in these proceedings.[7]

EXPERIMENTAL PROCEDURE AND BACKGROUND

Pentatelluride single crystals were grown in our laboratories under similar conditions to previously reported methods.[4,8] Ribbon-like crystals were obtained in excess of 1.5 mm long and 100 μm in diameter with the preferred direction of growth along the *a*-axis, as determined by face indexing. The pentatellurides are composed of long chain systems with an orthorhombic crystal structure.[9] They exhibit slightly anisotropic transport properties with the high conductivity axis along the growth, *a*, axis.

Initially the "Harman" technique was attempted in order to determine the figure of merit of samples of these dimensions.[10] Oriented ingots of Bi$_2$Te$_3$ materials were first employed, using different sample sizes, making them ever smaller by cleaving until the dimensions were comparable to that of the pentatellurides. Briefly, a current is passed through the sample to employ the Harman technique, which yields a measure of ZT of the sample. Then the thermal conductivity could be determined by measuring the resistivity and thermopower with separate measurements. It was discovered that the overall resistance of the sample became very important (recall R = ρL/A where A is the sample cross section) until Joule heating (I^2R) overshadowed the Peltier effect (αIT) necessary for the utilization of the Harman technique. These results will be presented in a short note elsewhere.[11] The possibility of this technique working for the pentatellurides is unlikely given the failure with the small cross section Bi$_2$Te$_3$ samples. Thus it was necessary to develop a technique which is appropriate for samples of these given dimensions. Subsequently, the thermal conductivity of the pentatellurides was measured using a Parallel Thermal Conductance (PTC) technique, which is the main thrust of this paper.[12] The PTC method requires the thermal conductance of the sample holder to be known (measured) and subtracted from the measured thermal conductance of the sample holder with the mounted sample. Sample conductance is measurable down to 10^{-6} W/K. Also, the accurate determination of sample dimensions is also a challenge and is still a source of large uncertainty. Usually, short fat samples have been preferred in order to minimize radiation losses. Usually the thermal conductivity is determined by the following relationship P = (λA/L)ΔT, by performing power vs. ΔT sweeps at fixed base temperatures. P is the power through the sample (P = I^2R_H), λ is the thermal conductivity of the sample, L is the length between thermocouple leads between which the temperature difference, ΔT, is determined and R_H is the resistance of the sample heater. Typically a standard steady state technique is used to measure the thermal conductivity of larger samples (3-10 mm diameter by approximately 5-10 mm long). The steady state technique is straightforward but much care must be taken to account for radiation, convection and heat conduction losses down the leads and through the any gas surrounding the sample. The determination of the thermal conductivity of a material to within an accuracy of 10% is considered a very good measurement, especially given the uncertainty in sample dimensions.

Parallel Thermal Conductance Method:

Due to the inability of the smaller pentatelluride crystals to support a heater and thermocouple, a sample holder has been developed. First, the characteristics of the sample holder itself must be measured, which determines the base line or background thermal conduction along with any losses associated with the sample stage. The second step consists of attaching the sample and measuring the new characteristics of the system. The concept of the sample stage with and without a sample is illustrated in Figure 1. By subtraction, the parallel thermal conductance (PTC) is calculated. This conductance is due to the sample, thermal contacts, and black body radiation from the sample. The thermal conductance due to the baseline and the baseline plus sample, and subsequently the deduced sample thermal conductance are shown in Figure 2. An advantage of this technique is the easiness and rapidity of the measurement. The entire measurement procedure, including the mounting process, cooling time, and 35 power-sweep points between 10 K and room temperature takes approximately 24 hours. Each power-sweep point at a stable temperature, typically consists of 3 stepped heater powers applied in order to determine the slope dP/d[ΔT]. Only one heater, one thermometer, and one differential thermocouple are necessary. This technique is also well adapted for a broad range of samples. Any sample of at least 1mm in length with a thermal conductance of at least one-tenth that of the baseline ($K_{BL} \approx 1\times10^{-4}$ W/K at room temperature) and (since this technique is a two-probe measurement) no more than one tenth that of the thermal contact conductance ($K_{Contact} \approx 5\times10^{-3}$ W/K at room temperature) are measurable. This technique is applicable for a broad range of materials of low thermal conductance, which is generally tied with small cross-area with respect to length.

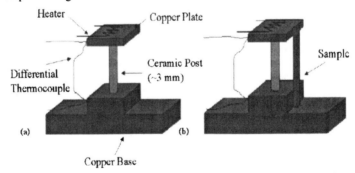

Fig. 1) Sample holder configuration: a) Sample holder alone, b) Sample holder with sample.

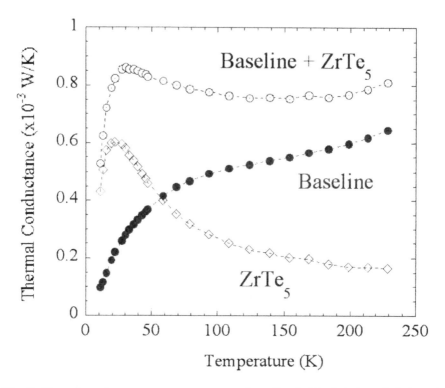

Fig. 2) Thermal conductance versus temperature: Base line (solid circles), typical thermal conductance of sample holder with sample (open circles), and sample thermal conductance (opened diamonds).

RESULTS

Validation of this system, as with any apparatus, requires evaluation of suitable standards. The thermal conductivity of standards as well as the reproducibility of the thermal contacts and base line stability were systematically and thoroughly evaluated. These results are presented in detail elsewhere.[12] With this technique we have measured, for the first time, the thermal conductivity of a pentatelluride sample in order to determine thermoelectric figure of merit for these materials. The thermal conductivity as a function of temperature of a number of pentatelluride samples is shown in Figure 3. The results show that the thermal conductivity exhibits a $\approx 1/T$ temperature dependence, which is typical of a single crystal material. Also evident is the "signature" peak in the thermal conductivity at lower temperatures, in this case around 20-40 K. This peak is due to the depletion of Umklapp scattering of the phonons (at T <

$\Theta_D/2$) and a crossover to other phonon scattering mechanisms, which are more predominant at lower temperatures, such as impurities, boundary scattering and size effects. As evident the room temperature thermal conductivity is between 4 - 8 W/m•K for these materials and rising to a thermal conductivity of 10 - 35 W/m•K at the peak at lower temperature, T ≈ 20 – 40 K. As with other materials, the magnitude and quality (sharpness) of the thermal conductivity peak is very dependent on sample perfection and quality. Accurate determination of the sample dimensions is still an issue and adds to the uncertainty of the overall determination of the thermal conductivity of these materials. A large number of samples of each composition have been measured and representative data of the average is shown in Figure 3. Not only is there uncertainty due to accurate determination of sample dimensions, but sample quality (uniformity of shape, magnitude of peak, etc.) is also an important factor, which affects the overall results. A detailed study of the thermal conductivity of these materials will be reported elsewhere.[13]

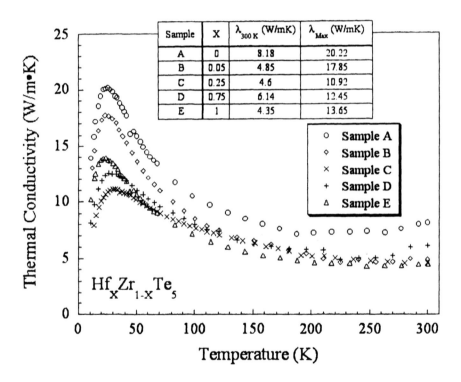

Fig. 3.) Thermal conductivity of $Hf_xZr_{1-x}Te_5$ for several x values versus temperature.

SUMMARY

A relatively simple (at least in principle) and efficient technique has been described, which allows the measurement of the thermal conductivity of small samples such as pentatellurides. The thermal conductivity of the pentatellurides is measured for the first time and is between 4 - 8 W/m·K at room temperature and rising to a thermal conductivity of 15 - 35 W/m·K at the peak at lower temperature, $T \approx 20 - 40$ K. There are still issues related to uncertainty due to the determination of sample dimensions. Future directions will incorporate studies not only to enhance the power factor (electronic properties) but also to reduce the thermal conductivity, now that it can be determined. Recent results on Th doped pentatellurides indicate that the thermal conductivity can be as low as 1-2 W/m·K, which is comparable to the best thermoelectric materials.

ACKNOWLEDGMENT

The authors would also like to acknowledge financial support from ARO/DARPA (grant #DAAG55-97-1-0267) for the research funds provided for this work. We would like to also acknowledge A. Pope for her assistance in the early stages of this work. We also would like to acknowledge Chris Gardner of Quantum Design for certain aspects that aided in these measurements.

REFERENCES

1.) F. J. DiSalvo, Science, **285**, 703-6 (1999)
2.) D. Y. Chung, T. Hogan, P. Brazis, M. Rocci-Lane, C. Kannewurf, M. Bastea, C. Uher, and M. G. Kanatzidis, Science, **287**, 1024-7 (2000)
3.) T. M. Tritt, Science, **283**, 804-5 (1999)
4.) R.T. Littleton, IV, J. W. Kolis, C. R. Feger and T.M. Tritt, MRS 1998: *Thermoelectric Materials 1998.* Edited by T. M. Tritt et. al. Vol. **545**, p 381-396
5.) R. T. Littleton IV, T. M. Tritt, C. R. Feger, J. Kolis, M. L. Wilson, M. Marone, J. Payne, D. Verebeli, and F. Levy, Appl. Phys. Lett., **72**, 2056-8 (1998)
6.) R. T. Littleton, T. M. Tritt, J. W. Kolis, and D. Ketchum, Phys. Rev. B, **60**, 13453-7 (1999)
7.) R.T. Littleton, IV, T. M. Tritt, B. T. Zawilski, D. R. Ketchum and J. W. Kolis, *Thermoelectric Materials 2000* (these proceedings) Edited by T. M. Tritt et. al.
8.) F. Levy and H. Berger, J. Cryst. Growth, **61**, 61-8 (1983)
9.) S. Furuseth, L. Brattas, and A. Kjekshus, Acta. Chem. Scand., **27**, 2367-2374 (1973)
10.) T. C. Harmann, J. Appl. Phys. **29**, 1373, (1958)
11.) B.M. Zawilski, R.T. Littleton IV, T.M. Tritt, unpublished
12.) B.M. Zawilski R.T. Littleton IV, and T.M. Tritt, to be submitted to Rev. Sci. Instrum.
13.) B.M. Zawilski, R.T. Littleton IV, and T.M. Tritt, unpublished

COMPOSITIONAL AND STRUCTURAL MODIFICATIONS IN TERNARY BISMUTH CHALCOGENIDES AND THEIR THERMOELECTRIC PROPERTIES

Duck-Young Chung[1], Melissa A. Lane[2], John R. Ireland[2], Paul W. Brazis[2], Carl R. Kannewurf[2], Mercouri G. Kanatzidis[1*]
[1]*Department of Chemistry and Center for Fundamental Materials Research, Michigan State University, East Lansing, MI 48824.* [2]*Department of Electrical and Computer Engineering, Northwestern University, Evanston, IL 60208*

ABSTRACT

Based on the versatile combination of PbQ- and Bi_2Q_3-type (Q = S, Se, Te) fragments, we explored new compounds in the Pb/Bi/Se ternary system. The new class of compounds, $Pb_5Bi_6Se_{14}$, $Pb_5Pb_{12}Se_{23}$, and $PbBi_8Se_{13}$ are homologues with different combination of alternating Bi_2Se_3- and PbSe-type layers. α- and β-$Pb_6Bi_2Se_9$ were obtained in different synthetic conditions and the former is isostructural to heyrovskyite ($Pb_6Bi_2S_9$) while the latter is a NaCl-type cubic phase. $Pb_5Bi_6Se_{14}$ shows a power factor of 11.2 $\mu W/cm \cdot K^2$ with electrical conductivity of 657 S/cm and thermopower of -131 $\mu V/K$ at 271 K. The most significant characteristic of this material is the extremely low thermal conductivity of less than 1.0 $W/m \cdot K$ at room temperature. On the basis of these properties, a preliminary doping study for $Pb_5Bi_6Se_{14}$ with Sn, Sb, and $SbBr_3$ as dopants was undertaken and the results are presented in this report.

INTRODUCTION

The need of high performance thermoelectric materials is now recognized and, in addition to further optimization of the $Bi_{2-x}Sb_xTe_{3-y}Se_3$ alloys, exploration of new materials is necessary to meet technological and societal demands. We have initiated exploratory research to identify new multinary systems, particularly, with Bi, Pb, S, Se and Te for thermoelectric applications. This class of compounds exhibit various combinations[1] of fragments which are essentially "excised" out of the structures of PbQ and Bi_2Q_3, as observed in the widely known mineral sulfo and selenosalts; $PbBi_2Q_4$(Q = S, Se, Te)[2, 3, 4], $PbBi_4Q_7$(Q = S, Se, Te)[4, 5, 6], $PbBi_4S_8$[7], $PbBi_6S_{10}$[8], $Pb_3Bi_2S_6$[9], $Pb_6Bi_2S_9$[10], $Pb_2Bi_2Se_5$[5], and $Pb_3Bi_4Se_9$[11]. This structural and compositional versatility of Pb/Bi/Q system motivated us to find new compounds with these elements. In addition, Pb/Bi/Q is a very interesting chemical system from the thermoelectric point of view. Binary bismuth and lead chalcogenides such as Bi_2Te_3, PbTe, and their solid solutions have been intensively studied for thermoelectric applications, however, the physical origin of their high thermoelectric figure of merit[12] of these compounds remains under investigation. Physical and chemical similarity of heavy Pb and Bi atoms creates widely disordered structures that may lead to low thermal conductivity by increasing phonon scattering. Therefore, this system should be explored for new thermoelectric materials. We have synthesized candidate materials such as $Pb_5Bi_6Se_{14}$, $PbBi_8Se_{13}$, $Pb_5Bi_{12}Se_{23}$, α- and β-$Pb_6Bi_2Se_9$ which are narrow gap semiconductors. Along with their thermoelectric and physicochemical properties, we present the fascinating structural features of these compounds and their interrelationships.

EXPERIMENTAL DETAILS

The starting materials, PbSe and Bi_2Se_3, were prepared using high purity (>99.999 %) elements by melting at high temperature. Stoichiometric amounts of PbSe and Bi_2Se_3 fine powder for the desired compound were thoroughly mixed and sealed in a silica tube under vacuum below 3×10^{-4} Torr. $Pb_5Bi_6Se_{14}$ and α-$Pb_6Bi_2Se_9$ were prepared by heating the mixture at 600 °C for 10 days and $Pb_5Bi_{12}Se_{23}$ at 750 °C for 1 h in a furnace. All compounds except α-$Pb_6Bi_2Se_9$ could also be obtained by melting the mixture in a flame for 2 min and quenching it in air. For $Pb_5Bi_{12}Se_{23}$, the yield was approximately 65 % and the major by-product turned out to be $Pb_5Bi_6Se_{14}$ and Bi_2Se_3. All other compounds were obtained in quantitative yield. The composition and purity of the products were identified by SEM/EDS and XRD.

All Pb/Bi/Se compounds obtained are air-stable and melt incongruently. The melting/decomposition temperatures were determined by Differential Thermal Analysis (DTA) as follows; 744 °C for $Pb_5Bi_6Se_{14}$, 689 °C for $PbBi_8Se_{13}$, 994 °C for α-$Pb_6Bi_2Se_9$, 867 °C for β-$Pb_6Bi_2Se_9$.

Charge transport measurements were done on pressed pellets prepared by pressing ~ 0.1 g of finely ground materials at ~0.85 ton/mm^2 for 20 min at room temperature. The obtained pellets in regular rectangular shape with a dimension of $1 \times 2 \times 6$ mm^3 were annealed at 500 °C for 3 days under vacuum.

RESULTS AND DISCUSSION

$Pb_5Bi_6Se_{14}$ has a three dimensional (3 D) structure composed of Bi_2Se_3-type and PbSe-type fragment layers alternating at 1:1 ratio, see Figure 1(a). This is similar to the sulfo-salt, so-called cannizzarite ($Pb_{46}Bi_{54}S_{127}$), with a unit cell of a = 189.8 Å, b = 4.09 Å, c = 74.06 Å, β = 11.93° and a space group $P2_1/m$. While this sulfo-salt mineral reveals incommensurate structural features, the seleno-salt did not appear to show this property probably because the misfit between -Pb-S-Pb-S- distance of cubic lattice and -S-S-S-S- distance of rhombohedral lattice is more prominent than the corresponding -Pb-Se-Pb-Se- and -Se-Se-Se-Se- distances. We note, however, that extensive investigations for the presence of a superstructure were not performed. $Pb_5Bi_{12}Se_{23}$ is composed of Bi_2Se_3- and PbSe-type layers alternating at 2:1 ratio, which produces a 2-D structure with a van der Waals gap between Bi_2Se_3-layers, see Figure 1(b). $PbBi_8Se_{13}$ consists of a 2-D structure with the 3:1 ratio of Bi_2Se_3- and PbSe-type layers which include a separated Bi_2Se_3-type layer by van der Waals gaps, see Figure 1(c). This structural evolution implies that more combination between a Bi_2Se_3-type and a PbSe-type single layer and/or even thicker PbSe-type layer could be possible. It is intriguing to consider these homologous members as "natural superlattices" in which one of the components is variable. Perhaps some of the beneficial effects expected from the thermoelectric superlattice structures currently investigated by other groups[13] may be found in the compounds reported here, only in our case the materials are actually bulk systems.

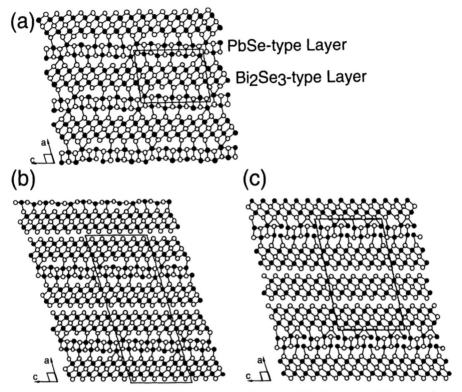

Figure 1. Structure of (a) $Pb_5Bi_6Se_{14}$, (b) $Pb_5Bi_{12}Se_{23}$, (c) $PbBi_8Se_{13}$ viewed down the b-axis. Open circles represent Se atoms and black and grey ones metal atoms.

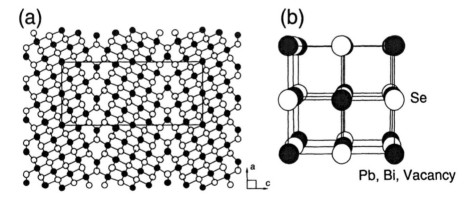

Figure 2. Structure of (a) α-$Pb_6Bi_2Se_9$ viewed down the b-axis and (b) cubic β-$Pb_6Bi_2Se_9$.

Two different structure types of $Pb_6Bi_2S_9$ were found in our investigation. The α-$Pb_6Bi_2Se_9$ possesses a 3-D structure isostructural to the well-known sulfo-salt heyrovskyite, $Pb_6Bi_2S_9$, which is one of the rarely found Pb-rich compounds. The structure of α-$Pb_6Bi_2Se_9$ belongs to a family of lillianite ($Pb_3Bi_2S_6$) homologs[14] which have a common structural motif consisting of the archetypal PbS layers cut by mirror planes, see Figure 2(a). The NaCl-type cubic β-$Pb_6Bi_2Se_9$ is composed of 6Pb, 2Bi, and one vacancy in cationic Na sites and 9Se in anionic Cl sites, see Figure 2(b).

Because all compounds are valence precise, they are expected to be semiconductors. Their energy gaps were observed at 0.53 eV for $Pb_5Bi_6Se_{14}$, 0.43 eV for $PbBi_8Se_{13}$, and 0.41 and 0.34 eV for α-$Pb_6Bi_2Se_9$.

The electrical conductivity and thermopower of these compounds were measured on single crystal samples, see Table 1. The compounds are all n-type which indicates the major charge carriers are electrons. $Pb_5Bi_6Se_{14}$ shows the most interesting properties with 657 S/cm and -131 μV/K at room temperature which corresponds to a power factor of 11.2 μW/cm·K^2. As shown in Figure 3, the most significant feature of $Pb_5Bi_6Se_{14}$ is extremely low thermal conductivity below 1.0 W/m·K at room temperature, even without considering radiative losses which are apparent in the data. This could be a common characteristic in this class of compounds probably because of wide range of disorder between the heavy Pb/Bi atoms involved in the structure.

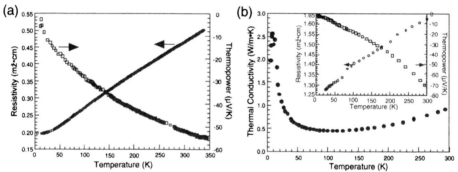

Figure 3. Variable temperature charge transport properties of (a) an undoped single crystal $Pb_5Bi_6Se_{14}$ and thermal conductivity data of (b) an ingot sample. The corresponding charge transport properties of the ingot sample is shown in the inset of (b).

Because of the highest power factor of $Pb_5Bi_6Se_{14}$ we selected it for further doping investigations. As a preliminary step for this study, relatively high concentrations of the dopants Sn, Sb and $SbBr_3$ were added. Due to the incongruent melting behavior of this class of compounds, the doped materials were obtained as polycrystalline agglomerates by the solid state diffusion of PbSe/Bi_2Se_3/dopant below the melting point of $Pb_5Bi_6Se_{14}$. As shown in Table 1, however, the resulting power factors of the doped materials are not very encouraging. Because these measurements were carried out on pressed pellet samples the conductivities are 3 to 5 orders of magnitude lower than the single crystal samples.

Table 1. Room Temperature Electrical Conductivity, Thermopower, and Power Factors Measured on Single crystals of α- and β-$Pb_6Bi_2Se_9$, $Pb_5Bi_{12}Se_{23}$, $PbBi_8Se_{13}$, $Pb_5Bi_6Se_{14}$, and on the Pressed Pellet Samples of $Pb_5Bi_6Se_{14}$ doped with Sn, Sb, and $SbBr_3$.

	Dopant	Conc. (mol.%)	Conductivity (S/cm)	Thermopower (μV/K)	Power Factor (μW/cm·K^2)
α-$Pb_6Bi_2Se_9$	–	–	502.4	-133.5	8.95
β-$Pb_6Bi_2Se_9$	–	–	889.1	-29.8	0.80
$Pb_5Bi_{12}Se_{23}$	–	–	1708.1	-52.1	4.64
$PbBi_8Se_{13}$	–	–	2165.3	-55.1	6.57
$Pb_5Bi_6Se_{14}$[a]	–	–	657.1	-130.7	11.23
$Pb_5Bi_6Se_{14}$	Sn	0.5	182.0	-144.6	3.81
		1.0	261.7	-104.6	2.86
		3.0	219.5	-116.2	2.96
		5.0	380.7	-41.6	0.66
$Pb_5Bi_6Se_{14}$	Sb	0.5	256.1	-98.1	2.46
		1.0	153.2	-129.5	2.57
		3.0	464.3	-55.9	1.45
		6.0	603.0	-44.7	1.20
$Pb_5Bi_6Se_{14}$	$SbBr_3$	0.05	256.1	-54.6	0.76
		0.10	317.6	-52.9	0.89
		0.30	304.4	-39.8	0.48
		1.00	318.1	-49.6	0.78

[a] The values at 271 K

Although there are significant changes in electrical conductivity and thermopower by doping $Pb_5Bi_6Se_{14}$, the magnitude of power factors are very low. This indicates that carrier concentrations are very high suggesting the amounts of Sn, Sb, and $SbBr_3$ used were large enough to "overdope" the material. We conclude that low doping levels appears to be more effective in improving the power factor of $Pb_5Bi_6Se_{14}$ and these levels should be examined in further studies.

CONCLUSIONS

We synthesized the new homologous compounds, $Pb_5Bi_6Se_{14}$, $Pb_5Bi_{12}Se_{23}$, $PbBi_8Se_{13}$, α- and β-$Pb_6Bi_2Se_9$, which are narrow-gap semiconductors in a range of 0.3 to 0.6 eV. A very interesting structural modulation by different combination of PbSe- and Bi_2Se_3-type layers is presented in the first three compounds. Wide range disorder of Pb/Bi involved in the structures may account for the extremely low thermal conductivity, particularly as shown in $Pb_5Bi_6Se_{14}$. $Pb_5Bi_6Se_{14}$ and α-$Pb_6Bi_2Se_9$ showed the most interesting charge transport properties and $Pb_5Bi_6Se_{14}$ was chosen as a host material for preliminary doping study with Sb, Sn, and $SbBr_3$. Although the results are not encouraging, this material showed sensitivity to doping agents at high doping levels. Based on the properties of the doped materials, lower doping levels should be more effective in improving the power factor. Other types of dopants should be investigated as well.

ACKNOWLEDGMENTS

This work was supported at NU and MSU by the Office of Naval Research grant no. N00014-98-1-0443. Work at NU made use of the Central Facilities supported by the National Science Foundation through the NU Materials Research Center (DMR-9632472).

REFERENCES

1. E. Makovicky, *EMU Notes in Mineralogy*, vol. 1, chap.5, 315-343 (1997)
2. (a) Iitaka, Y.; Nowacki, W. *Acta Cryst.*, **1962**, *15*, 691-698. (b) Takeuchi, Y.; Takagi, J. *Proc. Japan Acad.*, **1974**, *50*, 222-225.
3. Agaev, K. A.; Semiletov, S. A. *Soviet Phys. - Crystallogr.*, **1968**, *13*, 201-203.
4. Zhukova, T. B.; Zaslavskii, A. I. *Kristallografiya*, **1971**, *16*, 918-922.
5. Takeuchi, Y.; Takagi, J.; Yamanaka T. *Proc. Japan Acad.*, **1974**, *50*, 317-321.
6. Agaev, K. A.; Talybov, A. G.; Semiletov, S. A. *Kristallografiya*, **1966**, *11*, 736-740.
7. Takeuchi, Y.; Ozawa, T.; Takagi, J. *Z. Kristallograhie*, **1979**, *150*, 75-84.
8. Otto, H. H.; Strunz, H. *N. Jb. Miner. Abh.*, **1968**, *108*, 1-9.
9. Takagi, J.; Takeuchi, Y. *Acta Cryst.*, B, **1972**, *28*, 649-651.
10. Takeuchi, Y.; Takagi, J. *Proc. Japan Acad.*, **1974**, *50*, 76-79.
11. Liu, H.; Chang, L. L. Y. *Amer. Miner.*, **1994**, *79*, 1159-1166.
12. CRC Handbook of Thermoelectrics, Edited by D. M. Rowe, CRC Press, Boca Raton, 1995.
13. (a) Hicks, L. D.; Dresselhaus, M. S. *Phys. Rev. B*, **1993**. *47*, 12727-12731. (b) Venkatasubramanian, R.; Siivola, E.; Colpitts, T.; O'Quinn, B. *18th International Conference on Thermoelecrics,* Baltimore, USA, August 1999, p100-103. (c) Harman, T. C. et al, *16th International Conference on Thermoelecrics,* Dresden, Germany, August 1997, p416-423. (d) Harman, T. C.; Taylor, P. J.; Spears, D. L.; Walsh M. P. *18th International Conference on Thermoelecrics,* Baltimore, USA, August 1999, p280-284.
14. E. Makovicky, *EMU Notes in Mineralogy*, vol. 1, chap.3, 237-271 (1997)

Doping Studies of n-Type CsBi₄Te₆ Thermoelectric Materials

Melissa A. Lane[1], John R. Ireland[1], Paul W. Brazis[1], Theodora Kyratsi[2], Duck-Young Chung[2], Mercouri G. Kanatzidis[2], and Carl R. Kannewurf[1]
[1]Dept of Electrical and Computer Engineering, Northwestern University, Evanston, IL 60208.
[2]Department of Chemistry, Michigan State University, East Lansing, MI 48824.

ABSTRACT

We have previously reported the successful p-type doping of $CsBi_4Te_6$ which had a high figure of merit at temperatures below 300 K. In this study, several dopants were explored to make n-type $CsBi_4Te_6$. A program of measurements was performed to identify the optimum doping concentration for several series of dopants. The highest power factors occurred around 125 K for the 0.5% Sn doped $CsBi_4Te_6$ sample which had a power factor of 21.9 $\mu W/cm \cdot K^2$ and 1.0% Te doped $CsBi_4Te_6$ which had a power factor of 21.7 $\mu W/cm \cdot K^2$.

INTRODUCTION

Research on new thermoelectric materials for potential refrigeration applications focuses on discovering materials which function better than the $Bi_{2-x}Sb_xTe_{3-y}Se_y$ alloys at temperatures in the vicinity of and below 300 K [1,2,3]. We have previously reported a new p-type single crystal material, $CsBi_4Te_6$ doped with 0.05 mol% SbI_3, which outperforms the optimally doped $Bi_{2-x}Sb_xTe_3$ p-type alloy below room temperature [4]. This p-type material has a maximum ZT of ~ 0.8 at ~ 225 K [4]. It is necessary to find a suitable n-type dopant for $CsBi_4Te_6$ to develop an optimized thermoelectric device below room temperature. This paper reviews the initial results of testing various concentrations of several different dopants: Sn, Ge, Pr, and Te, in $CsBi_4Te_6$.

EXPERIMENTAL

Conductivity and dc Hall measurements were taken from 4.2 to 340 K using a computer controlled, five-probe technique [5]. The voltage sensing electrodes were 25 μm diameter gold wire, placed approximately 0.1 cm apart. The current electrodes were 60 μm diameter gold wire and were mounted across the ends of the samples. The electrode wires were attached to the sample with gold paste. For these materials, contacts made with gold paste were found to be superior to those made with silver paste. Hall measurements were performed with a 7.4 kG magnetic flux density and typically with a 1 mA applied current. All voltages were measured using a Keithley 181 nanovoltmeter. Hall measurements occurred in a reduced pressure (~10 mTorr) atmosphere of dry helium gas to improve thermal equilibrium. The relationship $R_H = -1/ne$ was used for determining carrier concentrations. The chemical synthesis procedure for these materials was performed as described previously [4,6].

Variable temperature thermopower measurements were taken on single crystal material using a computer controlled slow-ac technique from 4.2 to 295 K [7]. During the measurement procedure, the samples were kept in a 10^{-5} Torr vacuum. The samples were attached with gold paste to two 60 μm diameter gold wires which were in turn attached to separate quartz blocks with resistive heaters. Temperature gradients of 0.1 to 0.4 K were applied and measured with Au (0.07%Fe)/Chromel differential thermocouples. The voltage electrodes were 10 μm diameter gold wires. The sample and thermocouple voltages were measured with Keithley 2181 and Keithley 182 nanovoltmeters, respectively. Thermal conductivity data for some dopants were obtained by the computer controlled procedure described elsewhere [8,9].

RESULTS

The variable temperature electrical conductivity for Sn doped $CsBi_4Te_6$ is shown in Figure 1 (with the undoped $CsBi_4Te_6$ data from Ref. [10]). The samples show a weak metallic dependence in which the conductivity decreases as the temperature increases from 4 K to 340 K. The highest electrical conductivities at 295 K are for the 0.5% Sn and the 1.25% Sn samples with values of 1650 S/cm and 1380 S/cm respectively. The conductivity increases as the dopant is added until it peaks at 0.5% Sn, and then decreases to the conductivity value for 0.75% Sn. As doping increases from 0.75% to 1.25% Sn, the conductivity increases, until it reaches a maximum at 1.25% Sn. After reaching this second peak at 1.25% Sn, the conductivity begins to decrease.

The variable temperature thermopower data for the Sn doped $CsBi_4Te_6$ samples are shown in Figure 2 (with the undoped $CsBi_4Te_6$ data from Ref. [10]). The highest thermopower magnitude was for the 1.0% doped Sn which had a Seebeck coefficient of -79.6 µV/K at 163 K. As found in the conductivity study, there are two maximums as the concentration of the Sn dopant is varied. The Seebeck coefficient increases as the dopant is increased from 0.2% Sn to 0.4% Sn and then decreases as the dopant is increased above 0.5% Sn. As the concentration of Sn is increased to 1.0%, the thermopower again reaches a maximum and then steadily decreases.

The 0.5% Sn doped $CsBi_4Te_6$ sample was selected for Hall measurements based on the power factor performance. The carrier concentration and mobility are shown in Figure 3A. The carrier concentration of this sample at 295 K is 2.11×10^{19} cm^{-3} and the mobility is 3.25×10^2 $cm^2/V \cdot s$. The carrier concentration varies from 1.2×10^{19} cm^{-3} at 22.4 K to 2.38×10^{19} cm^{-3} at 312 K, which is close to the optimum value of 10^{19} cm^{-3} that is predicted for thermoelectric materials [11]. The mobility decreases as the temperature increases, changing from 4.6×10^3 $cm^2/V \cdot s$ at 22.4 K to 3.02×10^2 $cm^2/V \cdot s$ at 312 K.

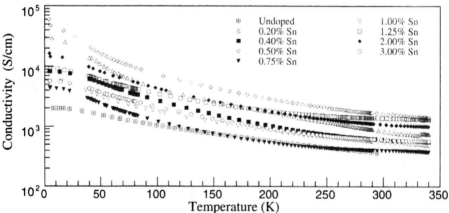

Figure 1. Temperature-dependent electrical conductivity for $CsBi_4Te_6$ doped with Sn.

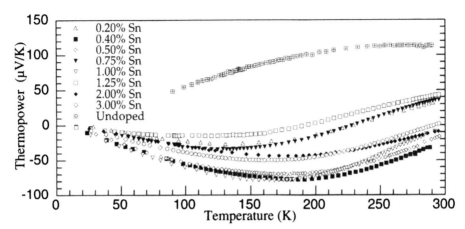

Figure 2. Temperature-dependent thermopower for Sn doped $CsBi_4Te_6$.

The transport properties of the n-type dopants, Ge, Pr, and Te, also have been studied in $CsBi_4Te_6$. Variable temperature conductivity and thermopower measurements were taken on 0.1% Ge, 0.5% Ge, 2.0% Ge, 0.2% Pr, 2.0% Pr, and 1.0% Te doped $CsBi_4Te_6$. These samples showed a weak metallic dependence in the electrical conductivity measurement and n-type behavior for the thermopower. In the case of Te, the substitution is on the Bi site; as yet the Ge and Pr positions are not known precisely. All samples show n-type thermopower behavior except the 0.1% Ge, which has a negative Seebeck coefficient below 172 K and a positive value from 172 K to 295 K, and the 2.0% Pr sample which became p-type at 285 K. Conductivity, thermopower and the temperature where the maximum thermopower occurs for these materials are shown in Table 1.

Hall measurements also were taken on 1.0% Te doped $CsBi_4Te_6$ samples as shown in Figure 3B. The mobility decreases as the temperature increases. At 27.3 K the carrier concentration is 1.16×10^{19} cm^{-3} and the mobility is 5.89×10^3 cm^2/V•s, and at 267 K the carrier concentration is 1.48×10^{19} cm^{-3} and the mobility is 4.15×10^2 cm^2/V•s.

Table 1. Conductivity and thermopower values for Ge, Pr and Te doped $CsBi_4Te_6$.

	0.1% Ge	0.5% Ge	2.0% Ge	0.2% Pr	2.0% Pr	1.0% Te
σ at 4.2 K (S/cm)	7.40×10^3	7.86×10^3	1.61×10^4	2.26×10^4	1.83×10^3	1.44×10^4
σ at 295 K (S/cm)	4.48×10^2	6.32×10^2	8.03×10^2	1.17×10^3	2.24×10^2	8.88×10^2
Max. S (µV/K)	46.9	-50.5	-80.0	-52.5	-60.3	-101
At Peak Temp. (K)	295	153	192	197	156	167
S at 295 K (µV/K)	46.9	-22.9	-42.8	-13.4	8.50	-27.3

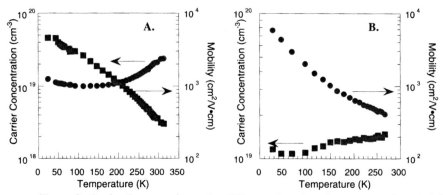

Figure 3. Carrier Concentration and mobility as a function of temperature for (A.) 0.5% Sn doped $CsBi_4Te_6$ and (B.) 1.0% Te doped $CsBi_4Te_6$.

DISCUSSION

In the initial study to make n-type $CsBi_4Te_6$, the 0.5% Sn and 1.0% Te doped samples showed the most promising results. The power factor, $S^2\sigma$, for Sn doped $CsBi_4Te_6$ is shown in Figure 4. The 0.5% Sn doped $CsBi_4Te_6$ has the highest power factor of 21.9 $\mu W/cm \cdot K^2$ at 127 K. The next best power factor is the 0.4% Sn doped $CsBi_4Te_6$ with a maximum of 8.94 $\mu W/cm \cdot K^2$ at 129 K. The peak of the power factors for the Sn doped $CsBi_4Te_6$ occurs at lower temperatures than the optimally doped p-type $CsBi_4Te_6$, which has a peak power factor of 51.5 $\mu W/cm \cdot K^2$ at 184 K [4]. As mentioned previously, the conductivity data for the Sn doped samples display no clear trend as the concentration of the dopant is varied. While annealing steps were carried out in sample preparation, it is believed that for some dopants additional annealing procedures should be performed.

The thermopower study with the Sn dopant suggests that this dopant does introduce excess carriers into the material. Although the 0.2%, 0.75%, and the 1.25% Sn doped samples show n-type behavior at low temperatures, they become p-type as they are warmed to higher temperatures. The 0.2% Sn may not have enough excess n-type carriers to compensate for the p-type carriers found in the undoped material. The 0.75% and 1.25% Sn may have too many defects which may also be improved by additional annealing steps. The carrier concentration of the 0.5% Sn is 1.01×10^{19} cm^{-3} and the mobility is 2.20×10^3 $cm^2/V \cdot s$ at the peak power factor temperature of 127 K. The carrier concentration is slightly higher and the mobilities are the same order of magnitude than those values previously reported for p-type $CsBi_4Te_6$, where the carrier concentration was found to be 3.01×10^{18} cm^{-3} to 10^{19} cm^{-3} and the mobilities 700 to 1000 $cm^2/V \cdot s$ at room temperature for 0.1% and 0.2% SbI_3 doped $CsBi_4Te_6$ [4].

The power factor for the Ge and Te doped $CsBi_4Te_6$ is shown in Figure 5. In the Ge series, the 2.0% Ge sample has the highest power factor with a maximum value of 10.3 $\mu W/cm \cdot K^2$ at 127 K. While the 2.0% dopant has the highest power factor, it is not clear that the optimum doping for this material has been achieved. The power factor for the 1.0% Te doped $CsBi_4Te_6$ sample shows a maximum power factor of 21.7 $\mu W/K^2 cm$ at 123 K. The carrier concentration at 123 K is 1.20×10^{19} cm^{-3} and the mobility is 1.26×10^{3} cm^2/V•s, which are in the same range as the p-type doped CsBi$_4$Te$_6$. It is interesting to note that the maximum power factors for Sn, Ge, and Te all occur at about the same temperature, 127 K, 127 K and 123 K respectively. This trend appears to be more related to the host material than to the species of the dopant.

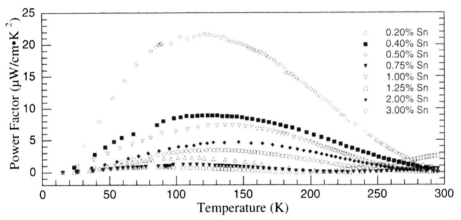

Figure 5. Power factors for the Sn doped CsBi$_4$Te$_6$ series as a function of temperature.

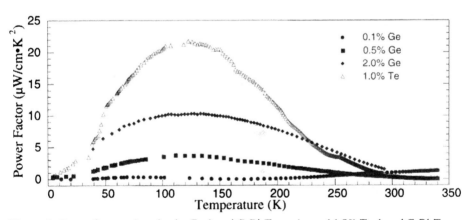

Figure 6. Power factor values for the Ge doped CsBi$_4$Te$_6$ series and 1.0% Te doped CsBi$_4$Te$_6$.

CONCLUSIONS

The dopants, Sn, Ge, Pr and Te were employed to make n-type $CsBi_4Te_6$. Of these dopants, 0.5% Sn and 1.0% Te have the highest power factors which occur in the vicinity of 125 K. All data were measured on single crystal material; future work will include preparing oriented ingot samples of these materials for thermal conductivity measurements. Preliminary thermal conductivity data and values calculated from the Wiedemann-Franz law, where appropriate, suggest that the figure of merit maxima will occur at slightly higher temperatures than those of the power factors. Not all of the data for every dopant and dopant concentration have been displayed here. This paper has presented, for those materials summarized here, typical results of what has been determined thus far. Continuing work will investigate more new n- and p-type dopants in the $CsBi_4Te_6$ system as well as optimizing the dopant concentrations which have already yielded the highest power factors or figure of merits.

ACKNOWLEDGMENTS

This work was supported at NU and MSU by the Office of Naval Research (N00014-98-1-0443) and by DARPA through ARO (DAAG55-97-1-0184). Work at NU made use of the Central Facilities supported by the National Science Foundation through the NU Materials Research Center (DMR-9632472).

REFERENCES

[1] M. Stordeur, *Phys. Status Solidi* vol. 161, 831, (1990).
[2] H. H. Jeon, H. P. Ha, D. B. Hyun, D. J. Shim, *J. Phys. Chem. Solids* vol. 4, 579, (1991).
[3] L. R. Testardi, J. N. Bierly, Jr., F. J Donahoe, *J. Phys. Chem. Solids* vol. 23, 1209, (1962).
[4] D. Y. Chung, T. Hogan, P. Brazis, M. Rocci-Lane, C. Kannewurf, M. Bastea, C. Uher, M. Kanatzidis, *Science* vol. 287, 1024, (2000).
[5] J. W. Lyding, H. O. Marcy, T. J. Marks, C. R. Kannewurf, *IEEE Trans. Instrum. Meas.* vol. 37, 756, (1988).
[6] M. G. Kanatzidis D. Y. Chung., L. Iordanidis, S. K. Choi, P. Brazis, M. Rocci, T. Hogan, C. R. Kannewurf, *Thermoelectric Materials 1998-The Next Generation Materials for Small-Scale Refrigeration and Power Generation Applications* edited by T. M. Tritt, M. G. Kanatzidi G. D. Mahan, H. B. Lyon Jr., (Mat. Res. Soc. Symp. Proc. **545**, Warrendale, PA. (1999) 233.
[7] H. O. Marcy, T. J. Marks, C. R. Kannewurf, *IEEE Trans. Instrum. Meas.* vol. 39, 756 (1990).
[8] O. Maldonado, *Cryogenics*, vol. 32, 908, (1992).
[9] T. P. Hogan, Ph.D. thesis, Northwestern University, 1996.
[10] J. L. Schindler, T. P. Hogan, P. W. Brazis, C. R. Kannewurf, D. Y. Chung, M. G. Kanatzidis, *Thermoelectric Materials-New Directions and Approaches* edited by T. M. Tritt, M. G. Kanatzidis, H. B. Lyon Jr., G. D. Mahan, (Mat. Res. Soc. Symp. Proc. **478**, Warrendale, PA) 327.
[11] A. F. Ioffe, *Semiconductor Thermoelements and Thermoelectric Cooling*, (Inforsearch Ltd, London, 1957).

EXPLORING COMPLEX CHALCOGENIDES FOR THERMOELECTRIC APPLICATIONS

Ying C. Wang and Francis J. DiSalvo
Department of Chemistry and Chemical Biology, Baker Lab
Cornell University, Ithaca, NY 14853, USA

ABSTRACT

Our research on ternary / quaternary chalcogenides for thermoelectric applications has lead to the identification of new interesting compounds and better understanding of the chemistry and physical properties of complex chalcogenides. The chemical, geometric, electronic diversity and flexibility has been well demonstrated in $BaBiSe_3$ and $Sr_4Bi_6Se_{13}$ type compounds. This presents both a challenge and more opportunity in controlling and optimizing the thermoelectric properties of these complex chalcogenides, compared with elemental and binary compounds. The importance of multivalley band structure in thermoelectric materials is emphasized. Only compounds with high crystal symmetry have the possibility of having a large number of degenerate valleys in the conduction bands or peaks in the valence bands, respectively. However, most of the complex chalcogenides crystallize in low crystal symmetry. An Edisonian method of exploratory synthesis and characterization may be the working approach to find good thermoelectric materials with ZT higher than 4.

INTRODUCTION

Thermoelectrics (TE) is a classical field discovered and investigated since the early nineteenth century. Thermoelectric phenomena are used in power generation and Peltier Cooling. TE devices have certain advantages over the conventional thermal-mechanical conversion power generator and compressor-based refrigeration devices. TE devices are light, compact, non-emissive, silent, simple, and very reliable. However, present TE devices have low efficiency typically 10% or less of Carnot efficiency [1-4].

Thermoelectric power generators convert heat or a temperature gradient into electricity. TE generators have found applications mainly in which the device reliability is paramount or the environment is very hostile. TE generators are well suited as the electrical power source in deep space missions where the surrounding can be extremely harsh with drastic temperature variation, asteroid debris, and high radiation fields. The Voyager 1 and 2 spacecraft launched in 1977 were each equipped with three multi-hundred Watt radioisotope thermoelectric generators (RTG). The RTGs have been operating without a single failure since [5-6].

TE cooling or Peltier cooling devices function like a heat pump, converting electricity to a temperature gradient. For decades, it has been the holy grail for scientists and engineers to design thermoelectric cooling devices with efficiency comparable to the conventional compressor-based refrigeration devices. While TE devices operate at about 10% of Carnot efficiency, the small conventional refrigeration devices operate at about 30% of Carnot efficiency. Thus the world market for TE cooling devices is small: about $80 million to $160 million per year. The use of thermoelectric devices at today's refrigeration market is limited by their low efficiencies [1].

Overall, the limit of the current TE devices is their low efficiency. The improvement of TE device efficiency depends on the availability of better TE materials. Currently, the best TE materials at room temperature are Bi_2Te_3 based alloys with a Figure of Merit $ZT=S^2T/\rho\kappa$ around 1, where S is Seebeck coefficient, ρ is electrical resistivity, κ is thermal conductivity, and ZT is a dimensionless unit. In order to increase the TE device performance to 30% of Carnot efficiency, we need a TE material with ZT higher than 4. After an examination of current literature [2-8], we concluded that a feasible strategy to find better TE materials is to investigate systems that have heavy elements, a large unit cell, high average coordination numbers, small electronegativity difference, high dielectric constant, a band gap around $10k_BT$, high band mass, high crystal symmetry, and multivalley band structure. If ease of fabrication is taken into account, congruently melting and good stability in air are desirable features in terms of applications.

Since Bi_2Te_3 based alloys are the best TE materials at room temperature, our exploratory synthesis and modification of known materials focus on those containing chalcogenides. This report will present a summary of our work on some bulk semiconductor materials, especially chalcogenides. The importance of the multivalley band structure feature in thermoelectrics will also be discussed.

GEOMETRIC AND ELECTRONIC FLEXIBILITY OF TERNARY AND QUATERNARY CHALCOGENIDES

Because $Bi_{2-x}Sb_xTe_{3-y}Se_y$ alloys are the best TE materials at room temperature, extensive research has focused on modifying the composition and structure of these alloys by incorporating electropositive elements such as alkali, alkaline earth, and rare earth metals. Many new compounds have been synthesized and characterized, for example, $APbBi_3Q_6$ (A = K, Rb; Q = S, Se) [9], β-$K_2Bi_8Se_{13}$ [10], $BaBiTe_3$ [11], $BaBiSe_3$ [12], $BaLaBi_2S_6$ and $KThSb_2Se_6$ [13], KBi_3S_5 [14], $K_2(RE)_{2-x}Sb_{4+x}Sb_4Se_{12}$ (RE = La, Ce, Pr, and Gd) [15], $CeBi_4Te_6$ [16], etc. A common and remarkable feature displayed by the plethora of complex bismuth / antimony chalcogenides is their chemical, geometric, electronic variety and flexibility. In the following, we will report our studies of $Sr_4Bi_6Se_{13}$ [17-18] and $BaBiSe_3$ type compounds [12,19].

We have synthesized compounds isotypic to $Sr_4Bi_6Se_{13}$: $Ba_3Bi_{6.67}Se_{13}$, $Ba_3Bi_6PbSe_{13}$, and $Ba_3Bi_6SnSe_{13}$ [17]. Some other isotypic compounds reported in the literature are β-$K_2Bi_8Se_{13}$ [10], $K_2Bi_8S_{13}$ [20], and $Ba_4Bi_6Se_{13}$ [21]. In $Sr_4Bi_6Se_{13}$ [17], there are 4 Sr sites, 6 Bi sites, and 13 Se sites. Three of the Sr sites can be replaced by Ba, K, or a mixture of Bi and K. The fourth Sr site can be replaced by 66.7% Bi, Pb, or a mixture of Sn and Bi. A comparison of $Ba_3Bi_{6.67}Se_{13}$, $Ba_3Bi_6PbSe_{13}$, and $Ba_3Bi_6SnSe_{13}$ will illustrate the chemical, geometric, and electronic flexibility demonstrated by complex chalcogenides. In $Ba_3Bi_{6.67}Se_{13}$, one of the Bi sites has a 66.7% occupancy due to the requirement of charge neutrality. If Pb^{2+} is used to substitute for this Bi^{3+}, then the site is fully occupied by Pb^{2+} as shown in $Ba_3Bi_6PbSe_{13}$ and as expected by oxidation state neutrality. If Sn^{2+} is used, not all the Sn atoms go to this site;

instead, some of the Sn atoms are mixed with Bi at this site and the rest are spread over several other sites. So Sn does not stay at one site as does Pb [18]. From the above substitution, mixed occupancy, and partial occupancy, it is clear that $Sr_4Bi_6Se_{13}$ structure type is very flexible in terms of chemistry and geometry.

BaBiSe$_3$ type compounds also display structural and electronic flexibility. We have synthesized $Ba_{1-x}K_xBiTe_3$ (x≤0.05), $BaCeBi_2Se_6$, $Ba_{1.5}Ce_{0.5}Bi_2Se_6$, $Ba_{1-x}K_xCeBi_2Se_6$ (x = 0.24), which are isotypic to BaBiSe$_3$ [12, 19]. Other isotypic chalcogenides reported in the literature are $BaSbQ_3$ (Q = Se, Te) [12], BaBiTe$_3$ [11], $KThSb_2Se_6$ and $BaLaBi_2S_6$ [13]. In BaBiSe$_3$ [12], there are 2 Ba sites, 2 Bi sites, and 6 Se sites. The Ba sites can be replaced by La, Ce, Th, a mixture of Ba and K, or a mixture of Ba and Ce [19]. The resulting compounds may have different electron concentrations. To maintain the charge neutrality, some structural modification occurs around covalently bonded zigzag Q chains (Q = S, Se, Te) in this structure and the compounds remain semiconducting by modifying the chain bonding as demonstrated by $KThSb_2Se_6$ and $BaLaBi_2S_6$ [13].

Overall, ternary / quaternary bismuth or antimony chalcogenides are much more flexible than binary chalcogenides, chemically, geometrically, and electronically. This presents both challenge and opportunity in doping the compounds n- or p-type to optimize their TE properties.

MULTIVALLEY BAND STRUCTURE IN TERMOELECTRICS

A microscopic analysis shows that Z is a function of $N_v \mu m^{3/2} / \kappa_{ph}$, where N_v is the number of multivalley or the number of k points with the same energy extrema near Fermi level, μ is the carrier mobility, m is the band mass, and κ_{ph} is the lattice thermal conductivity [1, 7]. Z increases as N_v increases. Although intervalley scattering resulting from a multivalley band structure may reduce Z somewhat from the simple model, the improvement of Z due to a larger N_v still dominates overall. A practical guideline to find better TE materials is to study materials with multivalley band structure features. Since there are so many compounds crystallizing in many different space groups, we should focus on those compounds that have as high an N_v as possible.

For semiconductor materials, a multivalley band structure means that the electronic band $E_j(k)$ should have multiple degenerate maxima at the top of valence band and multiple dgenerate minima at the bottom of conduction band. Extrema points must be zero-slope points, but zero-slope points are not necessarily extrema points. At the zero-slope points, $E_j(k)$ values are stationary. The existence of stationary values of $E_j(k)$ at certain k points may be essential due to requirements of crystal symmetry; or they may be accidental due to the particular crystal potential [22].

It is well established that there are 230 space groups, 24 Patterson groups, and 11 Laue groups in three dimensions. Every crystal can be described by one of the 230 space groups. For each space group in real space, there is a corresponding space group in reciprocal space, k space. There are 24 space groups in k space, which are called Patterson symmetry groups G'. A basic unit cell in reciprocal space, the Brillouin zone (BZ), is constructed from the reciprocal vectors **a***, **b***, **c***, which are purely derived from **a**, **b**, **c**, the translation vectors in real space. It should be noted that the contents of the primitive cell and therefore the full crystal symmetry are irrelevant to the size and shape of the BZ. Two different reciprocal lattices may have the same BZ shape but the electronic bands $E_j(k)$ can possess different symmetry, for example, in Im3 and Im3m [22-23].

The electronic bands $E_j(k)$ follow the symmetry of the reciprocal lattice **G'** which is determined by the crystal symmetry in real space. As a corollary, each constant-energy surface possesses all the symmetry of **G'**. The symmetry of $E_j(k)$ relative to the reciprocal lattice origin Γ or any other reciprocal lattice point $n_1a^*+n_2b^*+n_3c^*$ is the Laue point group **P'** associated with the crystal space group. The symmetry of $E_j(k)$ relative to other points k_0 are described by the point group **P'**(k_0) at k_0, called the little co-group of k_0 in **G'**. **P'**(k_0) is in general different from **P'** [22-23].

Based on the method and results in Cracknell's work [22], we investigated the occurrence of essential zero-slope points k_0 in k space, the corresponding **P'**(k_0), the maximum N_v satisfying the symmetry requirement, and the maximum N_v if accidental zero-slope points are allowed. The results are summarized in Table 1. The maximum N_v for accidental zero-slope points, denoted as $N_{v,max}$, is constrained only by the Laue point group. The maximum N_v for essential zero-slope points, denoted as $N_{v,max}(\mathbf{P'}(k_0))$, is determined by the Patterson symmetry.

Table 1. Summary of N_v and symmetry in solids.

Crystal System	Laue Group **P'**	Bravais Lattice	Reciprocal Lattice	Patterson Symmetry **G'**	$N_{v,max}(\mathbf{P'}(k_0))$ *	$N_{v,max}$
Triclinic	$\bar{1}$	P	P	P$\bar{1}$	1 ($\bar{1}$)	2
Monoclinic	2/m	P	P	P2/m	1 (2/m)	4
		B	B	B2/m	1 (2/m)	4
orthorhombic	mmm	P	P	Pmmm	1 (mmm)	8
		C	C	Cmmm	2 (2/m)	8
		I	F	Fmmm	2 (2/m, 222)	8
		F	I	Immm	4 ($\bar{1}$)	8
Tetragonal	4/m	P	P	P4/m	2 (2/m)	8
		I	I	I4/m	4 ($\bar{1}$)	8
	4/mmm	P	P	P4/mmm	2 (mmm)	16
		I	I	I4/mmm	4 (2/m)	16
Trigonal	$\bar{3}$	P	P	P$\bar{3}$	3 ($\bar{1}$)	6
		R	R	R$\bar{3}$	3 ($\bar{1}$)	6
	$\bar{3}$m	P	P	P$\bar{3}$1m	3 (2/m)	12
		P	P	P$\bar{3}$m1	3 (2/m)	12
		R	R	R$\bar{3}$m	3 (2/m)	12
Hexagonal	6/m	P	P	P6/m	3 (2/m)	12
	6/mmm	P	P	P6/mmm	3 (mmm)	24
Cubic	m3	P	P	Pm3	3 (mmm)	24
		I	F	Fm3	6 (2/m)	24
		F	I	Im3	4 ($\bar{3}$)	24
	m3m	P	P	Pm3m	3 (4/mmm)	48
		I	F	Fm3m	6 (mmm)	48
		F	I	Im3m	4 ($\bar{3}$m)	48

* **P'**(k_0) is the co-little group for the k_0 point where stationary values of $E_j(k)$ are required.

Table 1 reveals that the maximum N_v is limited by the crystal symmetry. Among the present TE materials, N_v is 4 or 6 [1, 7]. If a material with N_v higher than 6 exists, the degeneracy near the Fermi level must be accidental and result from the particular arrangement of atoms in the primitive cell. The maximum accidental degeneracy is attained if the extrema appear at a general k point. The occurrence of accidental degeneracy cannot be easily predicted or synthetically controlled. Only through the Edisonian methods of exploratory synthesis and characterization will we find materials with high N_v. Compounds with high crystal symmetry are the only ones that have the possibility of possessing a large N_v value.

Unfortunately, more than 50% of complex materials (ternary and up) crystallize in low symmetry structures (triclinic, monoclinic, and orthorhombic), whereas only one-third of elemental and binary inorganic compounds crystallize in low symmetry phases [24]. For chalcogenides containing heavy elements such as Sb/Bi/Pb, the lone pair effects of these heavy elements can cause the distortion of local polyhedral coordination environment and that often leads to low crystal symmetry. The majority of the reported ternary / quaternary chalcogenides containing heavy elements crystallize in monoclinic and orthorhombic systems; however, a few chalcogenides show higher crystal symmetry, for example, $CsPbBi_3S_6$ and β-$CsPbBi_3Se_6$ in $P6_3/mmc$ [9], $BaBi_2S_4$ [25], $BaBi_2Se_4$ and $SrBi_2Se_4$ in $P6_3/m$ [19a, 26]. Very few are cubic, but a few examples can be found: $Pb_3(PS_4)_2$ [27] and Pb_2GeQ_4 (Q = S, Se) [28]. These cubic phases themselves are not suitable for TE applications, since their carrier mobility is very low. Chemical and structural modifications are underway to find related high symmetry phases with high carrier mobility.

CONCLUSIONS

Better TE materials may be found in ternary / quaternary chalcogenides, as already demonstrated by some encouraging work in this area. They are much more complicated than binary chalcogenides. Ternary / quaternary chalcogenides display a higher chemical, geometrical, and electronic flexibility. This indicates more freedom in tuning their TE properties; on the other hand, it also suggests bigger challenge in controlling their TE properties. Another important note is that most of the complex chalcogenides crystallize in low symmetry space groups such as monoclinic and orthorhombic systems. This might be a hurdle toward increasing ZT significantly. But extensive exploratory synthesis and characterization will help produce complex chalcogenides with high symmetry and high ZT.

ACKNOWLEDGMENT

This project was funded by Office of Naval Research.

REFERENCES

1. F. J. DiSalvo, *Science* **285**, 703 (1999).
2. G. Mahan, B. Sales, J. Sharp, *Phys. Today* **50**, 42 (1997).
3. B. Sales, *Mater. Res. Soc. Bull.* **23**, 15 (1998).
4. T. M. Tritt, *Science* **272**, 1276 (1996).
5. *CRC Handbook of Thermoelectrics*, ed. D. M. Rowe (CRC Press, Boca Raton, FL, 1995).
6. C. Wood, *Rep. Prog. Phys.* **51**, 459 (1988).
7. G. D. Mahan, *Solid State Phys.* **51**, 82 (1998).
8. H. J. Goldsmid, *Electronic Refrigeration* (Pion, London, 1986).
9. D. -Y. Chung, L. Iordanidis, K. K. Rangan, P. W. Brazis, C. R. Kannewurf, and M. G. Kanatzidis, *Chem. Mater.* **11**, 1352 (1999).
10. D. -Y. Chung, K. -S. Choi, L. Iordanidis, J. L. Schindler, P. W. Brazis, C. R. Kannewurf, B. Chen, S. Hu, C. Uher, M. G. Kanatzidis, *Chem. Mater.* **9**, 3060 (1997).
11. D. -Y. Duck, S. Jobic, T. Hogan, C. R. Kannewurf, R. Brec, J. Rouxel, M. G. Kanatzidis, *J. Am. Chem. Soc.* **119**, 2505 (1997).
12. K. Volk, G. Cordier, R. Cook, H. Schäfer, Z. *Naturforsch.* **35B**, 136 (1980).
13. K. -S. Choi, L. Iordanidis, K. Chondroudis, M. G. Kanatzidis, *Inorg. Chem.* **36**, 3804 (1997).
14. T. J. McCarthy, T. A. Tanzer, M. G. Kanatzidis, *J. Am. Chem. Soc.* **117**, 1294 (1995).
15. J. H. Chen and P. K. Dorhout, *J. Alloys Comp.* **249**, 199 (1997).
16. D. -Y. Duck, T. Hogan, P. Brazis, M. Rocci-Lane, C. Kannewurf, M. Bastea, C. Uher, M. G. Kanatzidis, *Science* **287**, 1024 (2000).
17. G. Cordier, H. Schäfer, C. Schwidetzky, *Revue de Chimie minérale* **22**, 631 (1985).
18. Y. C. Wang and F. J. DiSalvo, *Chem. Mater.* **12**, 1011 (2000).
19. (a) Y. C. Wang, PhD dissertation, Cornell University, 2000, to be published. (b) Y. C. Wang, R. Hoffmann, F. J. DiSalvo, results to be published.
20. M. G. Kanatzidis, T. J. MaCarthy, T. A. Tanzer, L. -H. Chen, L. Iordanidis, T. Hogan, C. R. Kannewurf, C. Uher, B. Chen, *Chem. Mater.* **8**, 1465 (1996).
21. L. Iordanidis, P. W. Brazis, C. R. Kannewurf, M. G. Kanatzidis, *MRS 1998 Fall Meeting-SymposiumZ4.30*, Boston, MA, USA.
22. A. P. Cracknell, *J. Phys. C: Solid State Phys.* **6**, 826 (1973).
23. S. L. Altmann, *Band Theory of Solids: An Introduction from the Point of View of Symmetry* (Clarendon Press, Oxford, 1995), Chap 6.
24. ICSD/RETRIEVE 2.01, Gmelin Institute / Fiz Karlsruhe, Release February 1998.
25. B. Aurivillius, *Acta Chem. Scand.* **37A**, 399 (1983).
26. Y. C. Wang, R. Hoffmann, F. J. DiSalvo, *J. Solid State Chem.*, submitted.
27. E. Post and V. Krämer, *Mater. Res. Bull.* **19**, 1607 (1984).
28. K. M. Poduska and F. J. DiSalvo, results to be published.

Poster Session

Semiconductors With Tetrahedral Anions As Potential Thermoelectric Materials

Thomas P. Braun, Christopher B. Hoffman and Francis J. DiSalvo
Dept. of Chemistry and Chemical Biology
Cornell University, Ithaca, NY 14850, U.S.A.

ABSTRACT

In order to find the next generation thermoelectric (TE) material, we are focussing on the parameter N_v, the degeneracy of the band extrema in semiconductors near the Fermi energy. We attempt to synthesize 'multivalley' semiconductors by incorporating tetrahedral anions to introduce structural complexity while maintaining high crystallographic symmetry.

The synthesis and crystal structures of two new compounds that partially fulfill our requirements for potential TE materials are reported. $Pb_3(PS_4)Br_3$ is monoclinic, space group $P2_1/m$ with a=9.1531(1)Å, b=10.9508(3)Å, c=12.7953(1)Å and β=111.024(2)°. $Pb_3(PS_4)I_{1.75}Te_{0.625}$ crystallizes in the space group $R\bar{3}$ with a=9.4876(4)Å, c=46.189(3)Å. Both structures are built from alternating layers of lead halide and the thiophosphate ion $(PS_4)^{3-}$.

INTRODUCTION

Despite great efforts in recent years to find improved thermoelectric (TE) materials, no obvious candidates for the next-generation material are in sight. We strongly believe that these materials will have to be more than just minor variations of Bi_2Te_3 based current materials. Instead, both a deeper understanding of the underlying phenomena and the exploitation of additional features will be necessary. Different approaches have been pursued with varying success. [1,2] Among the more promising ones are the fields of superlattices and other low dimensionality structures or the use of high pressures or combinatorial chemistry to speed up the exploration of interesting phase spaces.[3-5] Our approach focuses on a critical parameter for the thermoelectric properties of semiconductors that has not been subject to systematic investigation before. It has been suggested that the figure of merit (ZT) for semiconductors should be directly proportional to the band degeneracy of (i.e. the number of extrema in) the conduction and valence bands. We are seeking to synthesize compounds with a high number of band extrema in order to verify and ultimately utilize these predictions.

Semiconductors containing tetrahedral anions are plausible candidates for these 'multivalley semiconductors'. In this article we describe the theoretical background of our approach and report our first synthetic results.

THEORY

It has been shown that for semiconductors the thermoelectric figure of merit (ZT) can be expressed as a function of two parameters: $ZT = f(E_g, B)$.[6] This function increases monotonically with both the band gap E_g and a materials parameter B. However, the increase of ZT is rather small for values of E_g above 10 kT which defines a lower limit for the band gap of semiconducting TE materials. We also seek the largest B parameter, which takes the following form: $B \propto N_v \, \mu \, (m^*)^{3/2}/\kappa_{ph}$.

N_v stands for the degeneracy of the band extrema near the Fermi level, μ is the carrier mobility and m^* is the effective band mass determined by the density of states. The need to minimize the phonon contribution of the thermal conductivity (κ_{ph}) has been noted before and great efforts have been undertaken to achieve progress in this respect. On the other hand little attention has been given to the potential improvement of ZT that is associated with the systematic optimization of N_v, the number of valleys or peaks in the conduction and valence band, respectively. Current TE materials already utilize this factor at least partially, as N_v is typically 6 or even 8. The crystalline symmetry of the material limits the maximum possible values for N_v. For cubic space groups it can reach 48, followed by 24 for hexagonal, 16 for tetragonal, 12 for trigonal, 8 for orthorhombic and even less for the remaining crystal classes. These numbers present the upper limits of N_v that are possible for the respective crystal class.

In order to reach these maxima a second condition has to be fulfilled: the band extrema have to be at general positions in the Brillouin zone. There seems to be no simple way to design structures that fulfill this requirement. The fundamental difficulty behind this problem is to formulate the conditions for an atomic arrangement (in real space) that results in a band structure (in reciprocal space) with the desired features.

In the absence of a general recipe to achieve this, we hope to force the band extrema to general positions by means of avoided crossings. In order to increase the likelihood for these to occur we need to increase the number of bands available, i.e. the number of atoms in the (primitive) unit cell. There is an added benefit of large unit cells, it is one factor exploited to lower κ_{ph}.

EXPERIMENTAL DETAILS

The intermediate starting material $Pb_3(PS_4)_2$ was synthesized from the elements in a stoichiometric ratio as reported elsewhere.[7] Pb (Seargent-Welch, 99.9%), P (Johnson Matthey, 99.999%) and S (Cominco American, 99.99%) were combined in an evacuated quartz tube and heated at 600 °C for 24 hours.

$Pb_3(PS_4)Br_3$ was synthesized by combining PbTe (prepared from the elements) and $Pb_3(PS_4)_2$ in a 3:1 mole ratio with a ten fold excess of a low melting flux. The flux used was a 8:1:1 mole mixture of $PbBr_2$ (Aldrich, 99.999%), KBr (Baker, '100.0%'), and LiBr (Aldrich, 99.9+%) with a melting temperature of 298 °C. The mixture was sealed in an evacuated quartz tube and heated at 500 °C for 16 hours, followed by slow cooling to room temperature over 100 hours. The excess flux was dissolved with hot water. The synthesis of $Pb_3(PS_4)I_{1.75}Te_{0.625}$ followed the same procedure as for $Pb_3(PS_4)Br_3$ with the exception of PbI_2 (Aldrich, 99%), KI (Baker, 99.6%), and LiI (Aldrich, 99.99%), being used in place of the bromide salts.

Single crystals of $Pb_3(PS_4)_2$, $Pb_3(PS_4)Br_3$ and $Pb_3(PS_4)I_{1.75}Te_{0.625}$ have been prepared and X-ray diffraction data have been collected on a Bruker CCD (SMART) system.

We confirmed that $Pb_3(PS_4)_2$ crystallizes in the cubic space group $P2_13$ with a=10.9247(2)Å (V= 1303.85(4)Å^3) but we find the opposite absolute configuration for this enantiomorphic space group. Since at the time of the original structure determination there were no simple tools available for the determination of the absolute configuration, we believe that $Pb_3(PS_4)_2$ actually adopts the configuration we found.

Pb$_3$(PS$_4$)Br$_3$ crystallizes in the monoclinic space group P2$_1$/m with a=9.1531(1)Å, b=10.9508(3)Å, c=12.7953(1)Å and β=111.024(2)°. There are 4 formula units (44 atoms) in the unit cell (V=1197.14(4)Å^3).

The mixed halide chalcogenide Pb$_3$(PS$_4$)I$_{1.75}$Te$_{0.625}$, space group R$\bar{3}$ with a=9.4876(4)Å, c=46.189(3)Å, V=3600.6(3)Å^3, contains twelve formula units (88.5 atoms) per unit cell. One anion position (Te2 in *(3b)*) is only half occupied and one lead atom is disordered on two positions (Pb2, Pb3 in *(18f)*) approximately 0.6Å apart. Details of the structure determination and a comparative discussion of the crystal structures will be published elsewhere.

RESULTS AND DISCUSSION

A database search for highly symmetric structures (tetragonal or higher) with a high number of atoms in the primitive unit cell yields a surprising result: A large fraction of the structures found contain tetrahedral complex anions like e.g. (SiO$_4$)$^{4-}$or (PO$_4$)$^{3-}$, but heavier analogs like the thiophosphate ion (PS$_4$)$^{3-}$ are found as well. One particularly exciting discovery seemed to be the compound lead thiophosphate (Pb$_3$(PS$_4$)$_2$) that crystallizes in a cubic structure with 52 atoms per unit cell. Unfortunately we found it to be a good insulator (low carrier mobility) and hence not a likely TE material. In our interpretation, this is due to the closed shell configuration of the tetrahedral anion and the cations. For high mobility a covalent network is necessary. Thus, we have to combine the advantages of the tetrahedral anions (structural complexity, larger number of atoms, potential for high N$_v$) with a covalent network built around these moieties. The mineral sillenite (Bi$_{12}$SiO$_{20}$) can be viewed as a prototype of this 'composite' of isolated, tetrahedral anions (SiO$_4$) and a three dimensional (Bi-O) matrix. Due to the large band gap of this transparent silicate-oxide, this specific compound is not a suitable TE candidate.

A more general strategy attempts to build these composite structures in a modular way by reacting pseudo-binary precursors with network-builders. The pseudo-binary compounds contain the tetrahedral anions and typically a heavy metal as the counter cation as, for example, the above mentioned lead thiophosphate Pb$_3$(PS$_4$)$_2$ or lead thiosilicate Pb$_2$(SiS$_4$). The 'network builders' are binary chalcogenides like e.g. PbTe, Bi$_2$Te$_3$ or SnSe$_2$. While some of the tetrahedral anions appear to be quite stable even at high temperatures, the thiophosphate anion (PS$_4$)$^{3-}$ shows a tendency to decompose (forming the thiohypodiphosphate ion $(P_2S_6)^{4-}$) in the presence of reducing agents like Te^{2-}. However if the temperature is kept below 550°C the $(PS_4)^{3-}$ anion can be preserved. We succeeded in the synthesis of two novel thiophosphate containing phases from halide flux reactions.

The initial reaction of $Pb_3(PS_4)_2$ with PbTe in a mixture of LiBr, KBr and $PbBr_2$ yielded a compound that proved to be tellurium free: $Pb_3(PS_4)Br_3$ contains $CdCl_2$ like layers of Pb-Br along [$\bar{1}$01] alternating with layers containing the thiophosphate ion $(PS_4)^{3-}$. The band gap of this composite halide thiophosphate is still too high and N_v cannot exceed 4 for the monoclinic crystal system. But the successful synthesis demonstrates that tetrahedral anions can be embedded in (in this case two dimensional) covalent networks.

The incorporation of tellurium in the network is possible if the bromide fluxes are replaced by the corresponding iodides. A reaction with iodides instead of bromides yields the compound $Pb_3(PS_4)I_{1.75}Te_{0.625}$. This compound crystallizes in a rhombohedral space group which fulfills more of our requirements outlined above: there are 88.5 atoms in the unit cell (one tellurium position is only partially occupied) and the trigonal symmetry allows for a maximum of 12 for N_v. Again we find alternating layers (in this case in the *ab* plane) containing the thiophosphate ion and lead iodide, respectively. In addition there are isolated tellurium ions present in both layers. The structures of both 'composite crystals' are shown in figure 1.

Figure 1: Crystal structures of $Pb_3(PS_4)Br_3$ (left) and $Pb_3(PS_4)I_{1.75}Te_{0.625}$ (right). Pb is represented by large open circles, Br/I by filled small circles, Te by open small circles and the $(PS_4)^{3-}$ ions as filled tetrahedra.

In order to further reduce the band gap of these compounds we are now attempting to increase the ratio of tellurium to iodine in these compounds. It should be possible to achieve this by simultaneous substitution of lead with bismuth and iodine with tellurium. Assuming full occupation of the anion sites in the mixed halide chalcogenide structure, a series of compounds with the general composition $Pb_{32-x}Bi_x(PS_4)_{12}I_{24-x}Te_{6+x}$ can be proposed. For the maximum value of x=24 the resulting composition would be $Pb_2Bi_4(PS_4)_2Te_5$. Experiments are currently under way to synthesize this and related compounds.

CONCLUSION

In our approach to identify the next generation thermoelectric material we have been focussing on the maximization of N_v, the degeneracy of the band extrema near the Fermi level. We attempt to synthesize these 'multivalley semiconductors' by incorporating complex anions such as the tetrahedral thiophosphate ion $(PS_4)^{3-}$. It has been established that this moiety is stable in molten halide flux systems which enables us to react pseudo binary precursors like $Pb_3(PS_4)_2$ or $BiPS_4$ with network builders such as e.g. Bi_2Te_3 or $PbTe$. We succeeded in the synthesis of the first composite materials that contain both halide/chalcogenide layers and layers of isolated tetrahedral anions. Once the band gap of these materials has been reduced further, we are looking forward to measuring their relevant properties in order to test, and hopefully confirm the underlying theoretical assumptions.

ACKNOWLEDGMENTS

Funding for this research was provided by ONR.

REFERENCES

1. T.M.Tritt, *Science*, **283**, 804 (1999)
2. F.J.DiSalvo, *Science*, **285**, 703 (1999)
3. L.D.Hicks, T.C.Harman, X.Sun and M.S.Dresselhaus, *Phys.Rev. B*, **53**, 10493 (1996)
4. M.S.Dresselhaus, G.Dresselhaus, X.Sun, Z.Zhang, S.B.Cronin and T.Koga, *Mat.Res.Soc.Symp.Proc.*, **478**, 55 (1997)
5. J.F.Meng, D.A.Polvani, C.D.W.Jones, F.J.DiSalvo, Y.Fei and J.V.Badding, *Chem.Mater.*, **12**, 197 (2000)
6. G.D.Mahan, *Solid State Phys.*, **51**, 82 (1998)
7. E.Post, V.Krämer, *Materials Research Bulletin*, **19**, 1607 (1984)

LATTICE DYNAMICS STUDY OF ANISOTROPIC HEAT CONDUCTION IN SUPERLATTICES

B. YANG, G. CHEN
Mechanical and Aerospace Engineering Department
University of California at Los Angeles,
Los Angeles, CA 90095-1597

ABSTRACT

Past studies on the thermal conductivity suggest that phonon confinement and the associated group velocity reduction are the causes of the observed drop in the cross-plane thermal conductivity of semiconductor superlattices. In this work, we investigate the contribution of phonon confinement to the in-plane thermal conductivity of superlattices and the anisotropic effects of phonon confinement on the thermal conductivity in different directions, using a lattice dynamics model. We find that the reduced phonon group velocity due to phonon confinement may account for the dramatic reduction in the cross-plane thermal conductivity, but the in-plane thermal conductivity drop, caused by the reduced group velocity, is much less than the reported experimental results. This suggests that the reduced relaxation time due to diffuse interface phonon scattering, dislocation scattering, etc, should make major contribution to the in-plane thermal conductivity reduction.

INTRODUCTION

Thermal transport in semiconductor superlattices has been the subject of numerous investigations because of the potential applications to thermoelectric devices and quantum well lasers.[1,2] Many groups have reported a significant reduction in the in-plane[3-5] and the cross-plane[6-12] thermal conductivities in semiconductor superlattices. Several theoretical studies, treating the lattice excitations as particles[13-15] or lattice wave[16-18], have been carried out to explain the experimental observation. Chen[15] and Chen and Neagu[14] attributed the large cross-plane thermal conductivity reduction in Si/Ge and AlAs/GaAs superlattices to the phonon interface reflection and diffuse scattering, based on the solution of the Boltzmann transport equation. Hyldgaard and Mahan[16] and Tamura et al.[17] took an alternative approach for explaining this reduction, based on the lattice dynamic calculation of the phonon dispersion in superlattices. For the in-plane direction, on the other hand, Chen[13,19] and Hyldgaard and Mahan[18] suggested that the diffuse interface scattering is still the main cause of the observed thermal conductivity reduction in GaAs/AlAs superlattices. These studies, however, are based on the phonon dispersion of bulk materials instead of superlattices and treat phonons as particles. It remains to be seen whether the group velocity reduction could also lead to a significant drop in the lattice thermal conductivity in the in-plane direction, as it is suggested for the cross-plane direction.

In this work, we investigate the effects of phonon confinement on both the in-plane and the cross-plane thermal conductivities, based on a face-centered cubic (fcc) lattice dynamics model.[17,20] We also calculate the dispersion relation, the density of sates (DOS) and the projection of the thermal energy propagation factor (TEPF), i.e. the squared group velocity weighted by the DOS, along the in-plane and cross-plane directions. Thermal conductivities of superlattices, based on the constant relaxation time approximation, are computed as a function of the period thickness for Si/Ge and AlAs/GaAs superlattices.

THEORY MODEL

Our method of studying the acoustic phonon dispersion in superlattices employs a fcc lattice dynamics model[17,20]. For Si/Ge and AlAs/GaAs superlattices, we only consider acoustic vibrations. So the two atoms in a unit cell are replaced by a heavier atom with an equivalent mass.[17] We define u_{lmn} to be the displacement of an atom from its equilibrium position $r_{lmn}=(x_{lm},z_n)=(la, ma, na)$ in the Nth period of the superlattices. To satisfy Bloch's theorem, the displacement vector can be rewritten in the form

$$u_{lmn} = \tilde{u}_n \exp[i(\mathbf{k}_{\parallel}\cdot \mathbf{x}_{lm} + qND - \omega t)] \quad (1)$$

where $\tilde{u}_n$ is the complex amplitude of the nth atom in the Nth period of the superlattice, $\mathbf{k}_{\parallel}$ the in-plane wavevector, q the cross-plane wavevector and D the length of a unit period in cross-plane direction. Given the number of monolayers n_D in each unit period, D is equal to the product of n_D and the lattice constant a, i.e. $n_D \cdot a$. Assuming perfect interfaces between individual layers in superlattices, the equation of motion for the atom u_{lmn} can be solved to determine the dispersion relation.

Based on the dispersion relation, the thermal conductivity is given by

$$\kappa_i = \sum_{\lambda} C_{ph}(\omega_{\lambda})(\partial \omega / \partial k_i)^2 \tau \quad (2)$$

where λ denotes a set of the superlattice modes at ($\mathbf{k}_{\parallel}$, q, j) with j the index specifying the polarization and the frequency band, $C_{ph}(\omega_{\lambda})$ represents specific heat related to the lattice vibration mode with a frequency of ω_{λ}, i identifies the thermal conduction direction and τ is phonon relaxation time, regarded as a frequency-independent parameter[16,17]. All the parameters used in the calculation are taken from Ref.17. We compare our results along the cross-plane direction with those in Ref.17 and they reach excellent agreements, as expected.

RESULTS AND DISCUSSION

Figures 1(a) and (b) show acoustic phonon dispersion for a (2X2) Si/Ge superlattice along the in-plane and cross-plane directions, respectively. The two transverse acoustic modes along the in-plane direction are no longer degenerate because the unit cell of superlattices becomes tetragonal with four atoms as a base, instead of an fcc. Because the large mass difference of constituent atoms exists along the cross-plane direction, and this difference doesn't exist along the in-plane direction, the band gaps at the folded Brillouin zone center and edge are much larger in the cross-plane direction. Consequently, the dispersion curves along the cross-plane direction, especially at high frequencies are much flatter.

Figure 2(a) shows the frequency dependence of DOS for a (5X5)Si/Ge superlattice(solid line), Si bulk solid (dotted line) and Ge bulk solid (dashed line). DOS per unit volume $D(\omega)$ [13,15] is given by

$$D(\omega) = \sum_{\lambda} \delta(\omega - \omega_{\lambda})/V \quad (3)$$

where V is the volume. As seen in this figure, DOS of superlattices takes approximately the average values of the corresponding bulk solids. The folded Brillouin zone has little effect on DOS of superlattices, without apparent high frequency suppression.

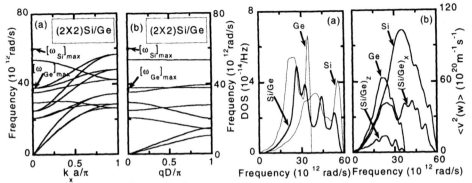

Fig.1 Dispersion curves of acoustic phonons in (a) in-plane direction and (b) cross-plane direction for a (2X2) Si/Ge superlattice. $[\omega_x]_{max}$ denotes the maximum acoustic phonon frequency of the bulk material.

Fig.2 (a) DOS and (b) TEPF for a (5X5) Si/Ge superlattice and the corresponding bulk materials. The subscripts x and z denote the phonon transport in the in-plane and cross-plane direction, respectively.

Under the constant relaxation time approximation, the thermal conductivity is proportional to TEPF,

$$<v_i^2(\omega)> = \frac{1}{V}\sum_\lambda \delta(\omega-\omega_\lambda)(\partial\omega/\partial k_i)^2 \qquad (4)$$

The in-plane and cross-plane TEPF's are plotted in Fig. 2(b) for a (5X5) Si/Ge superlattice. As seen in these figures, TEPF behaves differently in the in-plane and cross-plane directions. The in-plane acoustic phonon modes with the frequency higher than ω_{Ge} (shown in Fig.1) experience only slight confinement, but those along the cross-plane direction are significantly suppressed. Even at low frequencies, the magnitude of the in-plane TEPF is much larger than that along the cross-plane direction. This demonstrates that the superlattices act more like a phonon waveguide for the in-plane heat conduction. Also seen in Fig.2 are some dips in the curves for Si/Ge superlattices, caused by the mini-gaps in the phonon spectra shown in Fig.1, or the stop bands in the transmission curves.[19] It should be pointed out we also make similar calculation for AlAs/GaAs superlattices, but in order to save space, those figures are not listed here.

The variation of $\kappa` (=k/\tau)$ with the number of monolayers in each period is shown in Figs.3(a) and (b) for Si/Ge and AlAs/GaAs superlattices. The results we obtained for the cross-plane direction are consistent with those reported by Tamura et al..[17] As seen in these figures, $\kappa`_x$ and $\kappa`_z$ increase with decreasing layer thickness due to the effect of phonon tunneling.[19] This tunneling effect can explain the experimental results on Bi_2Te_3/Sb_2Te_3 superlattices.[10] The effect of phonon tunneling is much stronger in the cross-plane direction than that in the in-plane direction. As the layer becomes thicker, however, both $\kappa`_x$ and $\kappa`_z$ approach constants, in contrary to experimental results. It is unlikely that the frequency dependence of relaxation time[21,22] will explain the experimental data because the average group velocity doesn't change with thick period. In the superlattices with large period, phonons should be considered as particles due to the loss of coherence, which are described by Boltzmann transport equation.[13-15,18] It is also seen that the ratio κ/τ along the in-plane direction for Si/Ge superlattices, at room temperature, is reduced by a factor of 1.3 compared with the bulk material value if only the reduced group velocity due to phonon confinements is considered. In the same case, the corresponding reduction in the cross-plane direction is about an order of magnitude. The reasons for the different behavior of phonon confinement in the in-plane and cross-plane directions are as follows. The phonon confinements, caused by the phonon frequency mismatch between Si and Ge layers and the total internal reflection, limit the energy propagation in the cross-plane direction while the energy propagation along the film plane behaves like guided waves. The different effects of phonon confinement are not unique to Si/Ge superlattice; they are also apparent in AlAs/GaAs superlattices.

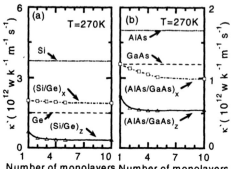

Fig.3 Ratio of thermal conductivity to relaxation time, k`, as a function of for (a) Si/Ge and (b) GaAs/AlAs superlattices as well as those of their bulk constituting materials.

Lee et al.[9] reported experimentally that the reduction in the cross-plane thermal conductivity is about a factor of 15 in the Si/Ge superlattices with the period comparable to our theoretical model, compared to their bulk values. Capinski and co-workers[7-8] observed that the corresponding reduction in AlAs/GaAs superlattices is about an order of magnitude. The phonon confinement effects can arguably explain the experimentally observed reduction in the cross-plane thermal conductivity.

In the in-plane direction, however, the reported experimental data show that the thermal conductivity is reduced by a factor of 6 in the AlAs/GaAs superlattices with the period comparable to our theoretical model,[1,3] and for Si/Ge superlattices the corresponding reduction is rather large[5], i.e. about a factor of 10. Clearly, the in-plane thermal conductivity reduction due to decreased phonon group velocity, cannot account for the large reduction reported in literature. This discrepancy suggests that the reduction in relaxation time due to diffuse interface phonon scattering[13,18], dislocation phonon scattering,[14] etc, should make the major contribution to the in-plane thermal conductivity reduction. This conclusion supports the results reported by Chen[13] and Hyldgaard and Mahan.[18]

CONCLUSIONS

In conclusion, we have investigated the effects of phonon confinement on the in-plane and cross-plane thermal conductivities of Si/Ge and AlAs/GaAs superlattices with the use of a fcc lattice dynamics model. We find that 1) the behavior of phonon confinement in the in-plane direction is different from that in the cross-plane direction. 2) the reduction in the in-plane thermal conductivity due to phonon confinement is much less than the experimental data. This suggests that the reduced relaxation time due to diffuse interface scattering, dislocation scattering, etc, should make the major contribution to the reduced in-plane thermal conductivity in superlattices.

ACKNOWLEDGMENTS

This work is supported by a DOD MURI grant on thermoelectrics (N00014-97-1-0516). G. Chen acknowledges the support from NSF through a young investigator award.

REFERENCES:

1. G. Chen, Ann. Rev. Heat Transf. **7**, 1 (1996).
2. G. Chen, S. G. Volz, T. Borca-Tasciuc, T. Zeng, D. Song, K. L. Wang, and M.S. Dresselhaus, MRS Proc. **545**, 357(1998).
3. T. Yao, Appl. Phys. Lett. **51**, 1798 (1987).
4. X. Y. Yu, G. Chen, A. Verma, and J. S. Smith, Appl. Phys. Lett. **67**, 3553 (1995).
5. E. Venkatasubramanian, E. Siivola, and T. S. Colpitts, ICT98 Proc. **17**, 191(1998)
6. G. Chen, C. L. Tien, X. Wu, and J. S. Smith, J. Heat Transf. **116**, 325 (1994).
7. W. S. Capinski, H. J. Maris, T. Ruf, M. Cardona, K. Ploog, and D. S. Katzer, Phys. Rev **B**. **59**, 8105(1999).
8. W. S. Capinski and H. J. Maris, Physica B **219**, 699 (1996).
9. S. M. Lee, D. Cahill, and R. Venkatasubramanian, Appl. Phys. Lett. **70**, 2957 (1997).
10. R. Venkatasubramanian, Naval Res. Rev. **58**, 44 (1996).
11. D.W. Song, C. Caylor, W.L. Liu, T. Zeng, T. Borca-Tasciuc, T.D. Sands, and G. Chen, ICT99 Proc., in press
12. T. Borca-Tasciuc, W.L. Liu, J.L Liu, T. Zeng, D.W. Song, C.D. Moore, G. Chen, K.L. Wang, M.S. Goorsky, T. Radetic, Ronald Gronsky, X. Sun and M. S. Dresselhaus, ICT99 Proc., in press.
13. G. Chen, J. of Heat Transf. **119**, 220 (1997).
14. G. Chen and M. Neagu, Appl. Phys. Lett. **71**, 2761 (1997).
15. G. Chen, Phys. Rev. B **57**, 14958 (1998).
16. P. Hyldgaard, and G. D. Mahan, Phys. Rev. B **56**, 10754 (1997).
17. S. Tamura, Y. Tanaka, and H. J. Maris, Phys. Rev. B **60**, 2627(1999).
18. P. Hyldgaard and G. D. Mahan, Thermal Conductivity (Technomic, Lancaster, PA, 1996) **23**, 172.
19. G. Chen, J. Heat Transf., **121**, 945 (1999).
20. D. A. Young and H. J. Maris, Phys. Rev. B **40**, 3685(1989).
21. A. Balandin and K. L. Wang, Phys. Rev. B **58**, 1544 (1998).
22. S.Y. Ren and J. D. Dow, Phys. Rev. B **25**, 3750 (1982).

Structure and Thermoelectric Properties of New Quaternary Tin and Lead Bismuth Selenides, $K_{1+x}M_{4-2x}Bi_{7+x}Se_{15}$ (M = Sn, Pb) and $K_{1-x}Sn_{5-x}Bi_{11+x}Se_{22}$

Antje Mrotzek[1], Kyoung-Shin Choi[1], Duck-Young Chung[1], Melissa A. Lane[2], John R. Ireland[2], Paul W. Brazis[2], Tim Hogan[3], Carl R. Kannewurf[2], and Mercouri G. Kanatzidis[1]

[1] Department of Chemistry, Michigan State University, East Lansing, MI 48824.
[2] Department of Electrical Engineering and Computer Science, Northwestern University, Evanston, IL. [3] Department of Electrical Engineering, Michigan State University, East Lansing, MI 48824.

Abstract
We present the structure and thermoelectric properties of the new quaternary selenides $K_{1+x}M_{4-2x}Bi_{7+x}Se_{15}$ (M = Sn, Pb) and $K_{1-x}Sn_{5-x}Bi_{11+x}Se_{22}$. The compounds $K_{1+x}M_{4-2x}Bi_{7+x}Se_{15}$ (M= Sn, Pb) crystallize isostructural to $A_{1+x}Pb_{4-2x}Sb_{7+x}Se_{15}$ with A = K, Rb, while $K_{1-x}Sn_{5-x}Bi_{11+x}Se_{22}$ reveals a new structure type. In both structure types fragments of the Bi_2Te_3-type and the NaCl-type are connected to a three-dimensional anionic framework with K^+ ions filled tunnels. The two structures vary by the size of the NaCl-type rods and are closely related to β-$K_2Bi_8Se_{13}$ and $K_{2.5}Bi_{8.5}Se_{14}$. The thermoelectric properties of $K_{1+x}M_{4-2x}Bi_{7+x}Se_{15}$ (M = Sn, Pb) and $K_{1-x}Sn_{5-x}Bi_{11+x}Se_{22}$ were explored on single crystal and ingot samples. These compounds are narrow gap semiconductors and show n-type behavior with moderate Seebeck coefficients. They have very low thermal conductivity due to an extensive disorder of the metal atoms and possible "rattling" K^+ ions.

Introduction
Currently, there is great interest in finding new materials [1] for thermoelectric applications [2] and also in developing new concepts for designing superior thermoelectric compounds [3]. The efficiency of a thermoelectric material is characterized by the figure of merit, ZT, with the absolute temperature T and $Z = \sigma \cdot S^2/\kappa$; where σ is the electrical conductivity, S is the thermopower and κ is the total thermal conductivity. Various alloys of Bi_2Te_3 are the best known thermoelectric materials so far and exhibit a figure of merit of about 1 [4], which is only 10% of the Carnot efficiency. Because increasing ZT to 1.5 or 2 would lead to many more technical applications, the motivation for exploring new materials is strong.

Our research is focussed on discovering new chalcogenide compounds with complex compositions and structures and heavy elements that possess high electrical conductivity, high Seebeck coefficient and low thermal conductivity for a ZT maximization [5]. The complex chalcogenides $BaBiTe_3$ [6], $K_2Bi_8Se_{13}$ [7, 8], $K_{2.5}Bi_{8.5}Se_{14}$ [8, 9], $CsBi_4Te_6$ [8, 10], $KBi_{6.33}S_{10}$ [7] exhibit promising thermoelectric properties. In these examples, the alkali and alkaline earth metals are incorporated in structurally diverse bismuth chalcogenides. The electropositive metal ions tend to fill tunnels in the anionic framework where they can function as "rattlers" to reduce the lattice thermal conductivity [3, 8]. Another way to maximize ZT is to increase the thermopower which is a measure for the asymmetry of the electronic structure near to the Fermi level according to the Mott formula [9]. If doped appropriately, a material with complex electrical band structure may give rise to more favorable thermoelectric properties. Therefore we extended our investigations to multinary systems. The attempt to incorporate the heavy metal Pb in the K/Bi/Se system led to the compound $K_{1+x}Pb_{4-2x}Bi_{7+x}Se_{15}$ [11] with interesting electrical properties. Because of the isovalent relationship between Pb and Sn we investigated the system K/Sn/Bi/Se as well. We report here the syntheses, structural and physicochemical characterization of the new compounds $K_{1.25}Pb_{3.5}Bi_{7.25}Se_{15}$, $K_{1.52}Sn_{2.96}Bi_{7.52}Se_{15}$ and $K_{0.66}Sn_{4.82}Bi_{11.18}Se_{22}$.

Results and Discussion
The compounds $K_{1-x}Sn_{5-x}Bi_{11+x}Se_{22}$ and $K_{1+x}Sn_{4-2x}Bi_{7+x}Se_{15}$ were prepared by stoichiometric reactions of K_2Se, Sn, Bi_2Se_3 and Se at 800°C and 900°C, respectively. $K_{1+x}Pb_{4-2x}Bi_{7+x}Se_{15}$ was prepared by the molten flux method at 800°C. The general formulas $K_{1+x}M_{4-2x}Bi_{7+x}Se_{15}$ (M = Sn, Pb) and $K_{1-x}Sn_{5-x}Bi_{11+x}Se_{22}$ express the variation of composition due to mixed occupancy by K^+, M^{2+} and Bi^{3+} ions. While Pb and Bi are not distinguishable by X-rays, the disorder of the metal atoms could be examined by the structure refinement of the tin compounds. The crystallographic data for the selenides are summarized in Table 1.

Table 1. Crystallographic data for $K_{1+x}M_{4-2x}Bi_{7+x}Se_{15}$ (M = Sn, Pb) and $K_{1-x}Sn_{5-x}Bi_{11+x}Se_{22}$

	$K_{1.25}Pb_{3.5}Bi_{7.25}Se_{15}$	$K_{1.52}Sn_{2.96}Bi_{7.52}Se_{15}$	$K_{0.66}Sn_{4.82}Bi_{11.18}Se_{22}$
Space Group	P 2$_1$/m	P 2$_1$/m	P 2$_1$/m
Crystal Habit	silver needle	silver needle	silver ribbon
a, Å	17.4481(8)	17.454(5)	15.777(2)
b, Å	4.1964(2)	4.201(1)	4.1669(6)
c, Å	21.695(1)	21.750(6)	17.358(2)
β, °	98.850(5)	98.550(5)	99.249(2)
V, Å^3 ; Z	1569.54; 2	1577.16; 2	1126.24; 1
R1/wR2 [I>2σ(I)]	5.37 / 11.26	7.75 / 19.36	5.35 / 9.32
GOF on F^2	0.921	1.018	2.101

$K_{1.52}Sn_{2.96}Bi_{7.52}Se_{15}$ and $K_{1.25}Pb_{3.5}Bi_{7.25}Se_{15}$ crystallize isostructural to $A_{1+x}Pb_{4-2x}Sb_{7+x}Se_{15}$ (A= K, Rb)[11] which is closely related to β-$K_2Bi_8Se_{13}$ [7, 8] and $K_{2.5}Bi_{8.5}Se_{14}$ [8, 9]. As in the ternary selenides the structure of $K_{1.52}Sn_{2.96}Bi_{7.52}Se_{15}$, shown in Figure 1A, is composed of building units of the Bi_2Te_3-type and the NaCl-type that are assembled by edge-sharing, distorted MSe$_6$-octahedra. We found mixed occupancy with Bi/Sn and Sn/K, respectively, for most metal sites. Five (Bi,Sn)-octahedra wide fragments of the Bi_2Te_3-type are rearranged to step-shaped layers in a trans-trans fashion. They are linked by NaCl-type rods that are three Bi octahedra wide in the direction parallel to the Bi_2Te_3-type layers and two Bi octahedra wide in the direction perpendicular to the Bi_2Te_3-type layers. The three-dimensional connection of the building units to an anionic framework creates tunnels that are filled by K$^+$ ions on the tri-capped trigonal prismatic sites. The ADP of K is that is 50 % higher than the average ADP for the atoms of the framework. This unusually high thermal displacement parameter for K indicates that the "rattling" of the K$^+$ ions in their crystallographic position.

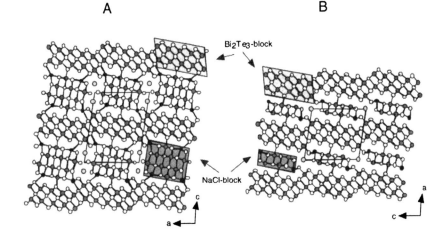

Figure 1. (a) The structure of $K_{1+x}Pb_{4-2x}Bi_{7+x}Se_{15}$ viewed down the b-axis. (b) The structure of $K_{1-x}Sn_{5-x}Bi_{11+x}Se_{22}$ viewed down the b-axis. The Bi_2Te_3-type and NaCl-type building units are highlighted in both structures. Small white spheres: Se, large light-gray spheres: K, middle-gray spheres: Bi, dark-gray spheres: Sn and Pb respectively.

$K_{0.66}Sn_{4.82}Bi_{11.18}Se_{22}$ crystallizes in a new structure type, that is closely related to the one of $A_{1+x}M_{4-2x}M'_{7+x}Se_{15}$ (A = K, Rb; M = Sn, Pb; M' = Sb, Bi) [11]. The structure, shown in Figure 1b, is composed of Bi_2Te_3- and NaCl-type units as well. Their three-dimensional linkage leads to tunnels that are filled with K^+ ions. The structure of $K_{0.66}Sn_{4.82}Bi_{11.18}Se_{22}$ contains the same step-shaped Bi_2Te_3-type layers of edge-sharing $(Bi,Sn)Se_6$-octahedra as in $K_{1.52}Sn_{2.96}Bi_{7.52}Se_{15}$. In contrast to $K_{1.52}Sn_{2.96}Bi_{7.52}Se_{15}$, they are joined by smaller NaCl-type units that extend three $(Bi,Sn)Se_6$-octahedra in the direction parallel to the Bi_2Te_3-type layers but only one $(Bi,Sn)Se_6$-octahedra in the direction perpendicular to the Bi_2Te_3-type layers. Bi and Sn are disordered over all the metal sites of the anionic framework. As in $K_{1.52}Sn_{2.96}Bi_{7.52}Se_{15}$, the K^+ ions filling the tunnels are "rattling" on their tri-capped trigonal prismatic sites as evidenced by the high thermal displacement parameter, five times higher than the average ADP for the atoms of the framework, although probably, static disorder is also present because the sites are only 33% occupied.

Properties

We measured the thermal conductivity, thermopower and electrical conductivity of the quaternary selenides on ingots and single crystals. $K_{1.25}Pb_{3.5}Bi_{7.25}Se_{15}$ exhibits interesting thermoelectric properties. The temperature dependence of the thermopower and electrical conductivity is shown in Figure 2 for single crystal and ingot measurements as well. Single crystals of $K_{1.25}Pb_{3.5}Bi_{7.25}Se_{15}$ show n-type behavior with a large Seebeck coefficient of - 150µV/K at 300K. The data exhibit an almost linear negative temperature dependence which indicates that the maximum thermopower lies at higher temperature and has not been reached yet. The temperature dependence of the thermopower of a polycrystalline ingot shows the same trend but much lower values (-25 µV/K at 300K) than the single crystal data. The single crystal electrical conductivity data show semiconducting behavior below 40 K and metallic behavior above. Corresponding to their thermopower values the electrical conductivity of the ingot (880 S/cm at 300K) is higher than the one of the crystal (260 S/cm at 300K). This difference can be explained by the difference in synthetic method that leads to a different extend of doping for the two different types of sample [12]. The thermal conductivity is not as sensitive to doping as thermopower and electrical conductivity and was found to be ~1.5 W/mK at room temperature for the ingot sample. The low value that is comparable to the one for optimized Bi_2Te_3 (~1.5 W/mK) is due to the dynamic motion of the K^+ ions and the mixed occupancies of most metal sites in the structure.

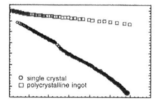 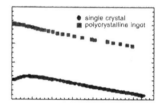

Figure 2. (a) Variable temperature thermopower for a single crystal and an ingot of $K_{1.25}Pb_{3.5}Bi_{7.25}Se_{15}$. (b) Variable temperature electrical conductivity for a single crystal and an ingot of $K_{1.25}Pb_{3.5}Bi_{7.25}Se_{15}$.

Thermopower and electrical conductivity of $K_{1.52}Sn_{2.96}Bi_{7.52}Se_{15}$ were measured on a single crystal, while the thermal conductivity was determined on a polycrystalline ingot. $K_{1.52}Sn_{2.96}Bi_{7.52}Se_{15}$ shows n-type behavior as well with a room temperature thermopower about -34 µV/K, see Figure 3. As in the Pb compound the absolute value of the thermopower increases linearly with the temperature indicating that maximum thermopower will be achieved at higher temperatures. The negative sign indicates that electrons are the main charge carriers. We will need more measurements on different samples to understand the influence of the substitution of Pb by Sn in the thermopower. The substitution of Pb by Sn leads to a higher absolute value of the thermopower compared to the ingot data of $K_{1.25}Pb_{3.5}Bi_{7.25}Se_{15}$ but reduces the thermopower in comparison with the single crystal data. Analogous to $K_{1.25}Pb_{3.5}Bi_{7.25}Se_{15}$, $K_{1.52}Sn_{2.96}Bi_{7.52}Se_{15}$ has a metal-like electrical conductivity that decreases with rising temperature. Its electrical conductivity is lower (540 S/cm) than the value for the polycrystalline ingot of $K_{1.25}Pb_{3.5}Bi_{7.25}Se_{15}$ but higher than the results for the single crystal. The thermal conductivity of $K_{1.52}Sn_{2.96}Bi_{7.52}Se_{15}$ (~1.5 W/m·K) is comparable to the one of the Pb compound. The favorable low thermal conductivities of $K_{1+x}M_{4-2x}Bi_{7+x}Se_{15}$ are due to the low symmetry of their structure as well as extensive disorder of the metal atoms and the presence of "rattling" atoms.

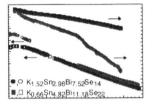 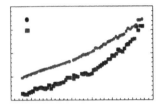

Figure 3. (a) Variable temperature electrical conductivity and temperature thermopower for single crystals of $K_{1.52}Sn_{2.96}Bi_{7.52}Se_{15}$ and $K_{0.66}Sn_{4.82}Bi_{11.18}Se_{22}$. (b) Variable temperature thermal conductivity for ingots of $K_{1.52}Sn_{2.96}Bi_{7.52}Se_{15}$ and $K_{0.66}Sn_{4.82}Bi_{11.18}Se_{22}$.

The thermopower and electrical conductivity of $K_{0.66}Sn_{4.82}Bi_{11.18}Se_{22}$ were measured on a single crystal while the thermal conductivity was measured on a polycrystalline ingot. $K_{0.66}Sn_{4.82}Bi_{11.18}Se_{22}$ shows a similar metal-like temperature dependence of the electrical conductivity, compare Figure 3. The room temperature value (550 S/cm) is slightly higher than for $K_{1.52}Sn_{2.96}Bi_{7.52}Se_{15}$. $K_{0.66}Sn_{4.82}Bi_{11.18}Se_{22}$ also has a negative thermopower that shows a negative linear temperature dependence. The maximum thermopower will be achieved higher than room temperature. Although the room temperature value for $K_{0.66}Sn_{4.82}Bi_{11.18}Se_{22}$ (-10 μV/K) is smaller than for $K_{1.52}Sn_{2.96}Bi_{7.52}Se_{15}$, we can not make a final statement yet how the size of the NaCl-type blocks affects the thermopower. All these materials are very sensitive to "self-doping" occuring during their preparation as it is evident by the conflicting results for $K_{1.25}Pb_{3.5}Bi_{7.25}Se_{15}$. The low thermopower of all ingot samples can be explained by being heavily doped. As in the selenides $K_{1+x}M_{4-2x}Bi_{7+x}Se_{15}$ (M = Sn, Pb), $K_{0.66}Sn_{4.82}Bi_{11.18}Se_{22}$ has a low thermal conductivity of ~1.5 W/m·K at room temperature due to the low symmetry of the crystal structure, extensive disorder of the metal atoms and "rattling" of K^+ ions.

Concluding Remarks

Synthetic investigations in the system K/M/Bi/Se (M = Sn, Pb) led to the new quaternary bismuth selenides $K_{1+x}M_{4-2x}Bi_{7+x}Se_{15}$ (M = Sn, Pb) and $K_{1-x}Sn_{5-x}Bi_{11+x}Se_{22}$. Their crystal structures are closely related and contain fragments of the Bi_2Te_3-type and NaCl-type lattices. The selenides are narrow band gap semiconductors with moderate electrical conductivities and negative Seebeck coefficients. Single crystals of $K_{1.25}Pb_{3.5}Bi_{7.25}Se_{15}$ exhibit the most promising thermoelectric properties while the results for the ingot samples of $K_{1+x}M_{4-2x}Bi_{7+x}Se_{15}$ (M = Sn, Pb) and $K_{1-x}Sn_{5-x}Bi_{11+x}Se_{22}$ are influenced by the heavy "self-doping" of the samples occuring during their preparation. Large, low symmetric unit cells, an extensive disorder of the metal atoms and the presence of "rattling" potassium atoms are responsible for the remarkable low thermal conductivity of the quaternary selenides. This work shows that the properties of these compounds are sensitive to dopants and that the thermopower might be improved by further progress in sample preparation processing and through appropriate doping.

Acknowledgements

Financial support from the Office of Naval Research (Grant No. N00014-98-1-0443) and the Deutsche Forschungsgemeinschaft is gratefully acknowledged. This work made use of the SEM facilities of the Center for Electron Optics at Michigan State University.

References

[1] (a) *Thermoelectric Materials 1998 - The Next Generation Materials for Small-Scale Refrigeration and Power Applications* edited by T. M. Tritt, M. G. Kanatzidis, G. D. Mahan, H. B. Lyon, *Mat. Res. Soc. Symp. Proc.* **1998**, *545*. (b) *Thermoelectric Materials – New Directions and Approaches* edited by T. M. Tritt, M. G. Kanatzidis, H. B. Lyon, G. D. Mahan, *Mat. Res. Soc. Symp. Proc.* **1998**, *478*.
[2] *CRC Handbook of Thermoelectrics*, edited by D. M. Rowe, CRC Press, Boca Raton **1995**
[3] (a) G. A. Slack in *CRC Handbook of Thermoelectrics*, edited by D. M. Rowe, CRC Press, Boca Raton **1995**, 407. (b) G. A. Slack in *Solid State Physics* edited by H. Ehrenreich, F. Seitz, D. Turnbull, Academic: New York **1997**, *34*, 1. (c) B. C. Sales, *Mater. Res. Bull.* **1998**, *23*, 15. (d) B. C. Sales, D. Mandrus, B. C. Chakoumakos, V. Keppens, J. R. Thompson, *Phys. Rev. B.* **1997**, *56*, 15081. (e) T. M. Tritt, *Science* **1996**, *272*, 1276. (f) M. G. Kanatzidis, F. J. DiSalvo, *ONR Quarterly Review* **1996**, *XXVII*, 14.
[4] (a) H.-H. Jeon, H.-P. Ha, D.-B. Hyun, J.-D. Shim *J. Phys. Chem. Solids* **1991**, *4*, 579. (b) L. R. Testardi, J. N. Jr. Bierly, F. J. Donahoe *J. Phys. Chem. Solids* **1962**, *23*, 1209. (c) C. H. Champness, P. T. Chiang, P. Parekh *Can. J. Phys.* **1965**, *43*, 653. (d) C. H. Champness, P. T. Chiang, P. Parekh *Can. J. Phys.* **1967**, *45*, 3611. (e) M. V. Vedernikov, V. A. Kutasov, L. N. Lukyanova, P. P. Konstantinov *Proc. Of the XVI*th *Int. Conf. On Thermoelectrics (ITC '97)*, Dresden, Germany **1997**, 56. (f) M. Stordeur in *CRC Handbook of Thermoelectrics*, edited by D. M. Rowe, CRC Press, Boca Raton **1995**, 239.
[5] D.-Y. Chung,, L. Iordanidis, K.-S. Choi, M. G. Kanatzidis, *Bull. Korean Chem. Soc.*, **1998**, *19*, 1283.
[6] D.-Y. Chung, S. Jobic, T. Hogan, C. R. Kannewurf, R. Brec, M. G. Kanatzidis, *J. Am. Chem. Soc.* **1997**, *119*, 2505.
[7] (a) M. G. Kanatzidis, T. J. McCarthy, T. A. Tanzer, L.-H. Chen, L. Iordanidis, T. Hogan, C. R. Kannewurf, C. Uher, B. Chen, *Chem. Mater.* **1996**, *8*, 1465. (b) B. Chen, C. Uher, L. Iordanidis, M. G. Kanatzidis, *Chem. Mater.* **1997**, *9*, 1655.
[8] M. G. Kanatzidis, D.-Y. Chung, L. Iordanidis, K.-S. Choi, P. Brazis, M. Rocci, T. Hogan, C. Kannewurf, *Mat. Res. Soc. Symp. Proc.* **1998**, *545*, 233
[9] D.-Y. Chung, K.-S. Choi, L. Iordanidis, J. L. Schindler, P. W. Brazis, C. R. Kannewurf, B. Chen, S. Hu, C. Uher, M. G. Kanatzidis, *Chem. Mater.* **1997**, *9*, 3060.
[10] D.-Y. Chung, T. Hogan, P. Brazis, M. Rocci-Lane, C. R. Kannewurf, M. Bastea, C. Uher, M. G. Kanatzidis, *Science* **2000**, *287*, 1024.
[11] D.-Y. Chung, K.-S. Choi, P. W. Brazis, C. R. Kannewurf, M. G. Kanatzidis, *Mat. Res, Soc. Symp. Proc* **1998**, *545*, 65.
[12] K.-S. Choi, D.-Y. Chung, A. Mrotzek, P. Brazis, C. R. Kannewurf, C. Uher, W. Chen, T. Hogan, M. G. Kanatzidis *Chem. Mater. submitted*.

Processing, characterization, and measurement of the Seebeck coefficient of Bismuth microwire array composites

T.E. Huber and P. Constant
Laser Research Laboratory, Howard University, Washington, DC 20059, USA.

ABSTRACT

We have fabricated Bi microwire array composites ranging in diameter from 10 to 50 micrometer using the method of high-pressure-injection (HPI) of the Bi melt into microchannel arrays (MCA) templates. The composites are dense, with Bi volume fraction in excess of 70 %. The parallel Bi nanowires, whose length appears to be limited only by the thickness of the host template (up to 2 mm), terminate at both sides of the composite in the Bi bulk. The individual Bi microwire crystal structure is rhombohedral, with the same lattice parameters as that of bulk Bi; the wires crystalline orientation is predominantly perpendicular to the (113) lattice plane. The transversal magnetoresistance and Seebeck effect of the wires has been measured in magnetic fields up to 0.8 Tesla and for temperatures ranging between 77 K and room temperature.

INTRODUCTION

The suitability of a material for thermoelectric applications is measured by the thermoelectric figure of merit $Z = S^2\sigma/\kappa$ where S is the Seebeck coefficient, σ is the electrical conductivity and κ is the thermal conductivity. Bulk Bi, a semimetal, and $Bi_{1-x}Sb_x$, have the highest Z at 100 K [1]. The changes in the thermoelectric properties of bismuth-antimony alloys as a function of magnetic field and various crystalline orientations has been studied extensively. The thermo magnetic figure of merit Z_M, which measures the cooling and heating efficiency of a given material when aided by an external magnetic field is as much as 0.5 for both pure Bi and some of its alloys with Sb at 0.75 Tesla and 77 K [2,3].

The properties of Bi and Bi-Sb are strongly anisotropic. For example the thermal conductivity λ of Bi at 100 K is 18 $Wm^{-1}K^{-1}$ along the trigonal direction and 13 $Wm^{-1}K^{-1}$ normal to the trigonal direction. The best thermoelectric performance is observed when the electrical current flows along the trigonal crystal axis. Since their lamellar structure confers to the single crystals an aptitude to cleavage along the trigonal planes. This lack of strength has largely prevented the use of these materials in thermoelectric devices [4]. Powder metallurgy may offer the opportunity to increase the mechanical strength of Bi and Bi-Sb alloys.

Microengineering Bi and Bi-Sb into composites may lead to a significant improvement in their thermoelectric performance, because of the reduction of phonon thermal conductivity from phonon scattering at the grain boundaries and interfaces. Devaux, Brochin, Lenoir, Martin-Lopez, Scherrer and Scherrer measured the figure of merit of pellets of polycrystalline materials, as a function of grain size (100 nm <d< 200μm) [5]. The samples were prepared by cold pressing and sintering of powders. They observed that there is a decrease of the figure of merit in this process brought about by a large decrease of the electrical conductivity. To increase the figure of merit it is important to introduce phonon scattering, so as to decrease the thermal conductivity, without decreasing the carrier's mobility. This seems to agree with theoretical predictions that by mixing together macroscopically a number of different thermoelectric components, the figure of merit can never be enhanced to a value greater than the largest among the various component values [6]. In order to understand the interplay between thermal and electrical conductivity, we need to study the simplest structure. A macroscopic wire array is a special case of a composite, where by neglecting the thermal conductivity of the insulating matrix, both the electrical and thermal conductivity are proportional to the filling fraction. $\kappa_{comp} = \phi \kappa_{Bi-Sb}$ and $\sigma_{comp} = \phi \sigma_{Bi-Sb}$, where ϕ is the conductor filling fraction. Therefore, the composite figure of merit of a wire array is not dependent upon the filling fraction.

Bi nanowire systems have recently attracted a great deal of interest, because of their potential applications to thermoelectric devices and their promise for studying 1D transport properties [7,8]. In fact, some new properties of Bi nanowires are actually the result of classical size effects that arise when the wire diameter d is smaller than the mean free path l.

EXPERIMENTAL DETAILS

Our samples were prepared using the high-pressure injection technique [9]. The pressure applied did not exceed 100 atmospheres. The   matrices are 10 μm and 50 μm channel diameter microchannel array (MCA) made of soda glass by Galileo Industries (Sturbridge, MA). Our samples crystalline structure was studied with the X-ray diffraction (XRD) method in the geometry where the wires are normal to the scattering surface. We find that the XRD peaks are very narrow, which indicates the long-range periodicity of the structure along the wire length. The XRD spectrum of the Bi nanowire arrays shows only a few strong peaks. The position of these peaks correspond within experimental resolution with those for the bulk, indicating that the rhombohedral crystal structure of bulk Bi is preserved in the nanowires. The appearance of only a few peaks of the rhombohedral structure suggests that individual wires are composed of crystalline grains that are highly oriented. The prominence of the (113) peaks in comparison to the (102) peak indicates that the wires crystalline orientation is predominantly perpendicular to the (113) lattice plane. We also observe the peak (102) with high intensity. However, since this peak is very strong in the bulk its observation could be interpreted as evidence of polishing residue or strained layer on the sample surface. As we explained above this crystalline orientation is not the most suitable for Peltier cooling.

We measured the transport properties of Bi microwire arrays with a conventional four-probe method. The MR was measured for magnetic fields B, up to 0.9 T, applied perpendicular to the wire length. As shown in Fig. 1, R(0.9 T) is several times larger than R(0) and thus large MR(H) = (R(H)-R(0))/R(0) are observed. The magnetoresistance size effects on Bi microwires can be understood in terms of the so-called ordinary MR, caused by the curving of trajectories in the magnetic field. The effect is determined by $\omega_c \tau$ where $\omega_c = eH/m^* c$ is the cyclotron frequency
and τ the relaxation time (e is the electron charge, m^* is the effective mass, and c is the velocity of light. At a given value of H, the value of ω_c is a intrinsic property of a given material. Because the Bi effective mass is very small, ω_c is very large. The mean free path of electrons in Bi is also,
characteristically, very long and consequently τ is large. The MR shown by our samples are larger than that exhibited by single crystal thin films for thickness t exceeding d, the wire diameter, at the same temperature. For example, Yang, Liu, Hong, Reich, Searson, and Chien [10] measured a magnetoresistance of 15% , at room temperature and for a 20 micrometer thick single crystal thin film in a magnetic field of 1 T. In comparison, we measured a MR of 20% for a 10 μm diameter wire.

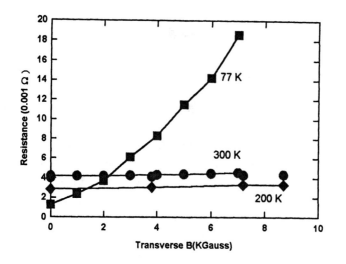

Fig. 1. Transverse (B⊥J) magnetoresistance of Bi microwire arrays (d= 10 μm).

We also measured the Seebeck coefficient of our wire arrays. The Seebeck coefficient is determined as the ratio of a potential difference to a temperature difference. The temperature differences are determined using copper-constantan thermocouples, the copper branches also being employed in obtaining the voltage differences. The reference junctions are maintained at room temperature. To minimize the heat conducted through the wires to the sample, which causes a temperature gradient within the junction of the thermocouple wires, we used 100 μm diameter wires. The scatter of the data is likely to be related to the small volume of the sample. The asymptotic value at high temperatures is 70±3 μV/K can be understood readily. The Seebeck coefficient of bulk Bi is anisotropic. For heat flow along the trigonal direction the Seebeck coefficient is approximately 95μV/K. For heat flow normal to the trigonal direction the coefficient is 55 μV/K. Therefore the observed value of the Seebeck coefficient is consistent with the intermediate orientation of our samples. Another very interesting feature of our samples is the decrease of Seebeck coefficient observed for low temperatures. In contrast, bulk samples display a much less noticeable effect [1]. Bodiul, D.V. Gitsu, Dolma, Miglei and Zegrya [11] studied the magnetoresistance and thermoelectric Seebeck coefficients of Bi microwires (0.1 μm < d < 10 μm). They observed a similar effect in the temperature dependence of the Seebeck coefficient and interpreted their results in terms of a classical size effect and enhancement of phonon drag effect in small diameter Bi wires. New measurements by Bodiul, Burchakov, Gitsu, and Nikolaeva are also consistent with this interpretation [12].

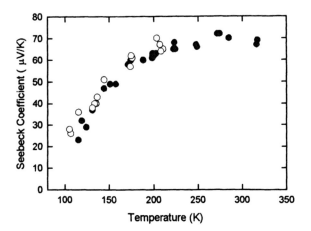

Fig. 2. Seebeck coefficient of Bi microwires. Full circles correspond to B=0. Empty circles are obtained for B=0.7 Tesla.

Another very interesting feature of the data is the absence of a magnetic field dependence. This contrasts with the large magnetic field dependences found for bulk Bi [13]. We have not been able to understand the reason for this property of microwires.

CONCLUSION

Microengineering these traditional thermoelectric materials into composites may lead to significant improvements in their mechanical strength and thermoelectric performance. We have fabricated dense composites consisting of arrays of Bi microwires ranging in diameter from 10 to 40 μm. We observe a large magnetoresistance that can be understood in terms of classical size effects. The Seebeck coefficient is found to be magnetic field independent up to 0.6 T.

ACKNOWLEDGEMENTS

Ansil Dyal and Ricky Calcao of Polytechnic University, Brooklyn, participated in the early stages of this work. Support by the Army Research Office through grant DAAD19-99-1-0282 and by the National Science Foundation through DMR-9632819 is acknowledged.

REFERENCES

1. H.J. Goldsmid in *"Electronic Refrigeration"* (Pion Limited, London, 1986), p. 7.
2. H.J. Goldsmid, "Thermomagnetic Phenomena" in "CRC Handbook of Thermoelectrics" edited by   D.M. Rowe (CRC Press, Boca Raton, 1994), p. 75.
3. G.E. Smith et al, J. Appl. Phys. **33** 841 (1962). R. Wolfe et al, Appl. Phys. Lett., **1**, 5, (1962).
4. H.J. Goldsmid and E.H. Volkmann, 16th ITC Proc., A. Heinrich, editor (IEEE, NJ, 1998), p.171.
5. X. Devaux, F. Brochin, A. Dauscher, B. Lenoir, R. Martin-Lopez, H. Scherrer, and S. Scherrer,16th ITC Proc, A. Heinrich, editor (IEEE Publications, Piscataway, NJ, 1998), p.199.
6. D.J. Bergman, 16th ITC Proc., A. Heinrich, editor (IEEE Pub., Piscataway, NJ, 1998), p. 401.
7. L.D. Hicks and M.S. Dresselhaus, Phys. Rev. **B47**, 16631 (1993).
8. D.A. Broido and T.L. Reinecke , Mat. Res. Soc. Symp. Proc. **545** 87 (1999), p. 485.
9   C.A. Huber and T.E. Huber, *J. Appl. Phys.* **64**, 6588 (1988).
10. F.Y.Yang, K. Liu, K. Hong, D.H. Reich, P.C. Searson, and C.L. Chien, Science **284** 1335 (1999).
11. P.P. Bodiul et al, Phys. Stat. Sol. (a) **53**, 87 (1979)
12. P.P. Bodiul, A.N. Burchakov, D.V. Gitsu, and A.A. Nikolaeva, to be published (2000).
13. G.E. Smith et al, Proc. 3th Int. Conf. of Phys.Semiconductors, Paris (Dunod, Paris, 1964), p. 399.

CHARACTERIZATION OF NEW MATERIALS IN A FOUR-SAMPLE THERMOELECTRIC MEASUREMENT SYSTEM

Nishant A. Ghelani[1], Sim Y. Loo[1], Duck-Young Chung[2], Sandrine Sportouch[2], Stephan de Nardi[2], Mercouri G. Kanatzidis[2], Timothy P. Hogan[2], George S. Nolas[3]
[1] Department of Electrical and Computer Engineering, Michigan State University, East Lansing, MI.
[2] Chemistry Department, Michigan State University, East Lansing, MI.
[3] R & D Division, Marlow Industries, Inc., Dallas, TX.

ABSTRACT

Several new materials in the $CsBi_4Te_6$, $A_2Bi_8Se_{13}$, (A = K, Rb, Cs), HoNiSb, Ba/Ge/B (B = In, Sn), and $AgPbBiQ_3$ (Q = S, Se, Te) systems have shown promising characteristics for thermoelectric applications. New synthesis techniques are able to produce samples at much higher rates than previously possible. This has led to a persistent challenge in thermoelectric materials research of rapid and comprehensive characterization of samples. This paper presents a description of a new 4-sample transport measurement system and the related measurement techniques. Special features of the system include fully computer-controlled operation (implemented in LabView™) for simultaneous measurement of electrical conductivity, thermo-electric power, and thermal conductivity. This system has been successfully used to characterize several new thermoelectric materials (including some of the above-mentioned compounds) and reference materials exhibiting a wide range of thermal conductivities.

INTRODUCTION

Recent advances in synthesis techniques have greatly increased the rate of fabrication of new electronic materials. The polychalcogenide flux technique [1], for instance, has produced a variety of new and interesting thermoelectric materials. Following the discovery of a new material, a fast preparation method is needed to produce a large number of doped samples. Such a method has been developed in our laboratories and we anticipate that a single researcher could produce approximately 80 - 100 samples per week using this procedure.

With existing charge transport characterization systems, it would be difficult to measure samples at such a rate, therefore a new facilities have been established to first screen through samples in a high throughput thermopower measurement system [2]. The most promising materials are then fully characterized in a 4-sample measurement system [3]. Special features of the 4-sample system include simultaneous measurement of electrical conductivity, thermoelectric power, and thermal conductivity. Typical run times to fully characterize four samples from 80 K to 400 K in 5 K steps are approximately 60 hours.

This paper presents a detailed description of the 4-sample system including recent upgrades in the mounting technique, followed by data recently collected on several of the above mentioned compounds.

EXPERIMENTAL SETUP

Thermoelectric power and thermal conductivity measurements were obtained using a pulse technique similar to that proposed by Maldonado [3]. The heater and thermocouple probes are mounted on the sample in the traditional four-probe steady state thermal conductivity configuration. Additional current and voltage leads connected for the simultaneous measurement of thermoelectric power and electrical conductivity as shown in Figure 1.

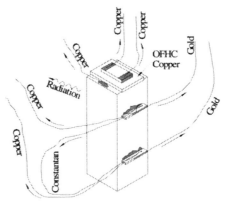

Figure 1. Sample mounting configuration for measuring electrical conductivity, thermopower, and thermal conductivity.

A copper-constantan differential thermocouple is used for measuring the temperature gradient, ΔT. The thermocouples are electrically isolated from the sample and thermally connected to the sample using Stycast™ epoxy to attach them on the surface of small copper plates, as shown in Figure 1. Heat losses through the electrical connections are minimized by using long leads (~ 15 cm) wrapped around a glass tube. One end of the sample is mounted to a gold plated OFHC copper stage using silver/gold paste and a heater resistor is attached to the other end with silver/gold paste. The voltage leads are then gold pasted to the sample along with the copper plates holding the thermocouples. This helps to assure that the voltage and temperature gradients are measured at the same locations on the sample for accurate thermoelectric power measurements. The next step includes loading the sample stage into the cryostat, evacuating the sample chamber at a slightly elevated temperature, testing contacts, and cooling the samples to the starting temperature. Most of the time is spent in step three, during data collection while under computer control. By incorporating multiple measurements in a single run, considerable time is saved by avoiding remounting, and recooling of the samples.

In this measurement technique the surface mount resistor (470 Ω) was used to periodically heat one end of the sample to establish a measured gradient of approximately 1 K. Typically a 12-13 minute period of oscillation was chosen and the temperature gradient was observed to reach steady state within half of the period, as shown for Bi_2Te_3 in Figure 2. In this condition, the peak-to-peak value was used in determining the thermal conductivity.

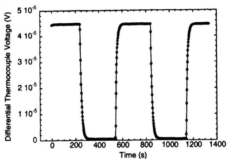

Figure 2. Typical ΔT data collected.

This slow ac technique gives the opportunity to determine the thermoelectric power by simultaneously measuring the voltage, ΔV, indicated in Figure 1 and finding the slope of the ΔV vs. ΔT curve. This further helps to avoid any offset voltages that might exist due to contacts or meter drift. After the thermal conductivity and thermoelectric power are measured, the heat pulse is turned off, and the four probe electrical conductivity is measured using both ac and dc techniques.

RESULTS

In testing the 4-sample system, a NIST 1461 stainless steel thermal conductivity reference was used. The sample was cut to a 3mm x 3mm x 10 mm geometry and mounted to the sample stage using the above mentioned procedure. Results of our measurements are shown in Figure 3 indicating good agreement with the NIST data. Some increase in the measured data at higher temperatures is assumed to be caused by radiation losses (also indicated in Figure 1).

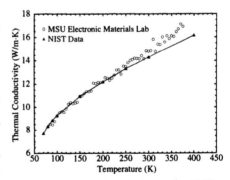

Figure 3. Thermal conductivity of a NIST stainless steel sample.

As a second reference a Corning Pyrex™ 7740 glass sample, cut to a 5mm x 5mm x 6mm geometry was mounted, and tested, with results as shown in Figure 4. In measuring the

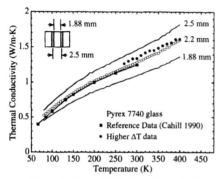

Figure 4. Pyrex 7740 glass. The solid lines indicate the variation in thermal conductivity due to change in the measured spacing of the thermocouple probes.

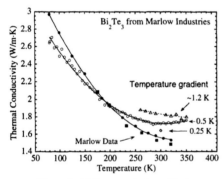

Figure 5. Bi$_2$Te$_3$ sample from Marlow Industries, Inc. Measurements at various temperature gradients are also indicated.

probe separation, it is interesting to note the variation in thermal conductivity as indicated by the solid lines in Figure 4, considering center-to-center and inside-to-inside edge spacing of the Cu plates, holding the differential thermocouple junctions. The plotted data however agree best with previously reported values [5], when the average of these probe separations or center-to-inside edge is used.

Figures 5 and 6 show measurements of a Bi$_2$Te$_3$ sample from Marlow Industries, Inc. in good agreement with Marlow data. The effects of radiative losses were experimentally investigated by measuring the thermal conductivity using different temperature gradients across the sample. These radiative contributions are clearly indicated in Figure 5 as seen by the increasing measured thermal conductivity with increases in the temperature gradient across the sample.

Figure 7 shows measurements of the cubic material AgPbSbTe$_3$. These samples were fabricated by direct combination of the elements in a suitable stoichiometry and heating to 800 °C for one hour. They were then cooled (100 °C per hour) to room temperature. The

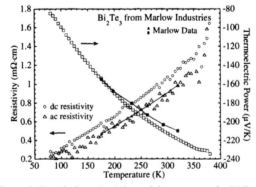

Figure 6. Electrical conductivity and thermopower for Bi$_2$Te$_3$.

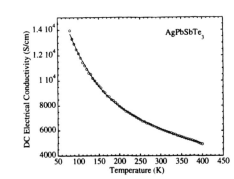

Figure 7. AgPbSbTe$_3$ sample.

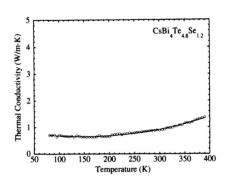

Figure 8. CsBi$_4$Te$_{4.8}$Se$_{1.2}$ sample.

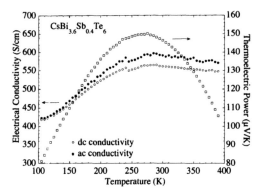

Figure 9. CsBi$_{3.6}$Sb$_{0.4}$Te$_6$ sample.

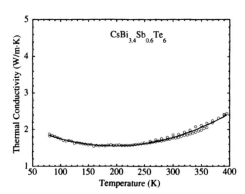

Figure 10. CsBi$_{3.4}$Sb$_{0.6}$Te$_6$ sample.

electrical conductivity can be very high in this material as shown in Figure 7. Measurements on similar compounds have shown low thermal conductivities (typically < 1 W/m·K).

Low thermal conductivity is shown for the CsBi$_4$Te$_{4.8}$Se$_{1.2}$ compound as indicated in Figure 8. Stoichiometric amounts of Cs metal and Bi$_2$Te$_3$-based Sb/Se solid solution loaded separately in two ends of H-shaped quartz tube under vacuum were heated to 600 °C for two days [6]. The product was then cooled to room temperature and washed with several portions of degassed methanol under a nitrogen atmosphere. After being dried, a chunk of well oriented silvery needles of $CsBi_{4-x}Sb_xTe_6$ or $CsBi_4Te_{6-y}Se_y$ solid solution was obtained quantitatively. The $CsBi_{3.6}Sb_{0.4}Te_6$ sample (Figure 9) shows p-type behavior with the maximum thermopower of 150 μV/K at 270 K. Thermal conductivity data for $CsBi_{3.4}Sb_{0.6}Te_6$ were collected from 80 K to 400 K and subsequently collected using a larger temperature gradient from 400 K to 250 K as shown in Figure 10.

CONCLUSIONS

A four-sample system for simultaneous measurement of electrical conductivity, thermoelectric power, and thermal conductivity has been developed and tested. The results obtained for several reference samples of wide ranging thermal conductivities are in good agreement with expected values. Thermoelectric power and electrical conductivity measurements as measured by different research groups have also shown good agreement. The system has been used to characterize several new materials exhibiting promising thermoelectric properties.

ACKNOWLEGEMENTS

This work has been supported by the ARO/DARPA Office (DAAG55-97-1-0184), DURIP award (ONR# N00014-98-0271), and the Office of Naval Research (ONR# N00014-98-1-0443). Special thanks to Dr. Ctirad Uher for many helpful suggestions.

REFERENCES:

1. Kanatzidis, M., DiSalvo, F. J., " Thermoelectric Materials: Solid State Synthesis," *Naval Research Reviews*, Vol. 48, No. 4 (1996).
2. Loo, S., Duck-Young, C., Kanatzidis, M., Hogan, T., "Thermoelectric Materials Measurement System for Doping and Alloying Trends," *Materials Research Society Symposium*, Spring 2000.
3. Hogan, T., Ghelani, N., Loo, S., Sportouch, S., Kim, S.-J., Duck-Young, C., Kanatzidis, M., " Measurement System for Doping and Alloying Trends In New Thermoelectric Materials," *Materials Research Society Symposium Proceedings*, Vol. 545 (1999).
4. Maldonado, O., "Pulsed Method for Simultaneous Measurement of Electric Thermopower and Heat Conductivity at Low Temperatures," *Cryogenics*, Vol. 32, No. 10 (1992), pp. 908-912.
5. Cahill, D.G., "Thermal-Conductivity Measurement from 30-K to 750-K – The 3-Omega Method," *Review of Scientific Instruments*, Vol. 61 (1990).
6. Duck-Young, C., Hogan, T., Ghelani, N., Brazis, P., Lane, M., Kannewurf, C., Kanatzidis, M., "Investigations of Solid Solutions of $CsBi_4Te_6$," *Materials Research Society Symposium Proceedings*, Spring 2000.

CRYSTAL GROWTH OF TERNARY AND QUATERNARY ALKALI METAL BISMUTH CHALCOGENIDES USING BRIDGMAN TECHNIQUE

Theodora Kyratsi[1], Duck-Young Chung[1], Kyoung-Shin Choi[1], Jeffrey S. Dick[2], Wei Chen[2], Ctirad Uher[2] and Mercouri Kanatzidis[1]
[1] Dept. of Chemistry and Center of Fundamental Materials Research, Michigan State University, East Lansing MI 48824-1322
[2] Dept of Physics, University of Michigan, Ann Arbor MI 48109

ABSTRACT

Our exploratory research in new thermoelectric materials has identified the ternary and quaternary bismuth chalcogenides β-$K_2Bi_8Se_{13}$, $K_{2.5}Bi_{8.5}Se_{14}$ and $K_{1+x}Pb_{4-2x}Bi_{7+x}Se_{15}$, to have promising properties for thermoelectric applications. These materials have needle-like morphology so they are highly anisotropic in their electrical and thermal properties. In order to achieve long and well-oriented needles for which, consequently, the best thermoelectric performance is expected, we developed a modified Bridgman technique for their bulk crystal growth. The preliminary results of our crystal growth experiments as well as electrical conductivity, Seebeck coefficient and thermal conductivity for the compounds obtained from this technique are presented.

INTRODUCTION

Exploratory work in the area of thermoelectric materials identified certain new systems as possible new candidates for thermoelectric applications, provided they could be optimized to give high figure of merit ZT. Our own investigations revealed β-$K_2Bi_8Se_{13}$ to be a promising material with very low thermal conductivity and high power factor [1-5]. Often this compound grows along with $K_{2.5}Bi_{8.5}Se_{14}$ the potential of which as a thermoelectric material has not been fully investigated yet. These compounds are closely related in composition and structure and are prepared by similar preparation methods. Another interesting new system is $K_{1+x}Pb_{4-2x}Bi_{7+x}Se_{15}$ [6], which is much less investigated, but it is nerveless promising because of its very low thermal conductivity. In order to learn more about stability of these compounds and their growth habits as well as to begin to evaluate their potential for thermoelectric applications, we decided to undertake large crystal growth studies. The initial results of these efforts are presented here.

Most of ternary and quaternary bismuth chalcogenides have needle-like morphology so they are highly anisotropic. Because of this anisotropy, it is important to be able to control the orientation of growing crystals. In this work, we developed a modified Bridgman technique to grow large oriented ingots of the compounds $K_2Bi_{6.4}Sb_{1.6}Se_{13}$, $K_{2.5}Bi_{8.5}Se_{14}$ and $K_{1.25}Pb_{3.5}Bi_{7.25}Se_{15}$, in order to be able to explore the thermoelectric properties in various crystal directions.

RESULTS AND DISCUSSION

$K_2Bi_8Se_{13}$ [4], $K_{2.5}Bi_{8.5}Se_{14}$ [4] and $K_{1+x}Pb_{4-2x}Bi_{7+x}Se_{15}$ [6] are anisotropic three-dimensional monoclinic structures, which propagate along the b-axis with very short repeating length (~4Å). This is a feature of almost all ternary and quaternary bismuth chalcogenide compounds and it is responsible for the needle-like crystal growth habits of these materials. Their structures consist of Bi_2Te_3-type and NaCl-type building units. The size of Bi_2Te_3-type building unit is different in each compound. The frameworks feature narrow tunnels filled with K^+ ions.

Solid solutions of the type $K_2Bi_{8-x}Sb_xSe_{13}$ (0<x<4) as well as $K_{2.5}Bi_{8.5}Se_{14}$ and $K_{1+x}Pb_{4-2x}Bi_{7+x}Se_{15}$ (x≈0.25) were grown with a vertical Bridgman technique. A vertical single-zone furnace, with temperature gradients 5-15°C/cm, was used. Each material was placed in a silica tube with rounded bottom and ~7mm ID. These tubes were carbon coated and sealed under high vacuum ($5 \cdot 10^{-5}$ mbar). They were lowered through the gradient temperature profile with dropping rates from 0.3 to 0.6 cm/h.

$K_2Bi_{8-x}Sb_xSe_{13}$

In the beginning of this study we attempted to grow large oriented crystals of β-$K_2Bi_8Se_{13}$ using vertical Bridgman growth but we discovered that the resulting ingots were mixed phase samples between β-$K_2Bi_8Se_{13}$ and $K_{2.5}Bi_{8.5}Se_{14}$. While the reasons for this co-crystallization are still being examined, we were able to grow single phase ingot samples of solid solution members such as $K_2Bi_{8-x}Sb_xSe_{13}$. Here the results of the system with x=1.6 are presented. One ingot of ~3cm length is shown in Figure 1. Specifically, the figure shows the ingot cleaved in two halves along the growth axis. It is clear that crystal growth starts from the bottom of the tube in a quasi-oriented fashion with the crystallites pointing upwards but with a considerable angular spread. This spread persists for about 0.5-1 cm and then ends giving rise to excellent parallel orientation of the crystals. Presumably this happens because certain crystals parallel to the growth axis dominate the growth at the expense of the other mis-oriented crystals.

Figure 1: $K_2Bi_{6.4}Sb_{1.6}Se_{13}$ *ingot*

$K_2Bi_{6.4}Sb_{1.6}Se_{13}$ has needle-like morphology composed of long dark metallic looking needles. Long ingots (3-4cm) of $K_2Bi_{6.4}Sb_{1.6}Se_{13}$ were grown with temperature gradient of about 14°C/cm and droping rate of about 0.4cm/h. The product was a pure phase, as concluded from powder X-ray Diffraction (XRD) studies.

The ingots were cut with a diamond saw to obtain specimens with parallel and perpendicular crystal growth exposure. The degree of the crystal orientation was assessed with X-ray diffraction parallel and perpendicular to the needle direction. Figures 2 and 3 show the diffraction patterns of these samples. It is readily apparent that strong preferential orientation of the needles has been achieved.

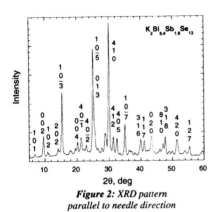

Figure 2: XRD pattern parallel to needle direction

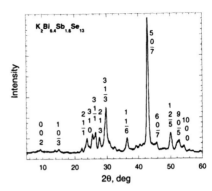

Figure 3: XRD pattern perpendicular to needle direction

Differential thermal analysis (DTA) measurements on ground crystals showed that $K_2Bi_{6.4}Sb_{1.6}Se_{13}$ melts at 674°C (Figure 4). By comparison β-$K_2Bi_8Se_{13}$ melts at 700°C. Infrared spectroscopy showed well-defined energy bandgap of 0.58eV (Figure 5), very similar to that of β-$K_2Bi_8Se_{13}$ at 0.59eV.

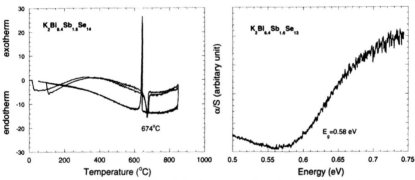

Figure 4: Differential Thermal Analysis of $K_2Bi_{6.4}Sb_{1.6}Se_{13}$. The crystallization point occurs at 644°C

Figure 5: Infrared absorption spectrum of $K_2Bi_{6.4}Sb_{1.6}Se_{13}$

Charge transport measurements were performed on samples, cut parallel and perpendicular to the needle direction. Figure 6 shows electrical conductivity measurements in both directions. Clearly the sample is very anisotropic with room temperature values of 292 S/cm and 59 S/cm respectively, as expected from its crystal structure. Parallel to the needle direction the electrical conductivity is ~5 times greater as expected.

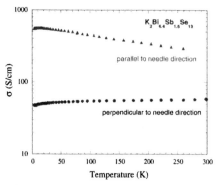
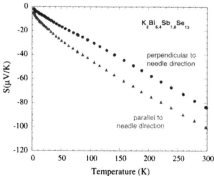

Figure 6: Electrical conductivity measurements of $K_2Bi_{6.4}Sb_{1.6}Se_{13}$ parallel and normal to the needle direction

Figure 7: Seebeck coefficient measurements of $K_2Bi_{6.4}Sb_{1.6}Se_{13}$ parallel and normal to the needle direction

Figure 7 shows Seebeck coefficient measurements parallel and normal to the needle direction. The room temperature values are -100 μV/K and -84 μV/K, respectively. The anisotropy in the Seebeck coefficient is much less compared to that of the electrical conductivity and this is attributed to the fact that thermopower is not as sensitive to carrier scattering mechanisms.

Thermal conductivity measurements parallel and normal to the needle direction also show substantial anisotropy as shown in Figure 9. Room temperature values are 1.2 W/mK and 0.64 W/mK, respectively. Along the needle direction the thermal conductivity is a factor of 2 greater than that perpendicular to it. Overall, both values are very low and smaller than those of the optimized $Bi_{2-x}Sb_xTe_3$ alloy [7].

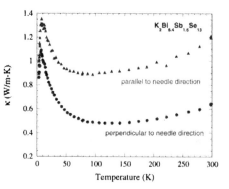

Figure 8: SEM photo of the $K_2Bi_{6.4}Sb_{1.6}Se_{13}$ ingot for the charge transport measurements.

Figure 9: Thermal conductivity measurements parallel and normal to needle direction

$K_{2.5}Bi_{8.5}Se_{14}$

$K_{2.5}Bi_{8.5}Se_{14}$ also has a needle-like morphology with long dark colored needles. It was grown with a temperature gradient of about 8 °C/cm and dropping rate of 0.5cm/h. Long and well-oriented needles were grown parallel to the length of ingot.

XRD studies on ground crystals and perpendicular to the growth direction were also performed and showed pure phase and strong preferential orientation. DTA measurements on ground crystals indicate that $K_{2.5}Bi_{8.5}Se_{14}$ melts at 699°C.

Seebeck coefficient measurements on oriented samples, revealed n-type charge transport behavior with room temperature value of ~ –11 µV/K, Figure 11. The room temperature electrical conductivity was measured to be ~740 S/cm.

$K_{1+x}Pb_{4-2x}Bi_{7+x}Se_{15}$

$K_{1+x}Pb_{4-2x}Bi_{7+x}Se_{15}$ (x≈0.25) has needle-like morphology with small dark needles with a metallic sheen. It was grown using a vertical Bridgman furnace with temperature gradient about 9.5°C/cm and dropping rate of about 0.4cm/h.

Oriented needles grow vertically, but this sample did not show the same degree of orientation as the other two samples described above. We need to improve the growth conditions in order to achieve better orientation. Figure 10 compares the microstructure of the obtained ingots for $K_{1.25}Pb_{3.5}Bi_{7.25}Se_{15}$, $K_2Bi_{6.4}Sb_{1.6}Se_{13}$ and $K_{2.5}Bi_{8.5}Se_{14}$. It is evident that at this stage the $K_2Bi_{6.4}Sb_{1.6}Se_{13}$ and $K_{2.5}Bi_{8.5}Se_{14}$ ingots show the best quality with greater orientation and fewer defects or microcracks.

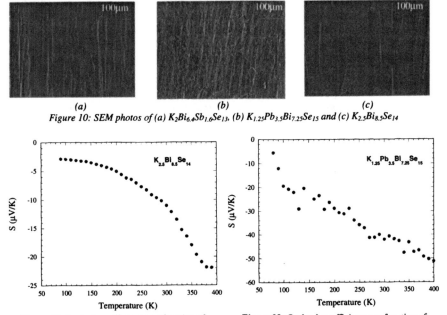

(a)   *(b)*   *(c)*

Figure 10: SEM photos of (a) $K_2Bi_{6.4}Sb_{1.6}Se_{13}$, (b) $K_{1.25}Pb_{3.5}Bi_{7.25}Se_{15}$ and (c) $K_{2.5}Bi_{8.5}Se_{14}$

Figure 11: Seebeck coefficient as a function of temperature of an oriented sample of $K_{2.5}Bi_{8.5}Se_{14}$

Figure 12: Seebeck coefficient as a function of temperature of an oriented sample of $K_{1.25}Pb_{3.5}Bi_{7.25}Se_{15}$

DTA measurements on ground crystals showed that $K_{1.25}Pb_{3.5}Bi_{7.25}Se_{15}$ melts at 734°C.

Seebeck coefficient measurements on oriented samples, showed n-type transport behavior with room temperature value ~–42 µV/K, Figure 12. The room temperature electrical conductivity was measured to be ~830 S/cm.

CONCLUSIONS

In this work, we demonstrate the ability to grow large oriented ingots of $K_2Bi_{6.4}Sb_{1.6}Se_{13}$, $K_{2.5}Bi_{8.5}Se_{14}$ and $K_{1.25}Pb_{3.5}Bi_{7.25}Se_{15}$ using a vertical Bridgman technique. The products showed very high degrees of orientation as concluded from XRD studies, charge transport and thermal transport studies.

$K_2Bi_{6.4}Sb_{1.6}Se_{13}$ and $K_{2.5}Bi_{8.5}Se_{14}$ show the highest degree of orientation whereas $K_{1.25}Pb_{3.5}Bi_{7.25}Se_{15}$ needs additional work to improve the growth conditions in order to achieve the best possible orientation. Charge transport measurements on oriented samples confirmed the high anisotropy of these materials and underscore the importance of growing well-oriented ingots. The best thermoelectric performance is observed along the needle direction.

ACKNOWLEDGEMENTS

We thank Prof. K.M. Paraskevopoulos and Dr. E. Hatzikraniotis (Aristotle University of Thessaloniki, Greece), for fruitful discussions.

At the University of Michigan and Michigan State University the work was supported by Office of Naval Research and DARPA. The work made use of the SEM facilities of the Center for Electron Optics at Michigan State University.

REFERENCES

1. D-Y Chung, L. Iordanidis, K-S Choi and M.G. Kanatzidis, *Bull. Korean Chem. Soc.*, **19**, 12 p. 1285 (1998)
2. P.W. Brazis, M.A. Rocci-Lane, J.R. Ireland, D-Y Chung, M.G. Kanatzidis and C.R. Kannewurf, *Proceedings of 18th International Conference on Thermoelectrics 1999*, p. 619.
3. M.G. Kanatzidis, D-Y Chung, L. Iordanidis, K-S Choi, P. Brazis, M. Rocci, T. Hogan and C.R. Kannewurf, *Mat. Res. Soc. Symp. Proc. 1998*, vol. 545, p. 233
4. D-Y Chung, K-S Choi, L. Iordanidis, J.L. Schindler, P.W. Brazis, C.R. Kannewurf, B. Chen, S. Hu, C. Uher and M.G. Kanatzidis, *Chem. Mat.* **9**, 12, 3060 (1997)
5. Th. Kyratsi, D-Y Chung, J.S. Dick, W. Chen, C. Uher, M.G. Kanatzidis, *work in progress*.
6. K-S Choi, D-Y Chung, A. Mrotzek, P. Brazis, C.R. Kannewurf, C. Uher, W. Chen, T. Hogan and M.G. Kanatzidis, *submitted*.
7. Encyclopedia of Materials Science and Engineering, Thermoelectric Semiconductors, MIT Press: Cambridge, MA, Pergamon Press: Oxford, 1986, p. 4968

THERMOELECTRIC PROPERTIES OF DOPED IRON DISILICIDE

Jun-ichi Tani, Hiroyasu Kido
Department of Inorganic Chemistry, Osaka Municipal Technical Research Institute,
1-6-50 Morinomiya, Joto-ku, Osaka 536-8553, Japan, tani@omtri.city.osaka.jp

ABSTRACT

In order to investigate the thermoelectric properties of Re-doped β-FeSi$_2$ (Fe$_{1-x}$Re$_x$Si$_2$), Ir-doped β-FeSi$_2$ (Fe$_{1-x}$Ir$_x$Si$_2$), and Pt-doped β-FeSi$_2$ (Fe$_{1-x}$Pt$_x$Si$_2$), the electrical resistivity, the Seebeck coefficient, and the thermal conductivity of these samples have been measured in the temperature range between 300 and 1150 K. Fe$_{1-x}$Re$_x$Si$_2$ is p-type, while Fe$_{1-x}$Ir$_x$Si$_2$ and Fe$_{1-x}$Pt$_x$Si$_2$ are n-type over the measured temperature range. The solubility limits of dopant are estimated to be 0.2at% for Fe$_{1-x}$Re$_x$Si$_2$, 0.5at% for Fe$_{1-x}$Ir$_x$Si$_2$, and 1.9at% for Fe$_{1-x}$Pt$_x$Si$_2$. A maximum ZT value of 0.14 was obtained for Fe$_{1-x}$Pt$_x$Si$_2$ (x=0.03) at the temperature 847 K.

INTRODUCTION

The transition metal silicides are of broad interest from the viewpoint of their structural and functional applications. The semiconducting phase of iron disilicide (β-FeSi$_2$) [1] has been studied as a candidate material for thermoelectric conversion application, because of its superior features such as a large Seebeck coefficient, low electrical resistivity, and chemical stability [2-5]. There have been many attempts to dope some additives into β-FeSi$_2$ to alter its semiconducting properties. The conduction types are p-type, produced by doping with V, Cr, Mn, and Al; and n-type, produced by doping with Co, Ni, Pt, and Pd [2-9].

Komabayashi et.al. [9] reported the electrical resistivity and Seebeck coefficient of Pt-doped β-FeSi$_2$ film at room temperature. They concluded that Pt is superior to Co as an additive for n-type β-FeSi$_2$ from the results of the power factor measurement.

Desirable qualities for thermoelectric materials include a large Seebeck coefficient S, a large electrical resistivity ρ, and a small thermal conductivity κ. These quantities determine the so-called thermoelectric figure of merit, $Z=S^2/\rho\kappa$. However, to our knowledge, there has been no investigation concerning the thermal conductivity and Hall effect of Fe$_{1-x}$Pt$_x$Si$_2$, and some of transport properties and figure of merit have not been determined.

Re and Ir are also expected to be attractive dopants of β-FeSi$_2$ because Re and Ir belong to VIIA and VIIIA group as the same as Mn and Co, which are well-known dopants of β-FeSi$_2$. In addition, the atomic radii of Re and Ir are larger than those of Mn and Co. Therefore, it is expected to have lower thermal conductivity than Mn-doped and Co-doped β-FeSi$_2$. However, there has no report concerning the thermoelectric properties of Re-doped and Ir-doped β-FeSi$_2$.

In this paper, the effects of doping on thermoelectric properties of β-FeSi$_2$ have been systematically investigated using Re, Ir, and Pt as impurities in the temperature range between 300 and 1150 K. We also present data for nondoped β-FeSi$_2$ for comparison.

EXPERIMENT

Powders of high purity, Fe (>99.9%), Si (>99.999%), Re (>99.9%), Ir (>99.9%), and Pt (>99.9%), were used as starting materials. They were ground together and heated in evacuated fused silica ampoules at 1373 K for 3 hours. The product was ground into fine powder, and then sintered in a graphite die (15mm in diameter) at 1003-1073 K for 5 min at 30 MPa by the spark plasma sintering (SPS) method. To convert the mixture of metallic α-FeSi$_2$ and ϵ-FeSi into the semiconducting β-FeSi$_2$, the as-sintered pellets were annealed at 1113 K for 168 h in a vacuum (<10^{-3} Torr). The density of the annealed samples was approximately 95% of the theoretical figure.

The Hall coefficient (R_H) and electrical resistivity (ρ) were measured using van der Pauw's technique [10] on 1.2-cm-diam, 0.2-cm-thick samples with Toyo Corporation Resitest 8320. The contacts between the samples and lead Au wires were formed by soldering with In. The Hall effect was measured at 300 K using an ac magnetic method, under an applied magnetic field of 0.39 T at a frequency of 200 mHz. The Seebeck coefficients (S) were measured by the standard four probe dc method in a He gas atmosphere in the temperature range from 300 to 1150 K using an ULVAC ZEM-1S. The temperature gradient across the length of the sample was about 5 K.

The thermal diffusion coefficients of the samples were measured by the usual laser flash method using a thermal constant analyzer (ULVAC TC-7000). The disk specimen was set in an electric furnace, and heated up to 1150 K under vacuum. After the temperature becomes stabilized, the front surface of the specimen was irradiated with a ruby laser pulse. The temperature variation at the surface was monitored with a Pt-Pt 13%Rh thermocouple and an InSb infrared detector. The specific heat of the samples was measured using a SEIKO DSC220. The thermal conductivity was calculated from the thermal diffusivity, specific heat capacity, and density.

RESULTS

Table I shows the results of the Hall effect and electrical resistivity measurements at 300 K of $Fe_{1-x}Re_xSi_2$, $Fe_{1-x}Ir_xSi_2$, and $Fe_{1-x}Pt_xSi_2$, as compared to those of nondoped β-$FeSi_2$.

Table I. The results of the Hall effect and electrical resistivity measurements at 300 K

Dopant	(at%)	Type	Hall coefficient (cm^3/C)	Carrier concentration (cm^{-3})	Mobility (cm^2/Vs)	Resistivity (Ωcm)
Nondoped	0.0	P	1.67×10^1	3.7×10^{17}	2.1	7.92×10^0
Re	0.1	P	5.31×10^0	1.2×10^{18}	1.8	2.96×10^0
Re	0.25	P	2.90×10^0	2.2×10^{18}	2.2	1.34×10^0
Re	0.5	P	2.50×10^0	2.5×10^{18}	2.8	8.86×10^{-1}
Re	1.0	P	2.92×10^0	2.1×10^{18}	5.4	5.46×10^{-1}
Ir	0.1	N	2.54×10^{-1}	2.5×10^{19}	0.19	1.31×10^0
Ir	0.25	N	1.19×10^{-1}	5.3×10^{19}	0.14	8.52×10^{-1}
Ir	0.5	N	5.68×10^{-2}	1.1×10^{20}	0.21	2.77×10^{-1}
Ir	1.0	N	5.35×10^{-2}	1.2×10^{20}	0.35	1.52×10^{-1}
Pt	0.1	N	8.44×10^{-2}	7.4×10^{19}	0.26	3.22×10^{-1}
Pt	0.25	N	4.64×10^{-2}	1.4×10^{20}	0.87	5.34×10^{-2}
Pt	0.5	N	2.31×10^{-2}	2.7×10^{20}	0.89	2.58×10^{-2}
Pt	1.0	N	1.32×10^{-2}	4.8×10^{20}	0.65	2.04×10^{-2}
Pt	3.0	N	6.80×10^{-3}	9.2×10^{20}	0.66	1.04×10^{-2}

The sign of R_H of $Fe_{1-x}Re_xSi_2$ is positive, indicating that the conductivity is mainly due to holes. However, the sign of R_H of $Fe_{1-x}Ir_xSi_2$ and $Fe_{1-x}Pt_xSi_2$ is negative, indicating that the conductivity is mainly due to electrons. These show that Re acts as an acceptor but Ir and Pt as donors. The carrier concentration of $Fe_{1-x}Ir_xSi_2$ at 300K is in good agreement with the Ir atom concentration within the composition range of $0.001 \leq x \leq 0.005$. On the other hand, the carrier concentration of $Fe_{1-x}Pt_xSi_2$ at 300K is about twice the Pt atom concentration within the composition range of $0.001 \leq x \leq 0.01$. These results suggest that Ir is a singly ionized donor but Pt is a doubly ionized donor at 300K. The solid solution is considered to exist within the composition range where the carrier concentration is almost proportional to the doping concentration (x). Indeed, the carrier concentrations are almost proportional to x at low carrier concentrations. With further increasing x, the carrier concentrations show a plateau. From the carrier concentration, the solubility limits of dopant are estimated to be 0.2at% for $Fe_{1-x}Re_xSi_2$, 0.5at% for $Fe_{1-x}Ir_xSi_2$, and 1.9at% for $Fe_{1-x}Pt_xSi_2$.

Figure 1 shows the temperature dependence of the electrical resistivity (ρ) of $Fe_{1-x}Re_xSi_2$ (x=0.0025), $Fe_{1-x}Ir_xSi_2$ (x=0.005), and $Fe_{1-x}Pt_xSi_2$ (x=0.01), as compared with nondoped β-$FeSi_2$.

The electrical resistivity of $Fe_{1-x}Pt_xSi_2$ (x=0.01) at 300 K is lowest because the carrier concentration of $Fe_{1-x}Pt_xSi_2$ is larger than that of other compounds. The electrical resistivity of $Fe_{1-x}Pt_xSi_2$ slightly increases with increasing temperature in the temperature range between 300 and 700 K.

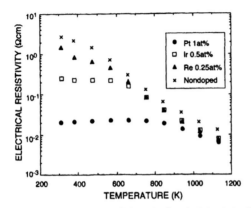

Figure 1. The temperature dependence of electrical resistivity (ρ) of $Fe_{1-x}Re_xSi_2$ (x=0.0025), $Fe_{1-x}Ir_xSi_2$ (x=0.005), and $Fe_{1-x}Pt_xSi_2$ (x=0.01), as compared with nondoped β-$FeSi_2$.

Figure 2 shows the temperature dependence of Seebeck coefficient (S) of $Fe_{1-x}Re_xSi_2$ (x=0.0025), $Fe_{1-y}Ir_ySi_2$ (x=0.005), and $Fe_{1-x}Pt_xSi_2$ (x=0.01), as compared with nondoped β-$FeSi_2$. The polarity of S of $Fe_{1-x}Ir_xSi_2$ and $Fe_{1-x}Pt_xSi_2$ is negative, indicating that the conductivity is mainly due to electrons. However, the polarity of S of $Fe_{1-x}Re_xSi_2$ is positive, indicating that the conductivity is mainly due to holes. The polarity of S is in good agreement with the sign of R_H. At low temperatures, the Seebeck coefficients of these samples greatly affected by the kinds of dopant. However, all of absolute Seebeck coefficients converge to a small value at high temperatures.

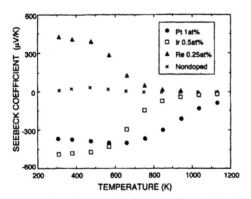

Figure 2. The temperature dependence of Seebeck coefficient (S) of $Fe_{1-x}Re_xSi_2$ (x=0.0025), $Fe_{1-x}Ir_xSi_2$ (x=0.005), and $Fe_{1-x}Pt_xSi_2$ (x=0.01), as compared with nondoped β-$FeSi_2$.

Table II shows the thermoelectric properties of of $Fe_{1-x}Re_xSi_2$, $Fe_{1-x}Ir_xSi_2$, and $Fe_{1-x}Pt_xSi_2$, as compared to those of nondoped β-$FeSi_2$.

Table II. Thermoelectric properties of of $Fe_{1-x}Re_xSi_2$, $Fe_{1-x}Ir_xSi_2$, and $Fe_{1-x}Pt_xSi_2$, as compared to those of nondoped β-$FeSi_2$.

Samples	Temperature* (K)	Resistivity (Ωcm)	Seebeck coefficient ($\mu V/K$)	Thermal Conductivity (W/cmK)	ZT
Nondoped	852	1.31×10^{-2}	-6	5.30×10^{-2}	1.21×10^{-4}
Re0.25at%	569	4.50×10^{-1}	+283	6.75×10^{-2}	1.50×10^{-3}
Re0.5at%	568	2.88×10^{-1}	+330	6.80×10^{-2}	3.16×10^{-3}
Ir0.5at%	568	2.24×10^{-1}	-428	6.80×10^{-2}	6.83×10^{-3}
Ir1at%	660	1.06×10^{-1}	-160	5.60×10^{-2}	2.85×10^{-3}
Pt0.5at%	755	2.75×10^{-2}	-376	4.59×10^{-2}	8.45×10^{-2}
Pt1at%	754	2.22×10^{-2}	-369	4.61×10^{-2}	1.00×10^{-1}
Pt3at%	847	1.11×10^{-2}	-283	4.36×10^{-2}	1.40×10^{-1}

(*) The temperature at which ZT shows a maximum value

The ZT value of $Fe_{1-x}Pt_xSi_2$ is larger than that of $Fe_{1-x}Re_xSi_2$ and $Fe_{1-x}Ir_xSi_2$. A maximum ZT value of 0.14 was obtained for $Fe_{1-x}Pt_xSi_2$ (x=0.03) at the temperature 847 K. The maximum ZT value of $Fe_{1-x}Pt_xSi_2$ is lower than that of Co-doped β-$FeSi_2$, which has ZT=0.19 at 923 K [11]. It is possible to control the carrier concentration in a wide range for Co-doped β-$FeSi_2$ sample because the Co atom is soluble in β-$FeSi_2$ up to 12at% [12]. Therefore, it is of great importance for $Fe_{1-x}Re_xSi_2$, $Fe_{1-x}Ir_xSi_2$, and $Fe_{1-x}Pt_xSi_2$ to increase the solubility limit of dopant in β-$FeSi_2$ by different process techniques, which improve thermoelectric properties.

CONCLUSIONS

In order to investigate the thermoelectric properties of $Fe_{1-x}Re_xSi_2$, $Fe_{1-x}Ir_xSi_2$, and $Fe_{1-x}Pt_xSi_2$, the electrical resistivity, the Seebeck coefficient, and the thermal conductivity of these samples have been measured in the temperature range between 300 and 1150 K. $Fe_{1-x}Re_xSi_2$ is p-type, while $Fe_{1-x}Ir_xSi_2$ and $Fe_{1-x}Pt_xSi_2$ are n-type over the measured temperature range. The solubility limits of dopant are estimated to be 0.2at% for $Fe_{1-x}Re_xSi_2$, 0.5at% for $Fe_{1-x}Ir_xSi_2$, and 1.9at% for $Fe_{1-x}Pt_xSi_2$. A maximum ZT value of 0.14 was obtained for $Fe_{1-x}Pt_xSi_2$ (x=0.03) at the temperature 847 K.

REFERENCES

1. N. Kh. Abrikosov, Bull. Acad. Sci. U.S.S.R. **20**, 37 (1956).
2. R. M. Ware and D. J. McNeill, Proc. IEEE **111**, 178 (1964).
3. U. Birkholz and J. Scelm, Phys. Status Solidi **27**, 413 (1968).
4. I. Nishida, Phys. Rev. B **7**, 2710 (1973).
5. T. Kojima, Phys. Status Solidi A **111**, 233 (1989).
6. J. Tani and H. Kido, J. Appl. Phys. **84**, 1408 (1998).
7. J. Tani and H. Kido, J. Appl. Phys. **86**, 464 (1999).
8. J. Tani and H. Kido, Jpn. J. Appl. Phys. **38**, 2717 (1999).
9. M. Komabayashi, K. Hijikata, and S. Ido, Jpn. J. Appl. Phys. **30**, 331 (1991).
10. L. J. van der Pauw, Philips Res. Rep. **13**, 1 (1958).
11. U. Stöhrer, Proc. 11th Int. Conf. on Thermoelectrics, 191 (1992).
12. J. Hesse and R. Bucksch, J. Mater. Sci. **5**, 272 (1970).

Transport Properties Of The Doped Thermoelectric Material $K_2Bi_{8-x}Sb_xSe_{13}$

Paul W. Brazis[1], John R. Ireland[1], Melissa A. Lane[1], Theodora Kyratsi[2], Duck-Young Chung[2], Mercouri G. Kanatzidis[2], and Carl. R. Kannewurf[1]
[1]Dept of Electrical and Computer Engineering, Northwestern University, Evanston, IL 60208-3118.
[2]Department of Chemistry, Michigan State University, East Lansing, MI 48824-1322.

ABSTRACT

The synthesis, physicochemical, spectroscopic, and structural characterization of the compound β-$K_2Bi_8Se_{13}$ has been previously reported. The results indicated that this material should be investigated further for possible thermoelectric applications. β-$K_2Bi_8Se_{13}$ exhibits excellent electrical conductivity values at room temperature while maintaining high Seebeck coefficients. In this work, the optimization of the compound β-$K_2Bi_8Se_{13}$ is continued by the introduction of varying concentrations of several different dopants. The value of x in $K_2Bi_{8-x}Sb_xSe_{13}$ was varied in order to find the composition with minimum thermal conductivity. Where possible, transport measurements were carried out on both single crystal and polycrystalline ingot material. From these data, the trends in the key parameters were identified for optimizing the power factor and figure of merit.

INTRODUCTION

The n-type material β-$K_2Bi_8Se_{13}$ was previously reported to have promising thermoelectric properties [1-3]. The average single-crystal room-temperature (295 K) electrical conductivity was 275 S/cm [1], the thermopower was –220 μV/K [1], and the thermal conductivity was 10 mW/cm·K [4]. These values give a power factor of 13 μW/cm·K^2 and a ZT of 0.39 at 295 K. These promising results suggested that the figure-of-merit may be improved by optimizing the carrier concentration through introducing dopants. It is also desirable to create both p-type and n-type materials, which are necessary for designing thermoelectric devices. This approach resulted in a threefold increase of the power factor to 38.5 μW/cm·K^2 for single crystals of β-$K_2Bi_8Se_{13}$ doped with 0.5% tin [2]. This suggests that a ZT of as high as 0.95 may be obtained with this composition [4].

Another approach to increasing ZT is to alter the crystal structure in order to reduce the lattice thermal conductivity without seriously reducing the thermopower or electrical conductivity. For example, Uher et al. reported a reduction of the thermal conductivity of the half-Heusler compound ZrNiSn by more than 20% when alloyed with HfNiSn [5]. This effect is also observed with solid solutions of Bi_2Te_3 with Bi_2Se_3 or Sb_2Te_3. Applying this technique for reducing thermal conductivity, the material β-$K_2Bi_8Se_{13}$ was alloyed with $K_2Sb_8Se_{13}$, which results in a series of compounds with the general formula $K_2Bi_{8-x}Sb_xSe_{13}$.

EXPERIMENTAL

The $K_2Bi_{8-x}Sb_xSe_{13}$ oriented ingot samples were synthesized using the Bridgeman technique [6]. The doping procedure was carried out using single-crystal samples with needle-like morphology [7]. Measurements for both oriented ingot and single-crystal samples were made along the *b*-axis of the crystal structure, which is the crystal growth direction. This direction exhibits the most interesting thermoelectric properties. Details of the crystal structure are described elsewhere [1].

The electrical conductivity of the oriented ingot and single-crystal samples was measured using a computer-controlled, four-probe technique. Electrical contacts consisting of fine gold wire (25 to 60 µm diameter) were attached with gold paste. Samples were placed under vacuum for at least 24 hours to allow the gold paste to dry completely, which improved contact performance. Excitation currents were kept as low as possible, typically below 1 mA, in order to minimize any non-ohmic voltage response and thermoelectric effects at the contact-sample interface.

Variable-temperature (4.2 K to 295 K) thermopower data were taken using a slow-ac measurement technique [8]. Samples were mounted on 60 µm-diameter gold wire using gold paste. Fine gold wire (10 µm in diameter) was used for sample voltage contacts, which were made as long as possible in order to minimize thermal conduction through the leads. The sample and thermocouple voltages were measured using Keithley Models 2182 and 182 nanovoltmeters, respectively. The applied temperature gradient was in the range of 0.1 to 0.4 K. Measurements were taken under a turbopumped vacuum maintained below 10^{-5} Torr.

Thermal conductivity data were obtained using an automated measurement technique [9] based on the heat pulse method described by Maldonado [10]. This measurement yields both thermal conductivity and thermopower data simultaneously. For comparison, the thermopower measurements were repeated using the thermopower measurement system described above. Samples were mounted using a four-probe measurement configuration. The length of the thermocouple leads was at least 20 cm and wrapped around a glass tube in order to minimize conduction losses. To minimize radiation losses, the peak temperature gradient was kept below 0.5 K. Vacuum levels were maintained at approximately 10^{-6} Torr throughout the experiment to minimize convection losses. To obtain the thermopower data, 10 µm-diameter gold wires were attached using gold paste at the same locations as the differential thermocouple junctions. Reference measurements on a commercially prepared Bi_2Te_3 sample using this technique differed by less than 10% from the manufacturer's data [11].

RESULTS

The electrical conductivity data for the $K_2Bi_{8-x}Sb_xSe_{13}$ alloys are shown in Figure 1a. The substitution of 10% of the bismuth with antimony ($x = 0.8$) nearly quadrupled the electrical conductivity, increasing from 244 S/cm to 884 S/cm at 295 K. A further increase in the concentration of antimony decreased the electrical conductivity, dropping to a room-temperature value of 139 S/cm when $x = 4.0$. The conductivity behavior became less metallic as x increased, except for $x = 4.0$, which showed a more metallic behavior than would be expected from the trend.

The thermopower data are shown in Figure 1b. The thermopower behavior for all samples showed a linear decrease toward zero as the temperature approached 0 K, which is characteristic of metallic materials. However, the magnitudes of the thermopower are more typical of semiconductors. The magnitude of the thermopower for β-$K_2Bi_8Se_{13}$ was reduced as the antimony concentration increased, dropping to a minimum value of -100 µV/K when $x = 1.6$. As the antimony concentration was further increased, the magnitude of the thermopower increased to -227 µV/K, which was similar to that of the pure β-$K_2Bi_8Se_{13}$ sample (-225 µV/K at 295 K).

The addition of 10% antimony ($x = 0.8$) to β-$K_2Bi_8Se_{13}$ increased the measured room-temperature thermal conductivity from 12.8 to 19.0 mW/cm·K. However, further increases in the concentration of antimony reduced the thermal conductivity. The lowest room-temperature thermal conductivity which was 12.0 mW/cm·K, was achieved for the $x = 1.6$ sample. However, a more apparent trend was observed for the thermal conductivity below 120 K, where the thermal conductivity was shown to reach a minimum value as x approached 3.2. This trend is less apparent at higher temperatures primarily because of the increased thermal radiation loss.

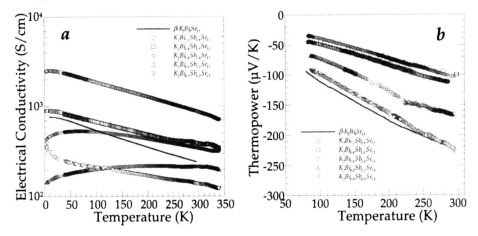

Figure 1. Temperature-dependent *a*) electrical conductivity and *b*) thermopower for $K_2Bi_{8-x}Sb_xSe_{13}$. Thermopower data for the $K_2Bi_{6.4}Sb_{1.6}Se_{13}$ composition as reported in ref. [6].

DISCUSSION

The series of alloyed materials was examined in order to reduce the lattice thermal conductivity of β-$K_2Bi_8Se_{13}$ by introducing increased disorder in the crystal lattice. Since the electrical conductivity is decreasing as x increases, the effect on the lattice thermal conductivity is not obvious from the experimental thermal conductivity data. In order to determine the effects of alloying on the lattice thermal conductivity, the thermal conductivity due to electron transport was first estimated. This is usually carried out by using the Wiedemann-Franz Law, which relates the electrical conductivity to the thermal conductivity due to the electrons:

$$\kappa_{el} = L_0 \sigma T \qquad (1)$$

where L_0 is the Lorenz number, which is equal to $2.44 \cdot 10^{-8}$ V^2/K^2. However, this value for L_0 is only appropriate for degenerate semiconductors or metals [12]. As discussed by Ioffe [12], the value of L_0 for semiconductors with an electrical conductivity of less than 2500 S/cm is equal to $1.49 \cdot 10^{-8}$ V^2/K^2. The lattice contribution shown in Figure 2 was determined by subtracting the electronic contribution calculated using Equation 1. It should be noted that the calculated lattice contribution shown in Figure 2 contains radiation losses, which are believed to be the most likely source of the increased thermal conductivity above 100 K.

It is observed that the lattice thermal conductivity of pure β-$K_2Bi_8Se_{13}$ is reduced by 36% for $x = 3.2$ at low temperatures. The lattice thermal conductivity is reduced through the entire temperature range for $x = 1.6$ and 2.4 by about 25%. For $x = 0.8$ and 4.0 the lattice thermal conductivity was higher than that for β-$K_2Bi_8Se_{13}$.

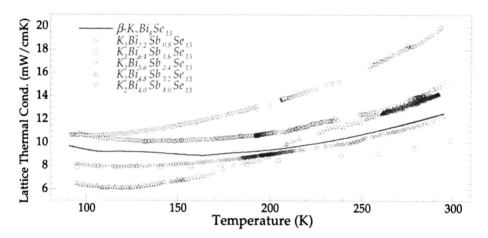

Figure 2. Lattice thermal conductivity, determined by subtracting the electronic contribution calculated from Equation 1. Data for the $K_2Bi_{6.4}Sb_{1.6}Se_{13}$ composition as reported in ref. [6].

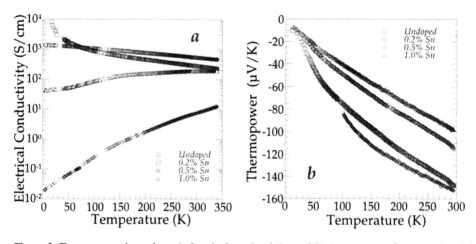

Figure 3. Temperature-dependent a) electrical conductivity and b) thermopower for the series of tin-doped single-crystal samples of $K_2Bi_{6.4}Sb_{1.6}Se_{13}$.

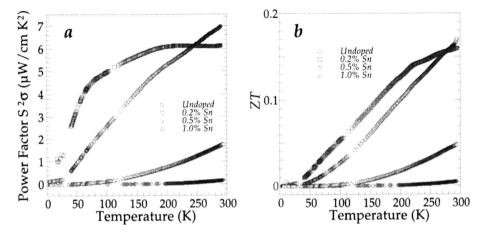

Figure 4. Temperature-dependent *a)* power factor and *b)* dimensionless figure-of-merit for the series of tin-doped single-crystal samples of $K_2Bi_{6.4}Sb_{1.6}Se_{13}$.

The alloy $K_2Bi_{6.4}Sb_{1.6}Se_{13}$ ($x = 1.6$), which showed the lowest overall lattice thermal conductivity, was used for doping studies in order to maximize ZT. Single-crystal samples of $K_2Bi_{6.4}Sb_{1.6}Se_{13}$ doped with tin were prepared and investigated as a function of the percent of tin present. The electrical conductivity and thermopower values are shown in Figure 3. Due to the low electrical conductivity of the 0.5% tin-doped sample below 100 K, the thermopower could not be measured because of the resulting increase in Johnson noise [4]. The electrical conductivity varied over two orders of magnitude, with the conductivity reaching a minimum at room-temperature for a doping of 0.5% tin. However, the thermopower increased by a factor of two over the undoped single crystal sample as the doping level approached 0.5% tin. As a result of the large change in the electrical conductivity, the power factor and ZT values (Figure 4) showed a strong dependence on doping concentration. The ZT values are calculated using the single-crystal power factor, the lattice thermal conductivity for the oriented ingot sample, and the electronic thermal conductivity calculated from the single-crystal electrical conductivity data.

CONCLUSIONS

The composition β-$K_2Bi_8Se_{13}$ has previously shown promising properties for thermoelectric applications. Alloys of this material were investigated in order to reduce the lattice thermal conductivity. The composition $K_2Bi_{6.4}Sb_{1.6}Se_{13}$ showed the best reduced thermal conductivity, but the overall ZT value was less than that of β-$K_2Bi_8Se_{13}$. However, doping the alloy with tin showed a significant improvement in the thermopower, resulting in an improved power factor and calculated ZT value. Although the improvements in overall ZT of $K_2Bi_{6.4}Sb_{1.6}Se_{13}$ by this initial doping study were not as large as expected, the transport properties were shown to be sensitive to small changes in the dopant concentration. Therefore a possible higher ZT value may be found if intermediate doping concentrations are investigated. This sensitivity to the doping concentration was observed in the doping studies of β-$K_2Bi_8Se_{13}$

[2], where the electrical conductivity was also significantly affected by small changes of the doping concentration. In this previous study, the best dopant concentrations showed a threefold improvement in the power factor and ZT as the dopant concentration changed by only 0.5% [4]. Therefore, with the reduced thermal conductivity of $K_2Bi_{6.4}Sb_{1.6}Se_{13}$, it is not unreasonable to expect a higher ZT when the optimal doping concentration for this composition is determined. It is expected that with further refinements of both alloy composition and doping concentration, the thermoelectric performance of the β-$K_2Bi_8Se_{13}$ system will be further enhanced.

ACKNOWLEDGMENTS

This work was supported at NU and MSU by the Office of Naval Research (N00014-98-1-0443) and by DARPA through ARO (DAAG55-97-1-0184). Work at NU made use of the Central Facilities supported by the National Science Foundation through the NU Materials Research Center (DMR-9632472).

REFERENCES

[1] D.-Y. Chung, K.-S. Choi, L. Iordanidis, J. L. Schindler, P. W. Brazis, C. R. Kannewurf, B. Chen, S. Hu, C. Uher, and M. G. Kanatzidis, *Chemistry of Materials*, vol. 9, pp. 3060-3071, (1997).
[2] P. W. Brazis, M. A. Rocci-Lane, J. R. Ireland, D.-Y. Chung, and M. G. Kanatzidis in *Proceedings, 18th International Conference on Thermoelectrics*, The International Thermoelectric Society (IEEE, Piscataway, NJ, 1999) pp. 619-622.
[3] J. L. Schindler, T. P. Hogan, P. W. Brazis, C. R. Kannewurf, D.-Y. Chung, and M. G. Kanatzidis, in *Thermoelectric Materials - New Directions and Approaches*, edited by M. G. Kanatzidis, H. Lyon, G. Mahan, and T. Tritt, (Mater. Res. Soc. Proc. **478**, Warrendale, PA, 1997) pp. 327-332.
[4] P. W. Brazis, "Electrical and Thermal Measurement Techniques for New Chalcogenide Thermoelectric and Mixed Conductor Materials," Ph.D. Dissertation, Northwestern University (2000).
[5] C. Uher, S. Hu, J. Yang, G. P. Meisner, and D. T. Morelli, in *Proceedings, 16th International Conference on Thermoelectrics*, The International Thermoelectric Society (IEEE, Piscataway, NJ, 1997).
[6] T. Kyratsi, D.-Y. Chung, K.-S. Choi, C. Uher, and M. G. Kanatzidis, in *Thermoelectric Materials 2000 - The Next Generation Materials for Small -Scale Refrigeration and Power-Generation Applications*, edited by T. Tritt, G. Nolas, G. Mahan, M. G. Kanatzidis, and D. Mandrus, (Mater. Res. Soc. Proc. **626**, Warrendale, PA, 2001) pp. Z8.8.1-Z8.8.6.
[7] T. Kyratsi, D.-Y. Chung, and M. G. Kanatzidis, (In preparation).
[8] H. O. Marcy, T. J. Marks, and C. R. Kannewurf, *IEEE Transactions on Instrumentation and Measurement*, vol. 39, pp. 756-760, (1990).
[9] T. P. Hogan, "Transport Measurements of New Ternary and Quaternary Chalcogenide and Ceramic Systems," Ph.D. Dissertation, Northwestern University, (1996).
[10] O. Maldonado, *Cryogenics*, vol. 32, pp. 908-912, (1992).
[11] Bi_2Te_3 n-type and p-type samples and data obtained from Marlow Industries Inc., Dallas, Texas
[12] A. F. Ioffe, *Semiconductor Thermoelements and Thermoelectric Cooling*. (Infosearch Ltd., London, England, 1957).

STRUCTURAL PROPERTIES OF STRAIN SYMMETRIZED SILICON / GERMANIUM (111) SUPERLATTICES

C. A. Kleint, A. Heinrich, T. Muehl, J. Schumann, and M. Hecker
Institute of Solid State and Materials Research Dresden
Helmholtzstr. 20, D-01069 Dresden, Germany

ABSTRACT

The use of Multi Quantum Well structures has been shown to provide a promising strategy for improving the thermoelectric figure of merit. In a recent paper the concept of carrier pocket engineering has been applied to strain symmetrized Si/Ge-superlattices leading to a ZT of 0.96 at room temperature for (111) orientation. Since the strain of the individual layers is crucial for the desired modification of their band structures, their experimental determination will be of importance. We have prepared a series of (111) oriented, 100 period (Si 2nm / Ge 2nm) superlattices on a graded $Si_{0.5}Ge_{0.5}$-buffer by sputter deposition. Deposition temperature and buffer thickness have been varied, the superlattices were characterized by AFM and XRD. The technique of XRD reciprocal space mappings of asymmetric reflections has been applied to describe the strain state of the superlattice. We found a buffer thickness of 1.1μm sufficient for more than 90% strain relaxation. XRD-data of 4nm-period superlattices are consistent with complete strain symmetrization.

INTRODUCTION

The use of low dimensional structures, like Multi Quantum Wells, quantum wires and quantum dot superlattices has been shown to provide a promising strategy for improving the thermoelectric figure of merit ZT. In a recent paper [1] the concept of carrier pocket engineering has been applied to Si / Ge superlattices. For (111) oriented, strain symmetrized Si / Ge-superlattices with 2nm individual layer thickness a room temperature ZT = 0.96 has been calculated. With increasing temperature ZT increases up to 1.8 at 700K. The existence of tensile and compressive stresses in the Si and Ge layers arising from pseudomorphic growth on a $Si_{0.5}Ge_{0.5}$ pseudo substrate leads to modifications of the band structures which are crucial for the significant enhancement of ZT. Provided that the individual layer thickness of the superlattice constituents is below their critical thickness for pseudomorphic growth (h_{c1}), strain symmetrized superlattices can be grown up to very high thicknesses, since the average superlattice strain is zero (h_{c2} = infinite). However, due to the large difference of the Si and Ge lattice parameters of about 4%, h_{c1} for the ±2% strained layers of a strain symmetrized superlattice is expected to be in the range of a few nanometers. For (100) oriented strain symmetrized superlattices values of 6 and 3 nm have been given for Ge and Si in [2]. For (111) oriented layers no data are available yet, however it is known, that $Si_{0.5}Ge_{0.5}$ on (111) Si has a h_{c1} significantly lower than on (100) Si due to the different orientation of the misfit dislocations with respect to the interface and the possibility of different Burgers vectors [3]. The question whether pseudomorphic growth can be achieved for (111) oriented structures is of fundamental importance for the realization of high ZT Si/Ge-systems. Therefore a major issue for the preparation of high ZT-Si/Ge structures will be the control of the superlattice strain state. We have prepared a series of (111) oriented 2nm Si / 2nm Ge superlattices on a graded $Si_{0.5}Ge_{0.5}$ buffer by magnetron sputter deposition at different substrate temperatures and with different buffer thickness and investigated their structural properties by XRD and AFM. We have applied the technique of (XRD) reciprocal space mappings of asymmetric reflections to investigate the strain state of the superlattice.

EXPERIMENTAL

The multilayers investigated in this study have been prepared in a fully automated DCA UHV-sputtering system consisting of a load lock and a deposition chamber, equipped with 3 conventional 4 inch magnetron sources and two Facing Target Sputtering stations, a 3 inch rotating sample holder / heater, a Knudsen cell for dopant evaporation and a 30kV RHEED-system for post deposition analysis. The base pressure of the system is in the mid 10^{-10} Torr range, the Ar pressure during deposition was 6mTorr. For the preparation of the multilayers we used the three confocal 4 inch magnetron sources operating in DC mode, equipped with undoped Si, Sb-doped Si and Sb-doped Ge single crystal targets. Before deposition the (111) Si wafers were Shiraki cleaned and held at 950°C for 5min to remove the oxide. The graded buffer was prepared by codeposition from the undoped Si and the Sb-doped Ge source. First a 40 step composition gradient was realized by varying the magnetron source powers from 4.1W/cm^2 (Si) and 0.14W/cm^2 (Ge) to 1.12W/cm^2 (Si) and 0.46W/cm^2 resulting in an average growth rate of 0.76Å/s followed by a $Si_{0.5}Ge_{0.5}$ layer at 0.45Å/s. The substrate temperature (400 to 650°C) and buffer thickness (0.6 to 1.1µm) have been varied to investigate strain relaxation and roughness. The 100 period superlattices have been grown at T_s between 400 and 500°C from the Sb-doped Ge and Si targets at growth rates of 0.33Å/s. As a final step a 30nm $Si_{0.5}Ge_{0.5}$ cap layer has been deposited.

XRD and reflectivity measurements were performed with an Philips Xpert diffractometer, with a curved Graphit monochromator in the diffracted beam using CuK$_\alpha$ radiation. A Topometrix AFM was used for studies of the layer surfaces.

RESULTS

Buffers

Since for thermoelectric applications a thin buffer is desired to minimize its contribution to thermoelectric transport properties we have prepared rather thin buffer layers compared to the literature on (100) oriented structures in order to find out a minimum thickness for complete strain relaxation. Our thinnest buffer with a total thickness of 0.6µm consisted of a 470nm gradient layer and 136nm of constant composition $Si_{0.5}Ge_{0.5}$. The total thickness has been increased up to 1.1µm (785nm gradient layer + 300nm constant composition layer). The growth temperature of the buffer has been varied from 400°C to 650°C. Since AFM roughness has been found to increase significantly above 450°C, we have restricted the strain investigations to buffer layers prepared at 450°C. Figure 1 shows an AFM image of the 1.1µm buffer prepared at 450°C. A hatch pattern is found for large parts of the investigated area with a surface roughness determined by the hatch undulations. Within a 10µm x 10µm area an rms-roughness of 3.6 nm was observed.

For the study of strain relaxation mappings of the reciprocal plane containing the growth direction [111] and the [112] in plane direction around the 422 lattice point were performed by a

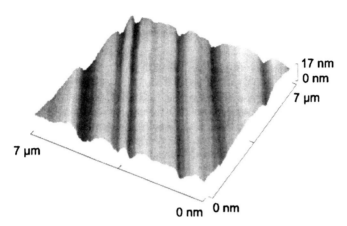

Figure 1. AFM image of a 1.1μm buffer prepared at 450°C. The hatching is oriented along [110]

series of ω/2Θ-scans with varying ω-offsets across the asymmetric 422 reflections of both buffer and substrate. Figure 2 shows the scan data for the thick buffer after transformation into reciprocal space with X parallel [112] and Y parallel [111] in units of $10^{-4} \cdot 2/\lambda$. Line (d) through the (unstrained, i.e. cubic) substrate CuKα_1 and CuKα_2 peaks points to the origin of reciprocal space. A fully relaxed, i.e. cubic buffer will be situated at this line, whereas a layer, grown pseudomorphically on Si will have its peak at line (a) with X=const through the Si-peak, i.e. will have the same in plane lattice parameter. As can be seen from figure 2, the thick, 1.1μm buffer is nearly completely relaxed, a detailed analysis gives 92% relaxation (b). The same degree of strain relaxation is observed for (100) oriented buffers with a thickness of 2.3μm, which showed a rms-

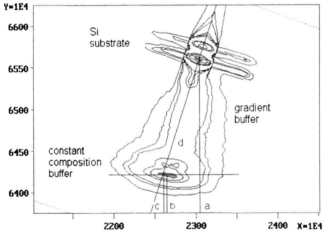

Figure 2. Reciprocal space mapping around the 422 lattice point of a 1.1μm buffer prepared at 450°C

roughness of 2nm [4]. From the X and Y coordinates a buffer composition of $Si_{0.51}Ge_{0.49}$ is obtained. For the thinnest, 0.6μm buffer, strain is released to 60% strain only.

Superlattices

The superlattices consisting of 100 periods (2nm Si/2 nm Ge) have been grown on top of 1.1μm buffers at temperatures of 400°C, 450°C and 500°C. Based on XRD investigations of the symmetrical (111) reflections (mosaic spreads and 2Θ-FWHMs), which will be not discussed in this paper, we found again a temperature of 450°C being optimal from the viewpoint of structural quality. Fig. 3 shows a scan along the growth direction through the (111) reflection. The intensity distribution of buffer plus substrate has been subtracted for clarity. The superlattice Bragg-peak is centered around 2Θ=27.85°, which gives an average superlattice composition of $Si_{0.5}Ge_{0.5}(\pm 0.01)$, showing that the individual Si and Ge layer thickness is equal. The satellites on the low angle (Ge) side of the Bragg peak are more intense than those on the high angle side, which due to the higher scattering amplitude of Ge compared to Si. A fit of the angular spacings of the satellites gives an superlattice period of 3.87nm close to the 4nm intended. In principal the strain state can be deduced from a simulation of the intensity distribution of the symmetrical reflection, since in-plane strain is connected with a change of the lattice spacings along the growth direction via the Poisson ratio. To demonstrate the effect on the intensity distribution, we have used a simulation procedure [5], which allows the incorporation of strain and a crystalline disorder concentrated at the interface (Figure 3). For 0.7 % strains with the correct sign the intensity of the satellites SL+1 and SL+2 is reduced compared to zero strain or strains of incorrect sign, whereas the low angle satellites remain unaffected, a behavior which qualitatively is observed for our samples. However the principal fit cannot be improved within the limits of the model used. Fluctuations of the lattice spacings and the introduction of an interface width allowing for some intermixing of the layers, have to be incorporated in the simulation, which should lead to a reduction of the satellite height and broadening of all peaks, as observed for our samples.

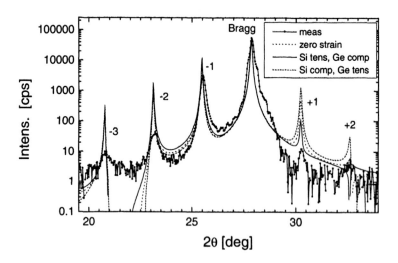

Figure 3. Influence of strain on superlattice (111) intensity distributions

Figure 4 shows reflectivity data of a superlattice with 3.87nm period. Whereas the superlattice period can be calculated from the Bragg peaks with high accuracy, again simple models allowing for the variation of layer thickness, composition, density and roughness fail to simulate the observed intensity distribution in a sufficient way. For developing strategies for improved simulation it might be interesting to notice that the AFM roughness of the superlattice structure is in general lower than the buffer roughness by some 20%, so cumulative roughness effects don't occur in the systems under consideration.

For investigations of the superlattice internal strain reciprocal space mappings of the asymmetrical 422 lattice point were carried out as described earlier for the buffer layers (Figure 5). The superlattice Bragg peak is found to be located at the position of the buffer peak on the dotted line connecting the substrate and the origin of the reciprocal lattice, indicating the correct average composition of the superlattice ($Si_{0.5}Ge_{0.5}$). The superlattice satellites and the Bragg peak are aligned along a line with x= const. The in plane lattice parameter of the superlattice is exactly halfway between Si and Ge, indicating zero average strain, as to be expected for strain symmetrized superlattices with pseudomorphic growth of the individual layers.

CONCLUSIONS

For the first time (111) oriented strain symmetrized Si/Ge-superlattices have been grown by magnetron sputter epitaxy. A buffer thickness of 1.1μm was found to be sufficient for strain relaxation of more than 90%. The buffer roughness of about 3.6nm can be considered as a very

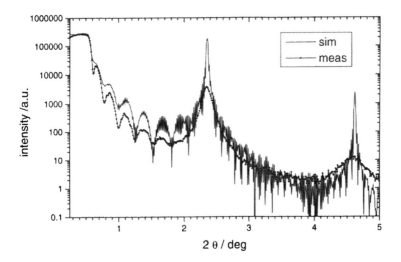

Figure 4. Reflectivity measurement of a superlattice prepared at 450°C. The short period oscillations are due to the 30nm cap layer.

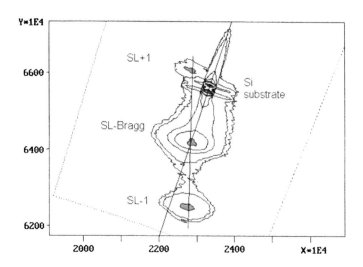

Figure 5. Reciprocal space mapping around the 422 lattice point of a 4.13nm period superlattice prepared at 450°C

good value for buffers with high final Ge concentration, a further reduction is however desirable. The X-ray data for superlattices with a Si and Ge individual layer thickness of 2nm on top of a $Si_{0.5}Ge_{0.5}$ pseudo-substrate show correct average composition and zero average strain, which is expected for strain symmetrized structures with pseudomorphic growth of the individual layers.

ACKNOWLEDGEMENTS

Research and one of the authors (CAK) are supported by BMBF under contract No. 03N2014B (NITHERMA-project). The authors would like to thank S. Krause for technical assistance.

REFERENCES

1. T. Koga, X. Sun, S. B. Cronin, and M. S. Dresselhaus, Appl. Phys. Lett., 75, 2438-2440 (1999)
2. S.-M. Lee, D. G. Cahill, and R. Venkatasubramanian, Appl. Phys. Lett., 70, 2957-2959 (1997)
3. R. Hull, Equilibrium theories of misfit dislocations in the SiGe/Si system, *Properties of strained and relaxed Silicon Germanium*, ed. E. Kasper, (INSPEC, 1995) pp. 17-27
4. J. L. Liu, C. D. Moore, G. D. U'Ren, Y. H. Luo, Y. Lu, G. Jin, S. G. Thomas, M. S. Goorsky, K. L. Wang, Appl. Phys. Lett., 75, 1586-1588 (1999)
5. J.-P. Loquet, D. Neerinck, L. Stockman, and Y. Bruynseraede, Phys. Rev. B, 38, 3572-3575 (1988)

Electric and Thermoelectric Properties of Quantum Wires Based on Bismuth Semimetal and its Alloys.

A.A.Nikolaeva[1], P.P.Bodiul, D.V.Gitsu, G. Para.
Institute of Applied Physics, Kishinev, Moldova.
[1] International Laboratory of High Magnetic Fields and Low Temperatures, Wroclav, Poland.

ABSTRACT

The resistance R(T), the magnetoresistance R(H) and the Seebeck coefficient S(T) of thin monocrystalline glass-coated wires of pure and doped by acceptor impurities (Sn) bismuth are studied. The measurements are carried out in the temperature range 4.2 - 300 K, and the magnetic fields up to 14 T. A significant dependence of the resistance R(T) and thermoemf S(T) on doping, sample diameter and surface state is revealed. By recrystallization, annealing and etching the samples a change in the surface scattering character of the charge carriers was achieved. The influence of the strong (up to 3-4%) elastic stretch on the band structure and kinetic coefficients of thin wires was studied. The thermoelectric figure of merit in doped bismuth wires is estimated, and possible ways of its increase in the temperature range 77 - 300 K are regarded.

INTRODUCTION

Bismuth and its alloys prepared in the form of nano- and submicron single crystal wires attract investigators dealing with thermoelectric cooling due to a number of objective reasons.

First of all, as result of large de Broglie wavelength (about 1000 A) a size quantization of the energy spectrum in the filamentary crystal of bismuth takes place already at thickness ~0,1 μm. In the latest theoretical and experimental works it is shown that at d < 50 nm in layers and wires of Bi due to the quantum size effect the transition semimetal-semiconductor is possible, leading to an increase of the thermoelectric figure of merit by an order [1]. Indeed, the effect of, weak localization of charge carriers on Bi nanowires was observed. [2] In the paper [3] a considerable drop of the heat conductivity is predicted when the diameter d decreases.

Considerable changes in absolute values of magnetothermoemf in doped Bi were found near the liquid-nitrogen temperature [4]. They are due to big values of partial thermopower and weak decompensation of electron-holes concentration at the doping. In order to realize above mentioned possibilities it is necessary to disbalance partial contributions to thermoemf of electrons and holes by different external influences (by magnetic field or elastic deformation). Effectively, this can be obtained by uniaxial stretch of the Bi wires where it reaches a record value – 3,5%.

EXPERIMENTAL DETAILS

Thin monocrystal wires of bismuth and its alloys with Sn were obtained by the liquid phase casting in a glass coating [5], with the diameters from 100 nm up to 5000 nm. The sample orientation was the same for all the diameters and compositions, the wire axis made up an angle of 20^0 with the bisector axis C_s in the bisector-trigonal plane.

For studying of kinetic effects under the influence of strong elastic stretches a displacement transformer having a certain elongation pitch was used. This allowed to measure the resistance and magnetoresistance both in the continuous or discrete regimes in the temperature range 4.2 - 300 K and in the magnetic fields up to 14 T. The measurements were carried out in the International Laboratory of Strong Magnetic Fields and Low Temperatures (Poland, Wrozlaw).

RESULTS AND DISCUSSION

The position of the Fermi level ε_F, (relative to the top of the valence band in T-point, calculated from the Shubnikov-de Haas (ShdH) oscillations) rise from $\varepsilon^h_F = 15$ meV (pure Bi) to 117 meV (Bi-0,07Sn). It should be noted that the period of the ShdH oscillations did not depend on diameter of wires.

Fig. 1 shows the temperature dependencies of the reduced resistance (R_T/R_{300}) of the doped bismuth wires of different diameters (d<0,5μm).

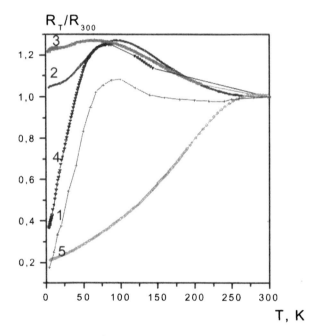

Figure 1 *The temperature dependence of the reduced resistance R_T/R_{300} in the Bi wires doped by Sn: 1. Bi, d=0,35μm, 2. 0,01at%Sn, d=0,45μm, 3. 0,02at%Sn, d=0,14μm, 4. 0,05at%Sn, d=0,35μm, 5. 0,07at%Sn, d=0,37μm*

Characteristic peculiarity of the resistance-temperature dependence in the pure and doped by acceptor impurity bismuth wires is the appearance of maximum, its value change with the wire diameter

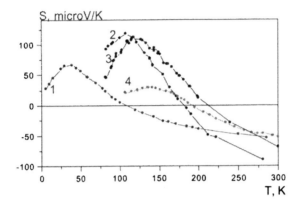

Figure 2 The temperature dependence of Seebeck coefficient of the doped Bi wires. 1. Bi, d=0,35µm, 2. Bi-0,01at%Sn, d=0,37µm, 3. Bi-0,02at%Sn, d=0,35µm, 4. Bi-0,07at%Sn, d=1.4µm.

and the doping degree. For pure bismuth wires the maximum is located in the temperature range of 70 - 90 K. As it was shown in [6], the dependence R(1/d) at 77 K obeys the classical formula of Foux-Sondheimer, and the peculiarity coefficient P determined from this dependence is equal to 0.6. At the liquid-helium temperature P~0,95 [6].

The temperature dependence of the thermoemf S(T) for wires of the above-mentioned alloys are given in Fig.2. At 300 K the thermoemf in wire does not depend practically on thickness and has a same value as in bulk crystals of the analogous crystallographic orientation.

In the low temperature region, the behavior of S(T) curves (pure Bi) shows a significant dependence on diameter. [8] The diminution of the wire diameter results in the increased of temperature when decrease of S (T) starts and its sign change occurs. For pure wires with d=0,1µm the sign inversion of thermoemf takes place at 120K. At a further decrease of temperature on S(T) dependence a positive maximum arises. Its position shifts into the region of higher temperatures, when d decreases, however the localization region does not exceed 40 K. Qualitatively analogous regularities are also observed in the doped Bi wires. Nevertheless, the range of sign change and maximal position on S(T) are displaced at higher temperatures.

In the Bi-0,01 at%Sn and Bi-0,02at%Sn alloys the maximum value of the positive thermoemf is 120 -140 µV/K at ≈100 K. (Fig.2, 3)

If one supposes that the thermal conductivity κ of the wires does not change with the diameter d (according to [3] when d drops the κ decreases) the thermoelectric efficiency in the thin wires of such composition at the range of 100K is nearly, 30 times larger than the one in the wire with d>1µm.

It is easy to see that the thermoemf sign change and maximum formation in the positive region of S(T) correlate with the maximum on the temperature dependencies of the resistance. For example, in the alloys Bi-0,07at%Sn - the maximum R(T) takes place at 210 - 250 K, and the thermoemf sign change occurs at 200 - 210 K. Qualitatively for the pure bismuth wires the thermoemf sign inversion can be explained using the expression for the bulk sample tensor S_{ii} [7, 8] :

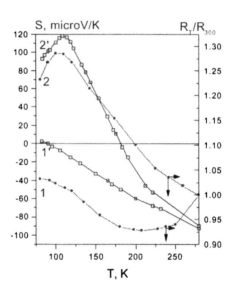

Figure 3 The temperature dependencies of Seebeck coefficient S in Bi-0,01at%Sn (1',2') and of R_T/R_{300} (1, 2) wires with different diameters: 1. $d=1,34\mu m$, 2. $d=0,37\mu m$.

$$S_{ii} = \frac{S_p^h - |S_p^e| \times b_{ii}}{1 + b_{ii}} \quad (1)$$

$$b_{ii} = \sigma_{ii}^e / \sigma_{ii}^n = \frac{\mu_{ii}}{\nu_{ii}}$$

where μ_{ii}, ν_{ii} are the mobilities of electrons and holes correspondingly. $S_p^{e,h}$ - the partial thermoemf.

In the wires with the diameter smaller than the free path length $l^{e,h}$ of the charge carriers, due to the surface scattering, a decrease of their average mobility takes place.

The thinner is the sample and the less is the surface specularity, the more significant is the mobility decrease. As far as in bismuth the free path length of electrons exceeds one of holes, the electron mobility in the thin wires will quickly decrease, therefore $b_{ii}= \mu_{ii}/\nu_{ii}$ decreases as well. So, according to (1), the fall of $b_{ii}(d)$ parameter in the Bi wires will lead to the thermoemf sign inversion at $b_{ii}<S_0^h/|S_0^e|$, and then to its positive value increase. However it is impossible to achieve a quantitative correlation, within

the limit of such simplification. It is clear that the real transport processes in the thin wires are more complex.

The surface scattering influence becomes obvious if, for example, one compares the dependencies R(T) and S(T) for Bi wire with d= 0,35μm before and after annealing, or for the same wire after recrystallization (Fig.4).

It should be noted that the crystalline orientation of the samples does not change.

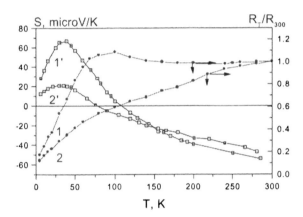

Figure 4 *The temperature dependencies of Seebeck coefficient S (1', 2') and of R_T/R_{300} (1,2) in Bi single crystal wires (d=0,35μm). (1) before and (2) after annealing.*

The annealing of the bismuth wires suppresses the maximum on dependence R(T) and displaces the region of the sign change of S(T) to lower temperatures (T<25K). At the same time the positive value of the S(T) decreases about 3-5 times and in general feature, approaches to its behavior in the bulk bismuth crystals of the same orientation.

In this aspect it would be of interest, in contrast to annealing and zone recrystallization, to obtain an impaired surface, in order to intensify the maximum on the temperature dependence R(T) of the pure and doped bismuth wires. This could be achieved by a strong stretching of the wire or by the wire surface etching. At 4.2 K a sharp increase of the resistance is observed, with the maximum formation critical to T, d, N_c (N_c is the doping impurity concentration). At a stretch of bismuth wires of the used crystallographic orientation, overlapping of L and T bands increases. This is probably the reason for the fact that at stretch on the temperature dependencies R(T) in the range 80 – 300 K, no significant change of the position and value of the maximum occurs.

The etching of the thinnest wires is a rather complicated procedure, because it was necessary to remove glass, this being a very difficult process when the diameters are 0,2÷0,5μ. However, the preliminary measurements of R(T) on the etched wire really promote an additional increase of the resistance when the temperature falls.

CONCLUSIONS

It is shown that in the thin filamentary crystals of bismuth and its alloys at temperatures below 200 K the values of the thermoelectric parameters can be altered in large limits by changing the crystals sizes and the surface scattering mechanism.

- The region of sign change of the thermoemf S(T) and its positive value in the pure and doped bismuth wires is a function of their diameter.
- By doping the bismuth wires by acceptor impurities up to 0,07at%Sn it is possible to displace the position of the positive maximum of S(T) from 40K up to 180K.
- It is established, that maximum value on the temperature dependencies of the resistance R(T) and in the positive range of S(T) are reciprocally related and determined by the character of surface scattering process of charge carriers. They can be controlled by usual technique procedures: annealing, zone recrystalization, etching, elastic deformation etc.

The elastic deformation achieving 3-3,5% in these wires shifts L-T bands relatively each other. In principle, having changed the crystallographic orientation of these wires, one could move apart L- and T- bands and obtain the transition semimetal-semiconductor in thicker wires (0,1- 1μ).

REFERENCES

1. L.D. Hicks, T.C. Harman, M.S. Dresselhaus, "Use of quantum-well superlattices to obtain a high figure of merit from nonconventional thermoelectric materials," *Appl. Phys. Lett.*, **63**, 3230-3232, (1993).
2. J.Heremaus, C.M. Trush, Z. Zhaug at all, "Magnetoresistance of bismuth nanowire arrays: A possible transition from one-dimensional to there- dimensional localization," *Phys. Rev. B*, **58**, No. 16, R10091-10095, (1998).
3. T.S. Tighe, J.M. Worlock and M.L. Roukes, "Direct thermal Conductance measurements on suspended monocrystalline nanostructures", *Appl. Phys. Lett.*, **70**, 2687, (1997).
4. N.A.Redko, V.I. Belitskii, V.V.Kosarev, I.A.Rodionov, "Anizotropia termoeds splavov BiSb v kvantuiushem magnitnom pole" [in Russian], *FTT*, **29**, No. 2, 463-466, (1986).
5. N.B. Brandt, D.V.Gitsu, A.M. Ioisher, B.P. Kotrubenko, A.A. Nikolaeva, "Poluchenie tonkih monokristallicheskih nitei vismuta v stekleannoi izoliatsii" [in Russian], *Prib. i Techn. Experiment.* (1976).
6. N.B. Brandt, D.V.Gitsu, A.A. Nikolaeva, Ya.G. Ponomaryov, "Investigation of size effects of thin cylindrical bismuth single crystals located in magnetic field," *J. Exp. Teor. Fiz,.* **72**, 2332-2344, (1997) [*Sov. Phys. JETP*, **45**, 1226, (1977)].
7. P.P. Bodiul, D.V. Gitsu, V.A. Dolma, G.G. Zegrya, N.F. Miglei, "The thermopower in bismuth whiskers," *Phys. Stat. Sol.*, **53**, 87, (1979).
8. P.P. Bodiul, A.N. Burchakov, D.V.Gitsu, A.A. Nikolaeva, "Thickness dependencies of kinetic properties of Quantum wires of pure and doped bismuth," *Plenum Press*587-592, (1999).

High-Z Lanthanum-Cerium Hexaborate Thin Films for Low-Temperature Applications

Armen Kuzanyan, George Badalyan, Sergey Harutyunyan, Ashot Gyulamiryan, Violetta Vartanyan, Silvia Petrosyan, Nicholas Giordano[1], Todd Jacobs[1], Kent Wood[2], Gilbert Fritz[2], Syed B. Qadri[2], James Horwitz[2], Huey-Dau Wu[2], Deborah Van Vechten[3], and Armen Gulian[4]

Institute for Physics Research, National Academy of Sciences, Ashtarak-2, 378410, Armenia.
[1]Physics Department, Purdue University, West Lafayette, IN 47907.
[2]Naval Research Laboratory, Washington, DC 20375.
[3]Office of Naval Research, Arlington, VA 22217
[4]Universities Space Research Association/ Naval Research Laboratory, Washington, DC 20375.

ABSTRACT

We have deposited and investigated thin films of lanthanum hexaborate with 1% of the lanthanum replaced by cerium. In bulk single-crystalline form, this material has, due to the Kondo-mechanism, the highest known Seebeck coefficient at sub-K temperatures. Thus it is a good candidate for several thermoelectric applications at very low temperatures. We are studying the kinetic properties of thin films such as the conductivity and Seebeck coefficient as a function of temperature and the dependence of these properties on film thickness, substrate material and deposition conditions. The consequent theoretical performance limits on the device applications of these films are considered with a focus on detectors and refrigerators.

INTRODUCTION

While most recent investigations of thermoelectric materials have focused near room temperatures, there are several reasons to investigate materials with large Seebeck coefficients at low temperatures. First of all, there is general interest in the relationship of the Kondo effect to thermoelectric properties. Secondly, there is interest in the fabrication of all-solid-state coolers. Finally, there is interest in the use of these materials for high energy resolution, single photon sensors for the UV and X-ray region of the EM spectrum. The basic concept here is that the thermoelectric materials can be used to read-out the impulsive temperature rise that results when a photon is absorbed. Thermoelectric sensors will not need to be voltage or current biased for read out [1,2]. This facilitates the construction of multi-pixel arrays and reduces the complexity of the thermal engineering of the sensors. The competing detectors are based on the temperature coefficient of resistance and measure the current through a voltage biased superconducting stripe biased on the sharply temperature dependent resistive transition. These sensors have demonstrated 6 eV resolution for 6 keV photons incident in single pixels. To beat this record the thermoelectric sensors will need to achieve a Seebeck coefficient of $S > 20$ μV/K at the expected bias temperature of <1K.

Only a few materials have had their thermoelectric properties investigated at low temperatures. We began our study by using trace impurities of iron in gold, the best studied system with Kondo effect peaks in $S \sim 10$ μV/K below 1K [3]. In a photon detector, we have already achieved energy resolution of about 1500 eV when 6 keV energy is deposited. We expect to be able to improve this to <100 eV with the same gold sensors by improving the devices and the readout electronics. However, large values for S have been reported in the lanthanum hexaborate system in which a small percentage of the lanthanum has been substituted by cerium [4,5]. We report here on the structural and electrical properties of thin films of $La_{0.99}Ce_{0.01}B_6$ composition.

EXPERIMENTAL DETAILS

We attempted to grow $La_{0.99}Ce_{0.01}B_6$ thin films using e-beam technique. Ceramic targets with 1% Ce (CERAC, Inc.) were used at the deposition. Thin films were prepared on a variety of substrates including Glass ceramics, YSZ, MgO, and Pyrex. The first run (depositions #1-12) revealed a strong variation of crystalline structure and resistivity for different thickness films (Figs. 1 and 2).

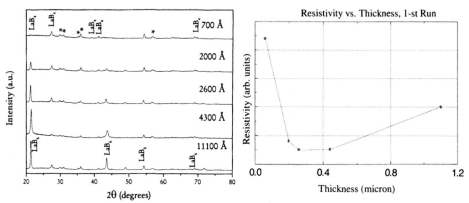

Figure 1. X-ray diffractograms of initial run (depositions #1-12). Mainly LaB_4 structure was observed in the thinnest films.

Figure 2. Resistivity measurements also reveal anomaly in thin films most likely related to the prevailing of the LaB_4 phase.

To characterize the crystalline structure, X-ray θ/2θ diffraction scans were taken using CuK_α radiation from Rigaku rotating anode X-ray source. All the films were measured under identical conditions. Figure 1 shows an overlay of the diffraction pattern observed as a function of film ggthickness. For 1.11 μm thick film, the majority of the peaks are identifiable with LaB_6 phase with a preferred orientation of (100). For 700 Å thick film, one can identify peaks only with LaB_4 phase with a (210) preferred orientation and peaks marked with an asterisk can not be identified with any of the known phases of La or Ce or their compounds with B. The peaks observed for 700 Å thick films are also present in all the scans for thicker films suggesting that these phases are formed during the initial stages of growth but as the film grows thicker the LaB_6 phase is formed. The lattice parameter of a = 4.1622±0.0114 Å for LaB_6 was obtained using a least square refinement routine and is in very good agreement for the bulk case (a = 4.153 Å).

The parameters of the second, more controlled set of depositions (#13-22) are given in Table 1 below.

In-situ annealing at the deposition temperatures in air at the pressures shown in Table 1 was used. Analysis of the X-ray diffraction data revealed a distinct (100) crystallographic texture

Table 1. (La,Ce)B6 Thin Film e-Beam Deposition

#dep	Pressure, MPa	Dep. rate**, Å/sec	Dep. Time, min	Dep. temp, °C	Annealing, min	Thickness***, µm
13	1.8	2.6	20	597	30	0.31
14	1.5	20	4	595	30	0.46
15	1.75	13.3	4	600	0	0.32
16	1.5	20.8	5	600	30	0.66
17	1.5	14.8	5	607	60	0.44
18	0.95	12.4	5	598	90	0.38
19*	1.5	9.7	5	600	30	0.3
20	1.5	15.8	8	610	30	0.78
21	0.2	7.8	6	613	10	0.29
22	2	-	2	585	3	0.07

*) At deposition #19 substrates were pre-heated 40 min.
**) Among other factors variation in the effective deposition rate is caused by the variation in tightness of the e-beam focusing.
***) Post deposition thickness measurements via profilometry.

in all the films. The mosaicity of films and c axis' tilt is characterized by measuring the ω-rocking curve. The texture coefficient TC=$\{I_{(hkl)i}/ I_{(hkl)i0}\}/(1/n)\Sigma^n_{j=1}\{I_{(hkl)j}/ I_{(hkl)j0}\}$ varies in the range of 1.3 to 1.99, when calculated using (100) and (210) XRD peaks. The main peak (100) has FWHM in the range 0.04-0.41 degrees, which corresponds to a grain size between ~200 and 2000 Å. A noticeable aspect of the depositions is that most of the films (as shown in Fig. 3 for films on the glass ceramics substrates) ended up displaying tensile stress, i.e. the lattice

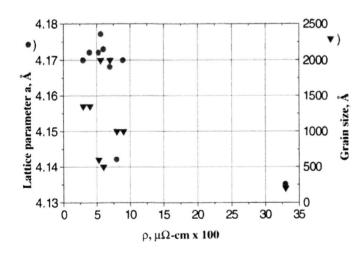

Figure 3. Lattice parameter and the grain size vs. film resistance (for films on glass ceramics substrates, similar results follow for films on other substrates: MgO and pyrex).

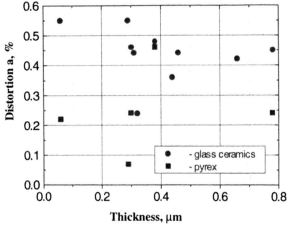

Figure 4. The magnitude of the distortion of the lattice parameter a of the films deposited on two classes of substrates.

constants are longer than the bulk value of 4.153Å. That two films displayed compressive stress indicates that stress free deposition conditions must exist. This feature was not indicative from the first run data. Another important consequence of Fig. 3 data is that the film resistivity (at these room temperature measurements) is not correlating with the grain sizes, so either the lattice imperfections or electron-phonon interaction yielded the dominant carrier scattering. An important improvement for this second run is that the dispersion of the values of the resistivity vs. film thickness is much smaller than for the previous one (the right bottom corner point in Fig. 3 belongs to the sample #13, which falls out of typical deposition conditions, see Table 1).

Measurements of the lattice distortion revealed that it is typically smaller than 1% (for films on all types of substrates) and can be as small as 0.07% (on a pyrex substrate).

SEEBECK COEFFICIENT

To measure the Seebeck coefficients between room temperature and 77K a mechanically chopped laser beam was directed through the optical port of a cryostat and illuminated the

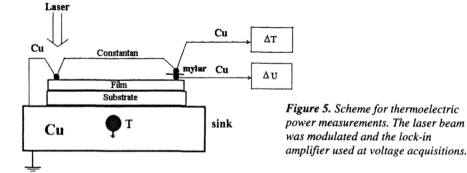

Figure 5. Scheme for thermoelectric power measurements. The laser beam was modulated and the lock-in amplifier used at voltage acquisitions.

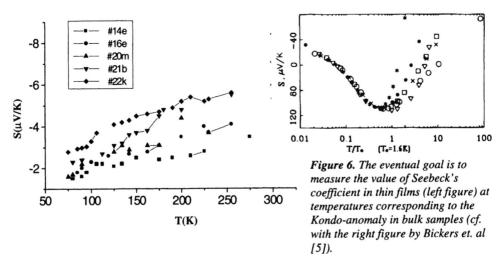

Figure 6. The eventual goal is to measure the value of Seebeck's coefficient in thin films (left figure) at temperatures corresponding to the Kondo-anomaly in bulk samples (cf. with the right figure by Bickers et. al [5]).

sample. A lock-in amplifier was used to read out the response at the same frequency. These measurements were taken by the differential technique described in Fig. 5.

Figure 6 (left) shows the results to date, which document the temperature dependence of Seebeck's coefficient as a function of temperature. The expected zero crossing of the Seebeck coefficient somewhere below the lowest measured point is plausible, especially for samples 14 and 20 which look likely to cross zero at about 50K. This is typically the case for bulk samples (Fig. 6, right). Clearly lower temperature measurements are required to determine which are the most promising deposition parameters and how the value of S in thin films relates to that in the bulk samples.

The same experimental scheme can be used to measure the expected low temperature (below 4K) Kondo anomalies in the Seebeck coefficient. Meanwhile, the resistivity measurements are much simpler. Our preliminary (only one sample was tested) results indicate Kondo-type anomaly at very low temperatures (Fig.7).

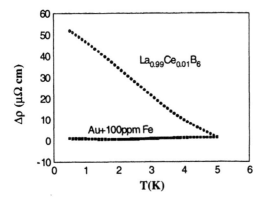

Figure 7. Preliminary data on Kondo-effect in lanthanum-cerium hexaborate films. For a comparison, similar effect in Au+100ppm Fe is shown. The magnitude of the effect is much higher in case the hexaborate film.

DISCUSSION

As was mentioned in the Introduction, thin hexaborate films are interesting for three groups of researchers. A) *Comparison with theory*. When the quality of the films becomes enough to be comparable with the bulk samples, one will be able to investigate the role of the film surface reflection on the kinetic properties – this is not yet done experimentally, though theoretical models exist. B) *Refrigeration*. It is interesting to devote some attention to the figure of merit of these materials. Taking into account that the best known thermoelectric materials have ZT~1 at 300 K, one can deduce their Z's in the range of 0.003 K^{-1}. At the same time, $(La,Ce)B_6$ at 0.3 K is expected to have ZT~0.1 (at these very low temperatures $ZT = S^2/L_0$, where $L_0 \approx 25$ nW-Ω/K^2). It yields Z~0.3 K^{-1}, i.e., about 100 times higher than that of the best known thermoelectric materials. The factor ZT determines the limitations on practical performance of the devices. In particular, the attainable limit of the refrigeration is given by $\delta T/T$ ~ ZT/2 [6], which means that for hexaborates materials one is limited by 0.05 K for single-stage process at T=1 K. Multi-stage schemes can be considered to get below 1 K. Thin film materials may become useful for designing these multistage solid-state refrigerators. C) *Detectors*. Application of these films to the development of the sensitive detectors is well described in [1,2], so we will not discuss further here.

CONCLUSIONS

In summary, this is the first (known) attempt to obtain Ce doped lanthanum hexaborate thin films. We hope to use them for thermoelectric sensors in efficient focal plane, multi-pixel, energy resolving detectors of individual UV and X-ray photons. While the process is not yet well controlled, we started to differentiate between yields on different substrates and deposition conditions from one side, and the physical properties from the other side. This work is continuing, and thin film samples are available for collaborative research.

ACKNOWLEDGEMENTS

This work is supported in part by funding from NRL, NASA, ONR and Ministry of Science and Education of Armenia. The work of T. Jacobs and N. Giordano was supported by NSF grant DMR-9970708.

REFERENCES

1. A. Gulian, K. Wood, G. Fritz, A. Gyulamiryan, V. Nikogosyan, N. Giordano, T. Jacobs, and D. Van Vechten, NIMA , **441**, No.3 (2000).
2. D. Van Vechten, K. Wood, G. Fritz, A. Gyulamiryan, V. Nikogosyan, N. Giordano, T. Jacobs, and A. Gulian in *Proc. Int. Conf. On Thermoelectricity* (ICT-99, Baltimore, MD, 1999) pp. 477-480.
3. D. K. C. MacDonald, W. B. Pearson, and I. M. Templeton, Proc. Roy. Soc. A **266**, 161 (1962).
4. K. Winzer, Solid State Comm., **16**, 521 (1975).
5. N. E. Bickers, D. L. Cox, and J. W. Wilkins, Phys.Rev. Lett., **54**, 230 (1985).
6. D. K. C. MacDonald, E. Mooser, W. B. Pearson, I. M. Templeton, and S. B. Woods, Phil. Mag. **4**, 433 (1959).

Thermionics

Thermal Conductivity Of Bi/Sb Superlattice

D. W. Song and G. Chen
Mechanical and Aerospace Engineering Department
University of California at Los Angeles, Los Angeles, CA 90025

S. Cho, Y. Kim, and J. B. Ketterson
Department of Physics and Astronomy
Northwestern University, Evanston, IL 60208

ABSTRACT

The temperature-dependent cross-plane thermal conductivity of a 1-μm thick 50Å Bi / 50Å Sb superlattice on a (111) CdTe substrate was measured, using a differential 3-ω method. This method uses the temperature difference between the superlattice sample and a reference sample to calculate its cross-plane thermal conductivity. However, the substrate thermal conductivity is comparable to or smaller than the superlattice thermal conductivity near room temperature. This results in a very small or negative temperature difference, making the existing data reduction method inapplicable. Based on an improved model, the temperature-dependent thermal conductivity of the Bi/Sb superlattice is obtained and is about half of the literature value of $Bi_{0.5}Sb_{0.5}$ bulk alloy.

INTRODUCTION

BiSb alloys have long been studied as a possible thermoelectric material.[1-10] Some have observed that high thermoelectric efficiencies can be obtained in a BiSb alloys, especially at low temperatures and in a magnetic field.[2-9] Since a reduction of the thermal conductivity, to below that of the corresponding alloy, has been reported in several superlattice systems, including GaAs/AlAs,[11-14] Si/Ge,[15] Bi_2Te_3/Sb_2Te_3,[16] and $CoSb_3/IrSb_3$,[17] it is of interest to investigate the thermal conductivity of Bi/Sb superlattices, particularly since both Bi and Sb are semimetals.

We report in this paper an experimental study on the thermal conductivity of a 50Å/50Å Bi/Sb superlattice grown on a (111) CdTe substrate using the 3-ω method. At near room temperature, the superlattice has a thermal conductivity value close to that of the substrate, rendering the usual way of obtaining the thermal conductivity based on one-dimensional steady-state heat conduction inapplicable. Based on a new data analysis scheme, we deduced the thermal conductivity of the superlattice and found that it is about half that of the alloy.

EXPERIMENTS

A 1-μm thick Bi/Sb superlattice film consisting of alternating layers of 50 Å Bi and 50 Å Sb was grown by molecular beam epitaxy[18] on a (111)B CdTe substrate. On top of the Bi/Sb superlattice film, an 800 Å layer of ZnTe was also deposited, to electrically insulate the Bi/Sb superlattice. However, the ZnTe film was not sufficiently insulating for the required measurements, and therefore an additional 1100 Å layer of Si_xN_y was deposited onto the ZnTe surface by plasma-enhanced chemical vapor deposition. A 10 μm wide, 2 mm long heater was then deposited on the Si_xN_y surface by conventional microfabrication technology. In addition to this "superlattice" sample, a "reference" sample was fabricated, which consists of the same collection of films as the former except for the Bi/Sb superlattice layer. The films common to the "superlattice" and "reference" samples were deposited in the same steps to ensure identical deposition conditions.

To measure the cross-plane thermal conductivity of the Bi/Sb superlattice, we employed a differential 3-ω technique, which uses the measured temperature differences between the sample and the reference to subtract the thermal conductivity of the film. The temperature rise of the heater is deduced from the third harmonic of the voltage drop across the heater, which is driven by a sinusoidal current. This measurements were done at temperatures from 80 K to 300 K inside a cryostat using flowing liquid N_2. Both the "superlattice" sample and the "reference" sample were measured. The thermal conductivity of the CdTe substrate was obtained by using Cahill's slope method.[19]

III. DATA REDUCTION AND DISCUSSION

In a typical differential 3-ω measurement, the cross-plane thermal conductivity of a thin film, k_F, of thickness d_F on a substrate is calculated using a simple one-dimensional approximation from Fourier's Law:

$$k_F = q'' \, d_F / (T_F - T_{REF}) \qquad (1)$$

where q" is the cross-plane heat flux at the metal wire, and T_F and T_{REF} are the amplitudes of the AC temperature rise of the wires on the "thin film" sample and the "reference", respectively. This approximation, however, is valid only when two conditions are satisfied: (1) the film thermal conductivity is much smaller than the substrate thermal conductivity, and (2) the heater width is much larger than the film thickness. The CdTe substrate has a very low lattice thermal conductivity, especially at higher temperatures. The literature value for the thermal conductivity of bulk single-crystal CdTe, ranging from 7.6 W/mK at 150 K to 3.5 at 300 K,[20] is lower than those of bulk Bi[21] and bulk Sb.[22] This presents a possibility that the thermal conductivity of the CdTe substrate is comparable to or lower than that of the Bi/Sb superlattice film, for which Eq. (1) can no longer be used for data analysis. Indeed, the experimental temperature rises of the sample and the reference at 280 K, as shown in Fig. 1, are very close to each other. Therefore, we have developed a new strategy for analyzing the experimental data.[23] Beginning with the 2-dimensional heat conduction solution for a film of anisotropic thermal property on a substrate,[24] the temperature difference between the film and the substrate can be expressed as

$$\Delta T = T_f - T_s = \frac{q'' d_F}{K_F} C F, \qquad (2)$$

where C is a constant that compensates for the contrast between the in-plane and cross-plane film thermal conductivities, $K_{F,x}$ and $K_{F,y}$, and the substrate thermal conductivity, K_S:

$$C = \left(1 - \frac{K_{F,x} K_{F,y}}{K_S^2}\right), \qquad (3)$$

The constant F in Eq. (2) compensates for the effect of lateral heat spreading in the film:

$$F = \frac{2}{\pi} \int_0^\infty \frac{\sin^2 x}{x^3} \frac{\tanh(rx)}{1+(K_{F,y}/K_s)\tanh(rx)} dx \quad (4)$$

where r is the "aspect ratio" factor that includes the geometrical aspect ratio of the wire width, 2b, to the film thickness, d_F, and the thermal conductivity anisotropy:

$$r = \frac{2b}{d_F} \sqrt{K_{F,x}/K_{F,y}} \quad (5)$$

From the modified approximation in Eq. (2), k_F, and y, the cross-plane thermal conductivity of a thin film on a substrate can be calculated for any given value of cross-plane temperature drop across the thin film, even when vanishingly small or negative.

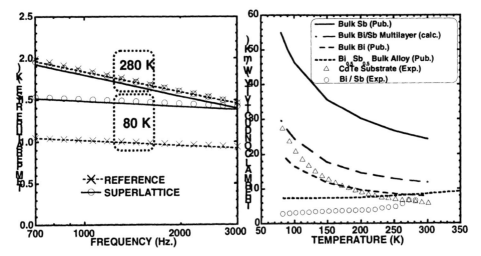

Figure 1. Temperature rise obtained by fitting the "superlattice" sample and the "reference" sample at 80 K and 280 K. The lines are simulated results, and o and x are the experimental data.

Figure 2. Experimental thermal conductivity of Bi/Sb superlattice and CdTe substrate, the published values for bulk Bi,[21] bulk Sb,[22] and bulk $Bi_{0.5}Sb_{0.5}$ alloy,[25] and the calculated value for a 50 % Bi / 50 % Sb multilayer.

Using the modified approximation in Eq. (2), the frequency-dependent temperature rise has been simulated to fit the experimental data for the "superlattice" and the "reference" samples. The lines in Fig. 1 show the calculated temperature rise for the "superlattice" and the "reference" samples at 80 K and 280 K. While the fit is very accurate for the "reference" sample, the simulated result slightly underpredicts the experimental temperature rise for the "superlattice" sample. This may be because Eq. (2) is a solution for only one film on a substrate, so that it works better for the two-film "reference" sample than for the three-film "superlattice" sample.

The experimental cross-plane thermal conductivity of the Bi/Sb superlattice as a function of temperature is plotted in Figure 2. Clearly, the thermal conductivity of the superlattice is not only much smaller than that of a bulk 50 % Bi / 50 % Sb multilayer, calculated as a parallel resistor array using the published values of the two bulk elements,[21,22] but also much smaller than the published value for bulk $Bi_{0.5}Sb_{0.5}$ alloy,[25] by a factor of ~2.

CONCLUSION

In this paper, the cross-plane temperature-dependent thermal conductivity of a 1-μm thick 50 Å Bi / 50 Å Sb superlattice film on a CdTe substrate has been measured. The small thermal conductivity contrast between the superlattice and the substrate leads to a failure of the one-dimensional steady-state approximation used in past work. This difficult is overcome with an improved model. The resulting thermal conductivity of the superlattice film was a factor ~2 lower than the literature value for the bulk $Bi_{0.5}$ / $Sb_{0.5}$ alloy.

ACKNOWLEDGMENT

This work was supported by DOD MURI on thermoelectrics (N00014-97-1-0516). The work at Northwestern University was supported by (DAAG55-97-1-0130).

REFERENCES

1. G. Gelhlhoff and F. Neumeier, Verhandl. Deut. Physik. Ges., **15**, 876, 1069 (1913).
2. W. C. White, Elec. Eng., **70**, 589 (1951).
3. M. Telkes, J. Appl. Phys. **25**, 765 (1954).
4. A. L. Jain, Phys. Rev., **114**, 1518 (1959).
5. S. Tanuma, J. Phys. Soc. Japan, **14**, 1246 (1959).
6. G. E. Smith and R. Wolfe, J. Appl. Phys., **33**, 841 (1962).
7. R. Wolfe and G. E. Smith, Appl. Phys. Lett., **1**, 5 (1962).
8. K. F. Cuff, R. B. Horst, J. L. Weaver, S. R. Hawkins, C. F. Kooi, and G. M. Enslow, Appl. Phys. Lett., **2**, 145 (1963).
9. Y. M. Yim and A. Amith, Solid State Electron. **15**, 1141 (1972).
10. S. Cho, I. Vurgaftman, A. B. Shick, A. DiVenere, Y. Kim, S. J. Youn, C. A. Hoffman, G. K. L. Wong, A. J. Freeman, J. R. Meyer, and J. B .Ketterson, Mat. Res. Soc. Symp. Proc. **545**, 283 (1999).
11. T. Yao, Appl. Phys. Lett., **51**, 1798 (1987).
12. G .Chen, C. L. Tien, X. Wu, and J. S. Smith, J. Heat Transfer, **116**, 325 (1994).
13. W. S. Capinski and H. J. Maris, Physica B, **219**, 699, (1996).
14. X. Y. Yu, G. Chen, A. Verma, and J. S. Smith, Appl. Phys. Lett., **67**, 3554 and **68**, 1303 (1995).
15. S.-M. Lee, D. G. Cahill, and R. Venkatasubramanian, Appl. Phys. Lett., **70**, 2957 (1997).
16. R. Venkatasubramanian, Naval Res. Rev., **58**, 44 (1996).
17. D.W. Song, C. Caylor, W.L. Liu, T. Zeng, T. Borca-Tasciuc, T.D. Sands, and G. Chen, Proc. 18th Int. Conf. Thermoelectrics, 679 (1999).

18. S. Cho, A. DiVenere, G. K. Wong, J. B. Ketterson, and J. R. Meyer, J. Vac. Sci. Technol. A (in press).
19. D. G. Cahill, Rev. Sci. Instrum., **61**, 802 (1990).
20. Y. S. Toulokian, Thermophysical Properties of Matter, IFI/Plenum, New York, NY, **1**, 1267 (1979).
21. Y. S. Toulokian, ibid **1**, 25.
22. Y. S. Toulokian, ibid **1**, 10.
23. G. Chen, S. Q. Zhou, D.-Y. Yao, C. J. Kim, X. Y. Zheng, Z. L. Liu, and K. L. Wang, Proc. 17th Int. Conf. Thermoelectrics, 1 (1998).
24. T. Borca-Tasciuc et al., unpublished.
25. Y. S. Toulokian, ibid **1**, 502.

Upper Limitation to the Performance of Single-Barrier Thermionic Emission Cooling

Marc D. Ulrich, Peter A. Barnes, and Cronin B. Vining[1]
Department of Physics, Auburn University, AL 36849
[1]ZT Services, Auburn, AL 36830

ABSTRACT

We have re-examined solid-state thermionic emission cooling from first principles and report two key results. First, electrical and heat currents over a semiconductor – semiconductor thermionic barrier are determined by the chemical potential measured from the conduction band edge, not the energy band offset between the two materials as is sometimes assumed. Second, we show the upper limit to the performance of thermionic emission cooling is equivalent to the performance of an optimized thermoelectric device made from the same material. An overview of this theory will be presented and instrumentation being developed to experimentally verify the theory will be discussed.

INTRODUCTION

Solid state thermionic emission cooling has received interest in the last decade as a possible alternative to standard thermoelectric cooling. It has been proposed that greater cooling power may be achieved with thermionic emission cooling [1,2]. Thermionic emission coolers comprised completely of semiconducting materials, such as in the diagram of figure 1, are also

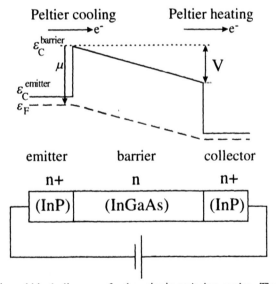

Figure 1. Schematic and block diagram of a thermionic emission cooler. The arrows show the direction of electron flow causing Peltier cooling on the left and Peltier heating on the right.

desirable because they can be monolithically integrated with solid state devices that require temperature control. The basic principle of cooling in thermionic emission devices, modeled in figure 1, is the transport of heat utilizing the Peltier effect.

The purpose of this paper is to present a first principles derivation of the electronic contributions to the cooling power for a semiconductor thermionic device and from this to determine the maximum cooling possible for both the ballistic and diffusive limits.

FIRST PRINCIPLES DERIVATION

The electrical current density, J_E, and heat current density, J_Q, over a semiconductor-semiconductor heterojunction barrier are derived from statistical mechanics [3]:

$$J_E = \int_{-\infty}^{\infty}\int_{-\infty}^{\infty}\int_{p_x^{free}}^{\infty} f(p)g(p)qv_x d^3p, \tag{1}$$

$$J_Q = \int_{-\infty}^{\infty}\int_{-\infty}^{\infty}\int_{p_x^{free}}^{\infty} f(p)g(p)(\varepsilon(p)-\varepsilon_F)v_x d^3p, \tag{2}$$

where $f(p)$ is the Fermi Dirac distribution function, $g(p)$ is the density of states, q is the elementary charge, v_x is the electron velocity in the x-direction, $\varepsilon(p)$ is the electron kinetic energy measured from the conduction band edge, ε_F is the Fermi level measured from the conduction band edge, and p_x^{free} is the momentum in the x-direction necessary to surmount the barrier. If a large barrier height is considered the results of these integrals are the historical Richardson equation for the electrical current density [4] and its equivalent for the heat current density. But, the Richardson approximation is the first term in a series expansion, so any arbitrary barrier height can be considered if all terms in the series are kept. This will also allow for fast numerical calculations at any barrier height. The complete series solutions for these integrals are:

$$J_E = A^*T^2 \sum_{n=1}^{\infty} \frac{(-1)^{n-1}}{n^2} \exp\left(n\frac{q\mu}{kT}\right), \tag{3}$$

$$J_Q = A^*T^2 \sum_{n=1}^{\infty} \frac{(-1)^{n-1}}{n^2}\left(-\mu + \frac{2kT}{nq}\right)\exp\left(n\frac{q\mu}{kT}\right), \tag{4}$$

where A^* is the effective Richardson constant, T is the temperature at the heterojunction, k is the Boltzmann constant, and μ is a chemical potential defined as the following, and is shown in figure 1:

$$\mu = \varepsilon_F - \varepsilon_C^{barrier}, \tag{5}$$

where $\varepsilon_C^{barrier}$ is the conduction band edge of the barrier and ε_F is the fermi level measured in the emitter. This solution is valid for negative or zero μ. We define μ outside the barrier to avoid problems with ballistic transport for which a chemical potential cannot be defined in the barrier. Equation 5 agrees with Wu and Yang's more detailed development of electrical current over a semiconductor heterojunction [5]. Our derivation extends the model to include the heat current density. Equation 5 is the correct barrier height for a semiconductor – semiconductor heterojunction and we have not seen it applied in the calculations of thermoelectric energy conversion. A chemical potential is defined as the energy necessary to add a particle to a system. Thus, the chemical potential governs the electrical and heat currents across a barrier as opposed to the energy difference between the adjacent conduction band edges, as is often assumed.

In order to apply equations 3 and 4 to thermionic emission cooling we consider ballistic transport across the barrier with an applied voltage and do not neglect the reverse current densities. For a thermionic emission cooler, the total current densities are:

$$J_E = A^* T_E^2 \sum_{n=1}^{\infty} \frac{(-1)^{n-1}}{n^2} \exp\left(n \frac{q\mu}{kT_E}\right) - A^* T_C^2 \sum_{n=1}^{\infty} \frac{(-1)^{n-1}}{n^2} \exp\left(n \frac{q(\mu-V)}{kT_C}\right), \quad (6)$$

$$J_Q = A^* T_E^2 \sum_{n=1}^{\infty} \frac{(-1)^{n-1}}{n^2} \left(\frac{2kT_E}{nq} - \mu\right) \exp\left(n \frac{q\mu}{kT_E}\right) - A^* T_C^2 \sum_{n=1}^{\infty} \frac{(-1)^{n-1}}{n^2} \left(\frac{2kT_C}{nq} - \mu\right) \exp\left(n \frac{q(\mu-V)}{kT_C}\right), \quad (7)$$

where T_E is the temperature at the emitter-barrier junction, T_C is the temperature at the barrier-collector junction, and V is the voltage across the barrier as shown in figure 1. To proceed analytically, we consider only the n = 1 term of equations 6 and 7 and combine the equations and eliminate the applied voltage. Thus, the heat current density removed from the emitter junction in terms of the electrical current density through the device is:

$$J_Q = \left(-\mu + \frac{2kT_E}{q}\right) J_E + \frac{2k}{q} J_E \Delta T - \frac{2k}{q} \left(A T_E^2\right) \exp\left(\frac{q\mu}{kT_E}\right) \Delta T$$
$$\equiv \Pi J_E + \frac{2k}{q} J_E \Delta T - \kappa_e \Delta T. \quad (8)$$

The first term is the Peltier effect where Π is the Peltier coefficient, the second looks like a Thomson effect, and the third is the electronic contribution to the thermal conductivity, κ_e. ΔT is the temperature difference between the emitter and collector and is positive if the collector is at a higher temperature. The close correspondence between equation 8 and conventional thermoelectric theory is reassuring.

TOTAL HEAT TRANSPORT

To examine a thermionic emission cooler, lattice effects are added directly into equation 8. In the ballistic limit, only lattice thermal conduction must be considered and the heat current density is:

$$J_Q = \Pi J_E + \frac{2k}{q} J_E \Delta T - \left(\kappa_e + \frac{\kappa}{d}\right) \Delta T, \quad (9)$$

where κ is the lattice thermal conductivity of the barrier and d is the barrier width. In the diffusive limit joule heating must also be considered. The solution to the one dimensional heat transport equation for electrical current through a resistive material with a temperature difference between the ends shows that half of the joule heat returns to the cold end [6]. Thus, in the diffusive limit the heat current density is:

$$J_Q = \Pi J_E + \frac{2k}{q} J_E \Delta T - \left(\kappa_e + \frac{\kappa}{d}\right) \Delta T - \frac{1}{2} J_E^2 \frac{d}{\sigma}, \qquad (10)$$

where σ is the electrical conductivity of the barrier. Equation 10 is exactly the equation for heat transport through bulk thermoelectric devices. This shows that a thermionic emission cooler in the diffusive limit will behave the same as a bulk thermoelectric device [1]. This provides a simple way of comparing a thermionic structure to a thermoelectric structure of the same material.

DEVICE OPTIMIZATION

We examine the cooling ability of a thermionic emission cooler by optimizing the maximum temperature difference that can be achieved. We consider the full series for the electrical and heat current densities and solve numerically. We use parameters for an InGaAs barrier because the InP-InGaAs family is an important semiconductor family in which many devices would benefit from integrated cooling. The mobility is taken to be 8000 cm^2-V/sec. The lattice thermal conductivity is taken to be 5 W/K-cm, which corresponds to the bulk value. There are two variables that can be optimized: the chemical potential and the barrier width.

Figure 2 shows the maximum temperature difference achievable as a function of the chemical potential for ballistic transport across a .2 micron barrier. The dashed line assumes saturation of the electrical current. The solid line is determined by limiting the electrical current density to be less than or equal to 100kA/cm^2. This limit is chosen because above this current density, joule heating in the contacts will be a significant issue [7] and electromigration may also threaten device stability [8]. With this self-imposed limit, the optimum chemical potential is approximately (-1.5kT/q).

Figure 3 shows the maximum temperature difference as a function of the barrier width in both the ballistic and diffusive limits for a chemical potential of (-1.5kT/q). Clearly there is an advantage to the cooling for the ballistic limit if relatively large barrier widths can be used. Using a mobility of 8000 cm^2-V/sec, the mean free path is only .1 micron. In this purely ballistic limit, no more than 2 degrees of cooling is possible, whereas a thick device in the diffusive regime could provide about 4.5 degress of cooling. The conclusion of this is two-fold. First, a single barrier ballistic InGaAs thermionic device will not be able to achieve better cooling than a standard InGaAs thermoelectric device. Second, between the ballistic and diffusive limits, it may be possible to achieve cooling greater than either limit. Because InGaAs is a good representation of the important thermoelectric parameters for ternary III-V materials, we expect similar results for other III-V material systems.

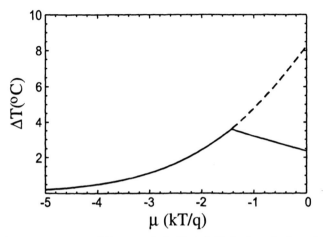

Figure 2. Maximum temperature difference of a ballistic InGaAs thermionic emission cooler as a function of the chemical potential for unlimited current (dotted line) and current limited to $100kA/cm^2$ (solid line).

CONCLUSIONS

We provide analytical solutions for the electrical and heat current densities over a heterojunction barrier that allow for fast numerical calculations for any chemical potential. The Peltier coefficient for a semiconductor – semiconductor heterojunction boundary must be defined

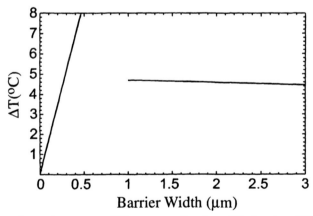

Figure 3. The maximum temperature difference as a function of the barrier width for an InGaAs device with $\mu = -1.5kT/q$, in both the ballistic limit (on the left) and diffusive limit (on the right).

using the chemical potential in equation 5 and shown in figure 1. The optimum chemical potential for thermionic emission cooling should be determined from a physically reasonable electrical saturation current density. Maximum cooling may occur between the ballistic and diffusive limits (see figure 3 for barrier widths from .5µm to 1µm) of a thermionic device and more investigation in this area is necessary.

ACKNOWLEDGEMENTS

This work was supported by the Army Research Office and monitored by Dr. Jack Rowe.

REFERENCES

1. G. D. Mahan and L. M. Woods, Phys. Rev. Lett. **80**, 4016, 1998.
2. A. Shakouri and J. E. Bowers, Appl. Phys. Lett. **71**, 1234, 1997.
3. N. W. Ashcroft and N. D. Mermin, *Solid State Physics*, (Harcourt Brace College Publishers, Fort Worth, 1976) pp. 253-254.
4. A. T. Fromhold, Jr., Quantum mechanics for applied physics and engineering, (Dover Publications, Inc., New York, 1981) pp.213-217.
5. C. M. Wu and E. S. Yang, Solid-State Electron. **22**, pp 241, 1979.
6. R. W. Ure, Jr., and R. R. Heikes, Theoretical calculation of device performance, *Thermoelectricity: science and engineering*, ed. R. R. Heikes and R. W. Ure, Jr. (Interscience Publishers, New York, 1961) pp. 458-517.
7. G. Y. Robinson, Schottky diodes and ohmic contacts for the III-V semiconductors, *Physics and chemistry of III-V compound semiconductor interfaces*, ed. C. W. Wilmsen (Plenum Press, New York, 1985) pp. 73-164.
8. *Electromigration and electronic device degradation*, ed. A. Christou (Wiley, New York, 1993).

Umklapp Scattering and Heat Conductivity of Superlattices

M.V. Simkin and G.D. Mahan
Department of Physics and Astronomy, University of Tennessee,
Knoxville, 37996-1200, and
Solid State Division, Oak Ridge National Laboratory,
P.O. Box 2008, Oak Ridge, TN, 37831

Abstract

The mean free path of phonons in superlattices is estimated. It is shown to be strongly dependent on the superlattice period due to the Umklapp scattering in subbands. It first falls with increasing the superlattice period until it becomes comparable with the latter after what it rises back to the bulk value. Similar behavior is expected of heat conductivity, which is proportional to the mean free path.

Superlattices offer an opportunity to control physical properties in unprecedented ways. Their thermal conductivity is of interest both for a fundamental understanding of these systems as well as in applications. Recently there has been a resurgence of interest in finding materials with improved thermoelectric transport properties for cooling and power generation. The quality of a material for such applications is given by the thermoelectric figure of merit, which is inversely proportional to the thermal conductivity κ. In materials of interest, such as semiconductors, the lattice contribution to κ dominates.

Experimental and theoretical works suggests that the thermal conductivity of superlattices is quite low, both for transport along the planes [1, 2, 10], or perpendicular to the planes [3, 4, 5, 6, 7, 8, 11].

The lattice heat conductivity κ is given approximately by an equation [12]:

$$\kappa \approx Cvl, \qquad (1)$$

where C is the lattice heat capacitance, v - the average phonon group velocity, and l - the mean free path. Recently we presented calculations of the thermal conductivity perpendicular to the layers [11] which were done in approximation which takes into account changes in phonon group velocities due to band folding, but neglects the dependence of the phonon mean free path on the superlattice period. The investigation of this dependence is the subject of the present work.

Three-phonon scattering due to anharmonicity is the dominant contribution to the lattice thermal resistivity. Umklapp processes, in which the net phonon momentum change by a reciprocal lattice vector, give the finite thermal conductivity [12]. Only phonons with energies of the order of Debye energy, Θ_D can participate in Umklapp scattering, giving a temperature dependence of the phonon mean free path l of the form [12]

$$l \approx \exp(\Theta_D/T). \qquad (2)$$

In superlattices, new mini-bands are introduced in the acoustic phonon dispersions along the growth direction, and they give rise to new Umklapp processes. The lowest phonon energy for Umklapp scattering in a superlattice of period L is of order Θ_D/L and phonon mean free path in a superlattice, L_{sl}, becomes:

$$l_{sl} \approx \exp(-\Theta_D/LT). \qquad (3)$$

The ratio of the mean free paths is:

$$l_{sl}/l \approx \exp(\Theta_D/T(1/L - 1)), \qquad (4)$$

which can be rather small for big L and small T.

According to Eq.4 l_{sl} decreases with L and eventually should become $l_{sl} = L$. This shall happen at the value of $L = L_c$ given by a solution to the equation:

$$L_c/l = \exp(\Theta_D/T(1/L_c - 1)) \tag{5}$$

What happens next? When L exceeds L_c according to Eq.4 it should become $l_{sl} < L$. However in this case Eq.4 is no longer applicable because when $l_{sl} < L$ superlattice effects should not matter and l_{sl} should assume the bulk value l. But as $L < l$ then Eq.4 should valid again. The only resolution of this contradiction is that l_{sl} starts to increase with L as $l_{sl} \approx L$ after reaching a minimum at $L = L_c$. It shall saturate, however when $L > l$ (in this case we are not bound to use Eq. 4 again ($L < l$!).

Eq.5 is not soluble analytically but assymptotics are easy to compute. For large l we get:

$$L_c/l = \exp(-\Theta_D/T), \tag{6}$$

which can be orders of magnitude small. When $l = 1$ we get $L_c = 1$.

The heat conductivity is proportional to l_{sl} (Eq.1) and should follow its behavior.

References

[1] T. Yao, Appl. Phys. Lett. **51**, 1798 (1987)

[2] P. Hyldgaard and G.D. Mahan, Proc.Int. Conf. on Thermal Conductivity, Nashville,(November, 1995)

[3] W.S. Capinski and H.J. Maris, Physica B **219-220**, 699 (1996)

[4] S.M. Lee, D.G. Cahill, R. Ventakasubramanian, Appl. Phys. Lett. **70**, 2957 (1997)

[5] R. Venkatasubramanian and T. Colpitts, in *Thermoelectric Materials–New Directions and Approaches*, ed. T.M. Tritt, M.G. Kanatzidis, H.B. Lyons Jr, and G.D. Mahan (Materials Research Society, 1997) Vol. 478, pg. 73; R. Ventakasubramanian, Phys. Rev. B **61**, 3091 (2000)

[6] G. Chen and M. Neagu, Appl. Phys. Lett. **71**, 2761 (1997)

[7] P. Hyldgaard and G.D. Mahan, Phys. Rev. B **36**, 10754 (1997)

[8] S. Tamura, Y. Tanaka, and H.J. Maris, Phys. Rev. B **60**, 2627 (1999)

[9] G. Chen, Phys. Rev. B **57**, 14958 (1998)

[10] S. G. Walkauskas, D. A. Broido, K. Kempa, and T. L. Reinecke, J. Appl. Phys **85**, 2579 (1999)

[11] M. V. Simkin and G. D. Mahan, Phys. Rev. Lett. **84**, 927 (2000).

[12] J. Ziman *Electrons and Phonons* (Cambridge University Press, 1960)

Skutterudites II

PARTIALLY-FILLED SKUTTERUDITES: OPTIMIZING THE THERMOELECTRIC PROPERTIES

[1]G.S. Nolas, [2]M. Kaeser, [2]R.T. Littleton IV, [2]T.M. Tritt, [3]H. Sellinschegg, [3]D.C. Johnson and [4]E. Nelson.
[1]R&D Division, Marlow Industries, Inc., 10451 Vista Park Road, Dallas, Texas 75238
[2]Department of Physics, Clemson University, Clemson, South Carolina 29634
[3]Department of Chemistry, University of Oregon, Eugene, Oregon 97403
[4]US Army Research Laboratory, Adelphi, Maryland 20783

ABSTRACT

The skutterudite family of compounds continues to be of interest for thermoelectric applications due to the low thermal conductivity obtained when filling the voids with small diameter, large mass interstitials such as trivalent rare-earth ions. In the last few years there has been a substantial experimental and theoretical effort in attempting to understand the transport properties of these compounds in order to optimize their thermoelectric properties. One such approach involves partially-filling the voids in attempting to optimize the power factor while maintaining low thermal conductivity. In this report experimental research on skutterudites with the voids partially filled with heavy mass lanthanide and alkaline-earth ions is reported.

INTRODUCTION

The continuing effort in improving the thermoelectric properties of skutterudite compounds has resulted in attempts to fill the voids in the crystal structure with ever differing atoms.[1] To this end new synthesis approaches are also underway in order to form ever more varied compounds in this diverse materials system. Novel approaches to skutterudite compound synthesis[2, 3] that can result in the preparation of many compounds that could not be successfully formed employing "traditional" synthesis techniques are also of interest. Whether employing these novel synthesis approaches or traditional ones, an approach that has been reported to be an optimization route is partial void filling.[4,5,6] The goal in this research is in obtaining compounds with low thermal conductivity while maximizing the electronic properties.

THERMOELECTRIC POTENTIAL

Figure 1 illustrates the maximum dimensionless figure of merit, $ZT=TS^2\sigma/\kappa$ (where S is the Seebeck coefficient, σ is the electrical conductivity and κ the thermal conductivity) for skutterudite compounds at room temperature. The plot was constructed by gathering and plotting experimental data of $CoSb_3$ at different doping levels from the literature[1] and inserting different values for the lattice thermal conductivity, κ_g. The different optimum carrier concentrations can be directly attributed to the difference in the effective masses of n (m_e) and p-type (m_p) specimens ($m_e \sim 10m_p$). The n-type material is therefore optimized at a higher carrier concentration. As seen in this figure it is clear that $\kappa \sim \kappa_g$ in order to obtain skutterudites with a higher ZT than that of Bi_2Te_3-alloy materials presently used in thermoelectric devices (with ZT ~ 1). One important note is that the ZT values at higher temperatures are higher than those shown in Figure 1, implying that optimization is more readily obtained at higher temperatures.

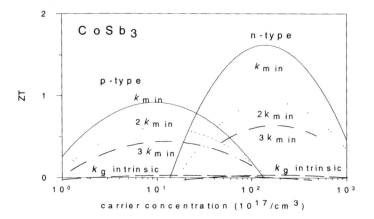

Figure 1. ZT vs. carrier concentration of n and p-type CoSb$_3$ at room temperature assuming an optimized power factor.

VOID FILLERS

Many different atoms have been introduced into the voids of skutterudites, including group-IV elements,[2,3] group-III elements,[2,6] alkali earth, lanthanide and actinide ions.[1] The chemical formula for a unit cell can be written as $\square_2X_8Y_{24}$ (typically X = Co, Rh or Ir and Y = P, As or Sb) illustrating the two voids per unit cell. These range from 1.763 Å for $\square_2Co_8P_{24}$ to 2.040 for Å for $\square_2Ir_8Sb_{24}$.[7] It is imperative that one take into account the void radius of the particular compound under investigation. Skutterudite antimonides possess the largest voids and are therefore of particular interest for thermoelectric applications. A partial list of different possible void fillers as well as an estimate of their ionic radii are listed in Table I. If we place smaller sized atoms in the voids, such as trivalent lanthinide ions in $\square Ir_4Sb_{12}$ (with the chemical formula indicating one-half the cubic unit cell) for example,[7] these rattle about in the larger diameter voids. More loosely bound rattlers produce local vibrational modes of lower frequency and are thus more effective in scattering the lower-frequency, heat-carrying phonons. The smaller the ion in the $\square Ir_4Sb_{12}$ voids, the larger the disorder that is produced and therefore the larger the reduction in κ_g. This concept is corroborated by the large atomic displacement parameters (ADPs) that have been observed in alkaline earth and lanthanide filled skutterudites.[8,9]

Table I. The average ionic radii, in Å, for ions in 12 coordinated, CN, sites taken from oxide and fluoride structural data. Note trend in ion sizes.

ion	Sr	Eu	La	Ce	Gd	Nd	Yb	Lu
charge	+2	+2	+3	+3	+3	+3	+3	+3
12 CN	1.58	~1.5	1.46	1.43	~1.4	1.40	~1.3	~1.3

RESULTS AND DISCUSSION

The specimens were prepared as described in References 4 and 7. X-ray diffraction, metallographic and electron-beam microprobe analyses were employed in structural and chemical characterization. Table II lists the physical properties of the specimens prepared for this report. Four-probe electrical resistivity (ρ), steady-state S and steady-state κ measurements were performed in a radiation-shielded vacuum probe inside a closed-cycle refrigerator. The κ_g values were estimated by subtracting the electronic contribution, κ_e, using the Weidemann-Franz relation ($\kappa_g = \kappa - \kappa_e$, with $\kappa_e = L_0 T/\rho$ where L_0 is the Lorentz constant).

Even a small concentration of interstitial void filler results in a large reduction in κ_g, as has been reported previously.[4,5,6] However the smaller and heavier the filler ion the lower the κ_g values. This is illustrated in Figure 2 where only a small amount of Gd or Yb in the voids of $\square Co_4Sb_{12}$, with a void radius of 1.892 Å,[7] produces a large reduction in κ_g. In particular the room temperature κ_g values are lower than that of the 5 % La-filled specimen. Data for the La-filled skutterudites with varying La concentration were taken from Reference 4 and are also shown in Figure 2. Of particular note is $La_{0.22}Gd_{0.02}Co_4Sb_{12}$ which possess a lower thermal conductivity than that of $La_{0.23}Co_4Sb_{12}$. It is evidence that only 2 % Gd filling, in addition to the La-filling, produces a relatively large affect on κ_g. The addition of two different void-fillers results in an additional phonon scattering mechanism due to the mass difference between the La, Gd and voids in this skutterudite. Such an affect is greatest at higher temperatures, as shown in Figure 2.

An interesting aspect of partial void filling is illustrated in Figure 3 where S is plotted for three of the skutterudites shown in Figure 2. The S values decrease with decreasing temperature in these n-type specimens and the larger the rare-earth filling fraction the lower the S values. The higher S values the higher the ρ values (Table II). These compounds illustrate properties indicative of semiconductors and may indicate that these n-type specimens display rigid band behavior where the lanthanide ions act as donor impurities. The electronic transport properties of these partially filled skutterudites are similar to that of Pt doped $CoSb_3$, for example.[10] These results illustrate the optimization approach for n-type skutterudites compounds.

Table II. Properties of the phase-pure specimens prepared for this report.

Specimen	density	a_0 (Å)	ρ (mohm-cm)	S (microV/cm)
$Br_{0.064}Co_4Sb_{12}$	95 %	9.043	0.27	7.3
$Gd_{0.021}Co_4Sb_{12}$	92 %	9.041	3.21	-274
$Yb_{0.066}Co_4Sb_{12}$	94 %	9.043	1.32	-229
$La_{0.22}Gd_{0.02}Co_4Sb_{12}$	92 %	9.073	0.35	-91
$Eu_{0.15}Co_4Sb_{12}$	93 %	9.053	1.09	-181
$Eu_{0.32}La_{0.02}Co_4Sb_{12}$	93 %	9.070	0.34	-91
$Sr_{0.32}Co_4Sb_{12}$	94 %	9.079	0.30	-84
$Yb_{0.19}Co_4Sb_{12}$	95 %	9.051	0.64	-146

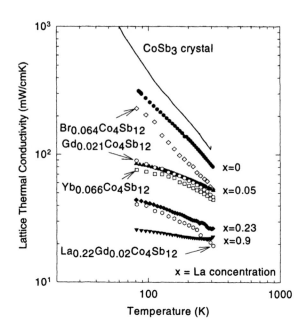

Figure 2. Lattice thermal conductivity of four partially filled skutterudites. Also shown are $La_xCo_4Sb_{12}$ for x=0, 0.5 and 0.23, $La_{0.9}Co_4Sn_3Sb_9$ and single crystal $CoSb_3$.

Figure 4 shows S, ρ and κ_g from 300 to 10 K for four partially filled skutterudites. The κ_g values are dependent on the filler-concentration and filler-ion, with the Yb-filled specimen showing the lowest κ_g values in this temperature range. The S values of these n-type compounds decrease with increasing temperature, typical of semiconductor behavior, with amplitudes that are dependent on the doping level. The ρ values also increase with increasing temperature indicative of metallic or heavily doped semiconductor behavior. In all cases an upturn in the ρ data is observed at the very lowest temperatures.

The data for the $Yb_{0.19}Co_4Sb_{12}$ compound is most interesting. It has been previously reported that Yb in $YbFe_4Sb_{12}$ possess intermediate valence.[11,12, 13] Magnetic susceptibility measurements on our $Yb_{0.19}Co_4Sb_{12}$ specimen also indicate that Yb possess intermediate valence between 2+ and 3+. This is of interest for thermoelectric applications since more Yb can be interstitially placed into the voids of the skutterudite structure, as compared to other trivalent lanthanide ions, while maintaining a similar carrier concentration. That is, a larger filling fraction in Yb-filled skutterudites as compared to Ce-filled skutterudites, which are trivalent in skutterudites above cryogenic temperatures,[14] will result in a similar power factor while possessing substantially lower κ_g, thus further optimizing ZT.

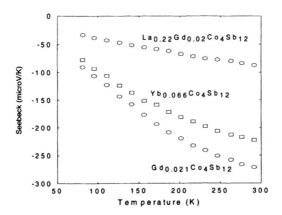

Figure 3. Seebeck Coefficient vs. temperature for three n-type partially filled skutterudites

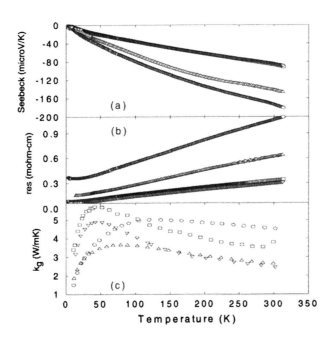

Figure 4. S, ρ and κ_g of $Eu_{0.15}Co_4Sb_{12}$ (circles), $Eu_{0.32}La_{0.02}Co_4Sb_{12}$ (squares), $Sr_{0.32}Co_4Sb_{12}$ (down rectangles) and $Yb_{0.19}Co_4Sb_{12}$ (up rectangles).

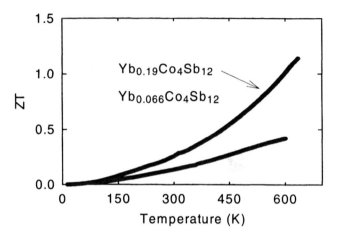

Figure 5. Figure of merit of the two Yb-filled skutterudites. The S and ρ used to calculate ZT are measured data, the κ_g data above room temperature was extrapolated from the measured data assuming a similar temperature dependence out to 600K.

Figure 5 shows ZT as a function of temperature for the two Yb-filled skutterudites. The $Yb_{0.19}Co_4Sb_{12}$ specimen has ZT = 0.3 at room temperature and ZT > 1 at T > 300 C, as reported previously. [15] In order to estimate ZT, S and ρ where measured to the temperatures shown in Figure 5 while the low temperature κ_g data was extrapolated above room temperature assuming a similar temperature dependence as from the peak to 300 K (see Figure 4). The room temperature power factor ($S^2\sigma$) for these skutterudites is the highest of all known thermoelectric materials, with the exception of Bi_2Te_3 alloy materials and optimized pentatellurides. [16] This relatively high ZT is a result of partial void filling, the phonon-electron scattering from the large effective mass of Yb-filled skutterudites [11,12,13] and the intermediate valence state of Yb in $CoSb_3$. Filling the voids with Yb allows for more interstitial Yb in voids before degradation of $S^2\sigma$, as compared to trivalent lanthanide ions, while also having lower κ values than in La or Ce-filled skutterudites. The large effective mass also allows for relatively high $S^2\sigma$ values for this highly doped compound. More research into Yb filling as well as bulk samples with other, heavier and smaller ions in a continuing effort to further optimize skutterudites for thermoelectric applications are currently under investigation by the authors. [17]

ACKNOWLEDGMENT

The authors are grateful for support from the U.S. Army Research Laboratory under contract number DAAD17-99-C-0006. GSN acknowledges useful conversations with B.C. Sales and T. Caillat.

REFERENCES
[1] G.S. Nolas, D.T. Morelli and T.M. Tritt, Annu. Rev. Mater. Sci. **29**, 89 (1999), and references therein.
[2] M.D. Hornbostel, E.J. Hyer, J. Thiel, J.H. Edvalson, and D.C. Johnson, Inorg. Chem. **36**, 4270 (1997), M.D. Hornbostel, E.J. Hyer, J. Thiel, and D.C. Johnson, Jour. Am. Chem Soc. **119**, 2665 5. (1997), and H. Sellinschegg, S.L. Stuckmeyer, J.D. Hornbostel, and D.C. Johnson, Chem. Mater. **10**, 1096 (1998)
[3] H. Takizawa, K. Miura, M. Ito, T. Suzuki and T. Endo, J. Alloys Comp **282**, 79 (1999).
[4] G.S. Nolas, J.L. Cohn and G.A. Slack, Phys Rev B **58**, 164 (1998).
[5] D.T. Morelli, G.P. Meisner, B. Chem, S. Hu and C. Uher, Phys. Rev. B **56**, 7376 (1997); G.P. Meiner, D.T. Morelli, s. Hu, J. Yang and C. Uher, Phys. Rev. Lett. **80**, 3551 (1998).
[6] B.C. Sales, B.C. Chakoumakos, D. Mandrus, Phys Rev B **61**, 2475 (2000).
[7] G.S. Nolas, G.A. Slack, D.T. Morelli, T.M. Tritt and A.C. Ehrlich, J. Appl. Phys. **79**, 4002 (1996).
[8] C.B.H. Evers, W. Jeitschko, L. Boonk, D.J. Braun, T. Ebel and U.D. Scholz, J. Alloys Comp. **224**, 184 (1995), and references therein.
[9] B.C. Chakoumakos, B.C. Sales, D. Mandrus and V. Keppens, Acta. Cryst. **B55**, 341, (1999), and references therein.
[10] T. Caillat, A. Borshchevsky and J.-P. Fleurial, J. Appl. Phys. **80**, 4442 (1996).
[11] N.R. Dilley, E.J. Freeman, E.D. Bauer and M.B. Maple, Phys. Rev. B **58**, 6287 (1998).
[12] A. Leithe-Jasper, D. Kaczorowski, P. Rogl, J. Bogner, M. Reissner, W. Steiner, G. Wiesinger and C. Godart, Solid State Comm. **109**, 395 (1999).
[13] N.R. Dilley, E.D. Bauer, M.B. Maple, S. Dordevic, D.N. Basov, F. Freibert, T.W. Darling, A. Migliori, B.C. Chakoumakos and B.C. Sales
[14] B. Chen, J. Xu, C. Uher, D.T. Morelli, G.P. Meisner, J.-P. Fleurial, T. Caillat and A. Borshchevsky, Phys. Rev. B **55**, 1476 (1997).
[15] M. A. Kaeser, T.M. Tritt, G. S. Nolas, R T. Littleton, IV, P. Alboni and A. L. Pope, Bull. Am. Phys. Soc. **44**, 48 (1999).
[16] R. T. Littleton, IV, T. M. Tritt, J. W. Kolis, and D. Ketchum, Phys. Rev. B **60**, 13453 (1999).
[17] see for example H. Sellingschegg, D.C. Johnson, G.S. Nolas, G.A. Slack, S.B. Schujman, F. Mohammed, T.M. Tritt and E. Nelson, Proceedings of the 17th International Conference on Thermoelectrics (IEEE, Catalog No. 98TH8365, Piscataway, NJ, 1998) p. 338, and other articles in this volume by the authors.

Bulk Synthesis of Completely and Partially Sn filled CoSb3 Using the Multilayer Repeat Method

Heike Sellinschegg and David C. Johnson[1]
[1]University of Oregon, Dept. of Chemistry and Materials Science Institute, Eugene, OR 97403

Michael Kaeser, Terry M. Tritt
Clemson University, Department of Physics, Clemson, SC 29634

George S. Nolas
R & D Division, Marlow Industries, Dallas, TX 75238

E. Nelson
U.S. Army Research Laboratory, Adelphi, ML 20783

ABSTRACT

Filled skutterudite compounds possess very low thermal conductivities due to the scattering of a wide range of phonon modes caused by a loosely bound cation incorporated in a cavity of the structure. The inclusion of such a filler cation causes several synthetic difficulties since the desired compounds are thermodynamically unstable with respect to disproportionation. Modulated elemental reactants were used in this study to circumvent these difficulties. $Sn_xCo_4Sb_{12}$ samples with x=0.5 and nearly 1.0 were synthesized using this method. To prevent nucleation of unwanted binary compounds, the repeat unit made up of elemental layers was less than 20 angstroms 500mg of each sample were produced, allowing for the samples to be hot pressed into a pellet. Structural analysis as well as measurements of the physical properties are presented.

INTRODUCTION

In the development of new materials with higher figures of merit much research has been focused on compounds with a large, complex unit cell and a misfit between a particular lattice site and a so-called filler - cation. The idea behind this concept, originally proposed by Slack [1], is to maintain good electrical conductivity by making the elements with states at the Fermi level part of a rigid covalent network while suppressing the lattice thermal conductivity by the random thermal vibrations of the filler cation which effectively scatters lattice phonons.

One class of compounds satisfying Slack's criteria that has sparked significant interest is the skutterudite compounds. The skutterudites are materials with a complex crystal structure with 32 atoms per cubic unit cell and have shown potential to exhibit extremely promising thermoelectric properties. Many different skutterudite compounds with different filler cations have been synthesized, however, synthetic problems have so far prevented the preparation of some of the targeted materials.

The synthesis of many of these materials is extremely challenging. The weak interactions between the filler cation and the lattice site causes many of the ternary skutterudite compounds to be thermodynamically unstable with respect to decomposition into binary compounds. The filling fraction of cations in the host lattice is usually quite limited, as for example in the case of Ce insertion into a Co_4Sb_{12} host lattice where the maximum filling is reported at 10% [2]. One compound of interest has been the Sn filled $CoSb_3$ structure. Takizawa et al have successfully demonstrated that high pressure conditions can result in skutterudites with a very high concentration of ternary cations [3]. The authors prepared these compounds by mixing fully crystallized $CoSb_3$ powder with tin powder in appropriate stoichiometric amounts and pressing the sample under a pressure of 5 GPa at a temperature of about 600C [3]. Skutterudite phases showing no contamination in their high angle x-ray pattern were obtained.

An alternate route to synthesizing $Sn_xCo_4Sb_{12}$ compounds is using modulated elemental reactants [4]. This technique is based on the design of reactants consisting of modulated elemental layers which are deposited sequentially. The thicknesses of the individual layers are chosen to yield the desired sample composition and are kept on the order of angstroms or tenths of angstroms to minimize solid state diffusion distances. The individual elements interdiffuse upon annealing at very low temperatures to form an amorphous intermediate at the desired composition. On further annealing at 250C, nucleation of the skutterudite structure occurs. Via this route, compounds that are thermodynamically unstable with respect to disproportionation, such as many of the skutterudites, can be synthesized as x-ray pure powders. The advantage of this technique is that it gives access to metastable compounds which are difficult to make otherwise. Some compounds such as a binary $FeSb_3$ compound have not been synthesized in any other way. While much of the initial synthesis was done in thin film form, our recent efforts have focused on scaling up this approach to make bulk samples. Pressed pellets are desired for the measurement of physical properties (Seebeck coefficient, electrical and thermal conductivity) because they approximate the form of the compounds which would be used in commercial devices.

In this paper we discuss the use of the multilayer repeat technique to prepare two bulk $Sn_xCo_4Sb_{12}$ samples where x is 0.5 and close to 1.0. We present structural data obtained via Rietveld refinements on these samples. The electrical and thermal conductivity as well as Seebeck coefficients will be reported and compared to the samples prepared using the high pressure method.

EXPERIMENT

Tin and cobalt are evaporated from Thermionics electron beam gun sources in a custom built high vacuum chamber. Antimony is deposited from a Knudsen cell. All elements are deposited at a rate of 0.3 angstrom per second - controlled by Leybold Inficon quartz crystal thickness monitors. The deposition and layer thicknesses are entered into a computer program that then executes the desired deposition sequence. The modulated elemental reactants are deposited simultaneously on an 8 inch diameter silicon substrate which has previously been coated with polymethylmethacralate (PMMA) and a zero background quartz plate. The multilayer repeat thickness (which includes one layer of each, Sb, Co and Sn) is approximately 16 Å; and about 2000 repeat layers are deposited. After deposition, the silicon wafer is removed from the chamber and annealed under inert atmosphere to 250C for one hour to interdiffuse the layers and crystallize the desired skutterudite. After annealing, the wafer is submerged in acetone dissolving the PMMA and separating the sample from the substrate. The suspended sample is collected on a Teflon filter paper, ground and then hot pressed into a pellet at a temperature of 400C and a pressure of 65,000 lbs/in^2. The off cut quartz plate is used for x-ray analysis. The as deposited sample is x-rayed using a Philips X'Pert MPD diffractometer. Low and high angle data are collected before the sample film on quartz is annealed in a vacuum furnace to 250C for 14 hours. High angle x-ray diffraction data is collected before and after the sample film on quartz is annealed for 14 hours in a vacuum furnace. High angle x-ray diffraction data is collected on the annealed film as well as on the finished pellet to establish that no contaminations were introduced into the sample pellet. Rietveld refinement [5] is used to determine the precise crystal structure. The composition of the sample is measured using Electron Probe Micro Analysis (EPMA). A portion of the free standing sample collected on the Teflon filter is pressed onto a conductive carbon tape and analyzed using a Cameca S-50 instrument with a setting of 10keV for accelerating voltage and 10nA beam current. The spot size was about 1 micrometer.

Differential Scanning Calorimetry (DSC) experiments are conducted to monitor the evolution of the sample as it is subjected to increasing temperature. A TA9000 calorimeter was used and approximately 1 mg of sample was put in an Al pan which was heated to 550C at a rate of 10 degrees per minute. The net heat flow associated with the irreversible changes occurring in the sample during this heating cycle were recorded.

Scanning Electron Microscopy (SEM) was used to inspect the pressed sample pellets. A basic JSM - 6300 system was used to collect data from a piece of the pellet cast in epoxy, polished and coated with carbon.

RESULTS AND DISCUSSION

Two tin - filled CoSb$_3$ samples with a composition corresponding to a 50% and a 100% filling were synthesized as shown in Table I.

Table I. Comparison of Samples.

	Sample 1	Sample 2
Lattice Parameter	9.114	9.083
Intended Sn – filling	1	0.5
Measured Sn - filling (EMPA)	0.98	0.58
Intended Multilayer Thickness	16.0	15.9
Measured Multilayer Thickness (x-ray)	16.8	16.5

Low angle diffraction of the as deposited sample as shown in Figure 1 confirms the layered structure of the starting materials. The Bragg peak at 5.3° 2θ corresponds to a repeat thickness of about 16.5 angstrom. Front to back reflections indicative of the total layer thickness are visible as well in Figure 1 up to approximately 5° 2θ, implying the total roughness of the film is on the order of ± 9 angstrom.

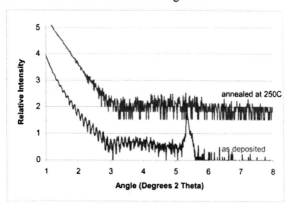

Figure 1. Low angle diffraction pattern of the $Sn_1Co_4Sb_{12}$ sample before and after annealing past the nucleation temperature. The Bragg peak at 5.3° 2θ in the as deposited scan shows that layering is present. The data collected after annealing shows no evidence of layering anymore.

Figure 2 contains a typical DSC scan of a Sn-Co-Sb sample. The sharp exotherm at 142°C signifies an irreversible change taking place as the sample is annealed. After annealing the sample past the nucleation temperature (as seen from the exotherm in the DSC scan) the Bragg peak disappears, denoting that the sample is no longer layered.

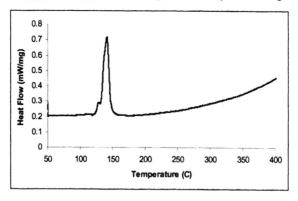

Figure 2. Differential scanning calorimetry collected for a Sn-Co-Sb sample. The exotherm at 142°C indicates an exothermic reaction taking place in the sample.

Figure 3 shows two high angle diffraction patterns – one collected before and one after heating past the exotherm. The data collected after heating past the exotherm shows that a crystalline skutterudite has formed. The x-ray pattern shows no contamination due to binary products or unreacted starting materials.

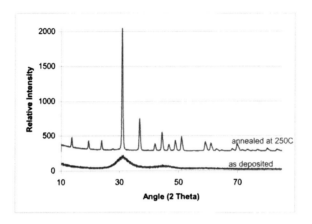

Figure 3. High angle x-ray diffraction data obtained before and after annealing the sample at 250C. The pattern of the sample after annealing is that of a crystalline skutterudite.

Rietveld refinement shows that the lattice parameter of both tin filled samples are larger than for the unfilled $CoSb_3$ structure. The lattice parameter of the sample with the higher tin content was larger than that of the sample with lower tin content, agreeing with previous studies examining the trend of lattice parameters with occupancy of the filler cation. This lattice expansion has been suggested to result from the thermal vibration of the incorporated atom.[6] The lattice parameters of the samples prepared in this study are similar to those reported by Takizawa et al for samples prepared using the high pressure technique.

Measurement of the thermoelectric properties of the bulk materials are currently in progress.

SUMMARY

The modulated elemental precursor techniques gives access new, metastable skutterudite compounds. Upscaling this synthesis method and producing bulk samples allows us to measure the physical properties and to compare them to compounds synthesized with different methods.

ACKNOWLEDGMENTS

Support of this research by the Office of Naval Research and the Defense Advanced Research Projects Agency through grant N00014-98-1-0447 and the support from the U.S. Army Research Laboratory under contract number DAAD17-99-C-0006 is greatly appreciated.

REFERENCES

[1] G. A. Slack, **CRC Handbook of Thermoelectrics**, ed. D. M. Rowe, CRC Press: Boca Raton, FL, (1995) pp 407-440.
[2] D. T. Morelli, G. P. Meisner, B. Chen, S. Hu, C. Uher, *Phys Rev B*, **56**, 7376 (1997).
[3] H. Takizawa, K. Miura, M. Ito, T. Suzuki, T. Endo, *Journal of Alloys and Compounds*, **282**, 79-83 (1999).
[4] M. D. Hornbostel, E. J. Hyer, J. Thiel and D. C. Johnson, *J. Am. Chem. Soc.*, **119** 2665-2668 (1997).
[5] "DBWS – 9411 An Upgrade for the DBWS Programs for Rietveld Refinement with PC and Mainframe Computers" *J. Appl. Cryst.*, **28**, 366-7 (1995).
[6] D. J. Braun, W. Jeitschko, *J. Less -Common Metals*, **72**, 147 (1980).

THE INFLUENCE OF Ni ON THE TRANSPORT PROPERTIES OF $CoSb_3$

CTIRAD UHER[1], JEFFREY S. DYCK[1], WEI CHEN[1], GREGORY P. MEISNER[2], JIHUI YANG[1,2]
[1]Department of Physics, University of Michigan, Ann Arbor, MI 48109, cuher@umich.edu
[2]Materials and Processes Laboratory, GM R&D and Planning, Warren, MI 48090

ABSTRACT

The effect of Ni doping on the Co site of the binary skutterudite $CoSb_3$ is investigated. We measured resistivity, Hall effect, magnetoresistance, thermopower, thermal conductivity, and magnetization of a series of samples of the form $Co_{1-x}Ni_xSb_3$ with x in the range x=0 to x=0.01. We find that Ni takes the tetravalent state Ni^{4+}, assumes the d^6 electronic configuration for the lower energy non-bonding orbitals, and gives an electron to the conduction band. Ni doping dramatically suppresses the thermal conductivity, changes the temperature dependence of the thermopower, and increases the carrier concentration. Low temperature anomalies in thermopower, Hall coefficient and magnetoresistance are found.

INTRODUCTION

Skutterudites have received considerable attention over the past several years as prospective novel thermoelectric materials [1,2]. The name skutterudite refers to the compounds designated as MX_3, where M represents Co, Rh, or Ir and X stands for a pnicogen atom such as P, As, or Sb. The skutterudite structure is characterized by two large voids (or cages) within the unit cell and these can accommodate certain rare earth elements [3], alkaline earths [4] or even monovalent thallium [5]. When the voids are filled the structure is referred to as a filled skutterudite. In this case, the appropriate chemical formula is RT_4X_{12}, where R stands for the electropositive filler atom and T represents a transition metal of the group eight elements (Fe, Ru, or Os), i.e., the elements one column to the left of the elements M that form the binary skutterudites MX_3. In the filled skutterudites the filler atom R supplies the missing electron (which arises when T replaces M) and provides enough electrons to saturate the bonds and stabilize the structure. The filler ion is weakly bonded to the neighboring pnicogen atoms and exhibits local oscillations that are very disruptive to the phonon transport. Consequently, lattice thermal conductivity is drastically diminished while the electronic properties are degraded to a much lesser extent. These are the general conditions one looks for in searching for good thermoelectric materials. The underlying premise—having a material with very poor heat conducting characteristics typical of amorphous or glassy solids while preserving reasonably robust electronic properties akin to those of crystalline solids—is referred to as the Phonon-Glass-Electron-Crystal concept first formulated by Slack [6].

Skutterudites that have attracted by far the greatest attention are compounds based on $CoSb_3$. They show not only some of the best thermoelectric characteristics, but the constituent elements are also abundant and markedly less expensive than most of the alternative skutterudite structures. $CoSb_3$ is a small gap semiconductor, and transport properties of this material (single crystals as well as polycrystalline specimens) have been studied extensively [7-12]. Doping is an effective way to alter the electronic properties of this skutterudite. Some forty years ago, Dudkin and his colleagues [13,14] established that one could dope (substitute) on both the transition metal site and on the pnicogen lattice. We and others [15-18] have shown that replacing a small fraction of Co atoms with Fe has a remarkably strong influence on the nature of the charge carrier transport and on the propagation of phonons. Anno *et al.* [19] have published recently an interesting study that indicated a strong influence of Ni on the electronic properties of $CoSb_3$. In their work they used Ni concentrations above 3 at%, a rather high level of substitution that resulted in 2-3 orders of magnitude higher free carrier concentration than that observed in pure $CoSb_3$. We were interested in exploring the influence of much lower doping levels of Ni, and in this paper we present our findings on the influence of Ni at concentrations below 1 at%. We show that even these low doping levels have a spectacular effect on the carrier spectrum and on the overall transport behavior of the compound.

SAMPLES AND EXPERIMENTAL TECHNIQUE

Polycrystalline $CoSb_3$ and its Ni-doped forms were prepared from high purity starting materials (99.995% purity for Co, 99.999% purity for Sb, and 99.999% purity for Ni) by induction melting the constituents at 1400°C for a brief period of time. This assures very good mixing, near-theoretical density, and good homogeneity of the ingot. However, because skutterudites undergo peritectic decomposition (the peritectic temperature of $CoSb_3$ is near 873°C), the ingot must be annealed below the peritectic temperature in order to achieve the desired skutterudite phase. We accomplish this by annealing at 700°C under argon for 20 hours. Sample stoichiometry is checked by electron probe microanalysis. The skutterudite phase is confirmed by x-ray powder diffraction. Samples are cut from the ingots with typical dimensions $3 \times 3 \times 8$ mm^3.

Galvanomagnetic measurements are performed using a 16Hz Linear Research bridge in conjunction with a cryostat equipped with a 5T superconducting magnet. Hall effect studies are done in both positive and negative magnetic field to correct for any misalignment of the Hall probes. Thermal transport studies are made with the aid of a longitudinal steady-state technique in the range 2-300K. We use copper-constantan thermocouples made from very thin wires (25μm diameter) to measure a thermal gradient set up by a small strain gauge heater attached to a free end of the sample. The copper legs of the thermocouple serve as voltage probes for thermopower measurements. Correction for the thermopower of the copper legs is based on independent measurements of their thermopower against a high-T_c sample and against a Pb standard [20]. Magnetization measurements are done in a Quantum Design magnetometer between 10-300K in applied magnetic fields up to 5T. For each sample the background magnetic moment of the plastic sample container was subtracted from the measured magnetic moment at each temperature and at each magnetic field.

THEORETICAL CONSIDERATIONS

Band structure calculations [21] indicate that $CoSb_3$ is a very narrow direct gap semiconductor ($E_g \approx 0.05$eV). The gap occurs at the Γ-point between the conduction band minimum dominated by the transition metal d-like character and a <u>single</u> valence band arising from hybridized transition metal d- and pnicogen p-orbitals. The valence band is unusual in the sense that, apart from a very small region near the Γ-point, it has a distinctly nonparabolic dispersion. If not for this single band, $CoSb_3$ would be a medium-size-gap semiconductor with the conduction and valence band manifolds separated by 0.57eV. More recent band structure calculations [22] confirm the above general features except that the direct gap is four times larger and the "unusual" linear dispersion is viewed as typical of narrow-gap semiconductors. The two-band Kane model seems to describe the bands near the Fermi level quite well. The transport properties of the usual p-type forms of CoSb$_3$ (undoped or intentionally doped) are much influenced by the linear dispersion of the valence band as has been amply demonstrated [7,9].

CoSb$_3$ is a diamagnetic solid, and the bonding configuration that takes into account its transport and magnetic characteristics was first postulated by Dudkin [23]. In this scheme each Sb atom bonds with two nearest neighbor Sb atoms (pnicogens form a nearly square planar 4-membered rings) and with two nearest cobalt atoms. Co atoms (d^7s^2) are octahedrally coordinated by the Sb atoms and, because of a large separation between the Co atoms, there is no direct metallic bond between the transition metal atoms. Three of the nine Co electrons bond with the pnicogens while the remaining six nonbonding electrons adopt the maximum spin-pairing configuration, the zero-spin d^6 state. Attempting to substitute either Fe or Ni for Co, one runs into an obvious problem of what to do with a missing electron in the case of Fe, and an extra electron in the case of Ni. Constraints imposed by structural and electrostatic considerations are reflected in the limited solubility of Fe and Ni in CoSb$_3$. The structure can accommodate no more than 10 at% of Ni and no more than about 25 at% of Fe. A somewhat wider boundary is possible if one allows for a slight deviation in pnicogen stoichiometry.

Apart from the number of d-electrons, iron and nickel have valence electron configurations similar to that of cobalt. Both Ni and Fe readily form octahedral bonds and, to substitute for Co, they must match closely the radius of the Co^{3+} ion (1.22Å). In the case of Ni, an excellent fit is achieved with its tetravalent state Ni^{4+} (1.21Å) that assures the d^6 configuration and promotes a single electron into the conduction band. Nickel thus acts as an electron donor. In the case of Fe, there are two valence states that have radii comparable to trivalent cobalt: the divalent Fe^{2+} (1.23Å) and the trivalent Fe^{3+} (1.22Å). The divalent iron does not provide enough electrons for bonding. The trivalent iron, on the other hand, supplies the necessary electrons to form the bond, and it also matches the size of the trivalent cobalt exactly. Based on these simple bonding considerations, the trivalent iron would seem to be the most likely configuration upon the iron substituting Co in CoSb$_3$. However, this leaves only five electrons among the non-bonding d-electrons, and it is impossible to pair all spins. The Fe-doped CoSb$_3$ is thus expected to display paramagnetism. The paramagnetic state indeed seems to develop in all Fe-doped samples. The effective magnetic moment increases with Fe concentration and asymptotes to a value of about 1.7μ_B per Fe atom. It is interesting to note that the spin-only value of the magnetic moment expected for the Fe^{3+} ion is $\mu_{eff} = 2[s(s+1)]^{1/2}\mu_B = 1.73\mu_B$ /Fe. This would seem to confirm the predominance of the trivalent iron with its low-spin d^5 electronic configuration in the non-bonding d-orbitals. Because the bonding configuration and, therefore, the valence band structure of Co^{3+} and Fe^{3+} are similar, the presence of trivalent iron should not lead to significant changes in the transport behavior between the pure CoSb$_3$ and its Fe-doped forms.

In reality, Fe substitution leads to unexpected and dramatic changes in the character of transport [15]. Even at concentrations on the order of 1 at% or less, the structure exhibits a strongly metallic features in its electrical resistivity, and the carrier density (holes) progressively rises from 1×10^{19}cm^{-3} for the undoped sample to near 10^{20}cm^{-3} at 10 at% Fe. The thermopower reflects the increasing hole concentration, and its magnitude decreases with Fe addition. In accord with the assumption of a nonparabolic valence band, the carrier density dependence of the thermopower is proportional to $p^{-1/3}$. A crossover from a semiconducting to a metallic domain of conduction upon doping with Fe is robust and counters the assumption of Fe "inactivity" assumed in the model of Dudkin.

The presence of Fe also has a strong influence on heat transport [16]. Addition of 10 at% Fe to $CoSb_3$ reduces the lattice thermal conductivity at 300K by a factor of four and, at its dielectric peak near 30K, the thermal conductivity is suppressed by an order of magnitude in comparison to its value for pure $CoSb_3$. From a detailed analysis of the thermal conductivity of $Co_{1-x}Fe_xSb_3$ it follows that the concentration of point defects increases with increasing Fe doping and the most likely point defects are the vacancies on the Co sites. Fe substitution promotes a higher density of vacancies, and it is these structural defects that give rise to strong reduction in the heat conduction.

RESULTS AND DISCUSSION

Figure 1 shows the x-ray powder diffraction patterns for the $Co_{1-x}Ni_xSb_3$ samples. There is no secondary phase, and this is further confirmed by electron probe microanalysis. Table I lists the nominal compositions and the compositions determined by electron probe microanalysis. All samples are very close to stoichiometry.

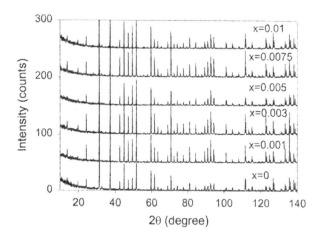

Figure 1. X-ray powder diffraction spectra for $Co_{1-x}Ni_xSb_3$. Successive spectra are shifted by 50 counts for clarity.

Table I. The nominal compositions and the compositions determined by electron probe microanalysis for the series of $Co_{1-x}Ni_xSb_3$ samples.

Nominal Composition	Composition by EPMA
$CoSb_3$	$CoSb_{3.0225}$
$Co_{0.999}Ni_{0.001}Sb_3$	$Co_{0.999}Ni_{0.0009}Sb_{3.0102}$
$Co_{0.997}Ni_{0.003}Sb_3$	$Co_{0.997}Ni_{0.0031}Sb_{3.0559}$
$Co_{0.995}Ni_{0.005}Sb_3$	$Co_{0.995}Ni_{0.0049}Sb_{2.9862}$
$Co_{0.9925}Ni_{0.0075}Sb_3$	$Co_{0.9925}Ni_{0.0076}Sb_{2.9987}$
$Co_{0.99}Ni_{0.01}Sb_3$	$Co_{0.99}Ni_{0.01}Sb_{3.0309}$

In Figure 2 we plot the temperature dependence of the electrical resistivity. At room temperature, the resistivity first increases with the addition of Ni, reflecting the change from the high mobility p-type to low mobility n-type conduction. Further increase of the Ni concentration results in a reduction of $\rho(300K)$ up to $x = 0.0075$ where it begins to increase slightly. Over the entire temperature range all samples display an activated behavior. While the low-temperature semiconducting behavior has been seen before [10, 12], the increase in resistivity over several orders of magnitude below about 50 to 100 K is a testament to the high purity and crystalline quality of the samples. In comparison to the Fe-doped samples (see Figure 3) we previously studied [15], very small amounts of Fe in $CoSb_3$ changed the semiconducting behavior of the resistivity to a metallic-like one. A plot of $\log(\rho)$ vs. $1000/T$ for the undoped $CoSb_3$ sample

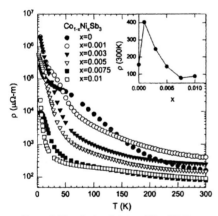

Figure 2. Electrical resistivity of $Co_{1-x}Ni_xSb_3$ versus temperature from 2 K to 300 K. The inset is a plot of resistivity at 300K versus Ni concentration, x.

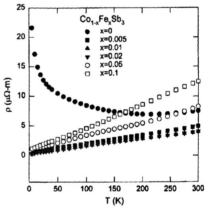

Figure 3. Electrical resistivity of $Co_{1-x}Fe_xSb_3$ versus temperature from 2 K to 300 K.

(same data as in Fig. 2) is presented in Figure 4. The dashed lines are fits to the equation $\rho=\rho_o\exp[E_a/(k_BT)]$, where E_a is the activation energy and k_B is Boltzmann's constant. Two distinct regions characterized by different activation energies can be seen: (1) below 20 K, $E_{a1} \approx$ 0.6 meV, and (2) above 100 K, $E_{a2} \approx$ 51 meV. A gentle increase in resistivity with decreasing temperature such as in region (1) has been previously attributed to impurity band conduction (also with a 0.6 meV activation energy) [12]. The activation energy of region (2) suggests an intrinsic semiconducting energy gap of approximately 102 meV. This result is in reasonable agreement with previous experiments and calculations [10, 21, and 22]. The curves of $\log(\rho)$ vs. $1000/T$ for the Ni-doped samples are not as simple a picture as that of the undoped $CoSb_3$ sample because of the n-type doping.

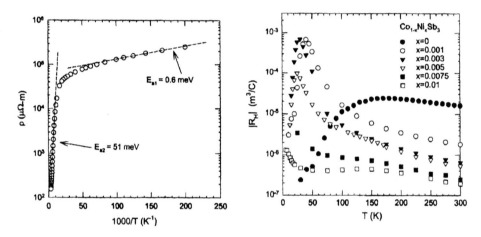

Figure 4. Electrical resistivity of $CoSb_3$ versus inverse temperature from 2 K to 300 K. Activation energies of 0.6 meV and 51 meV are found for T < 20 K and T > 100 K, respectively.

Figure 5. Hall coefficient for Ni-doped samples between 2 K and 300 K. For x=0, R_H is positive; for x≠0, R_H is negative.

The Hall coefficient, R_H, as a function of temperature was also investigated and is shown in Figure 5. It should be noted that R_H showed some variation with magnetic field strength at lower temperatures which suggests the participation of charge carriers from more than one energy band. Our pure $CoSb_3$ has a room temperature hole concentration of 3.7×10^{17} cm^{-3} which is low compared to that for other polycrystalline $CoSb_3$ [12, 18, and 24]. Rather, this carrier concentration is typical of that for single crystal $CoSb_3$ [7, 10, and 11] which indicates very good stoichiometry. All of the Ni-doped samples had negative R_H, indicating dominant n-type conduction. Figure 6 plots the room temperature carrier concentration obtained by using $n=1/(R_H e)$ where e is the electron charge. This equation assumes a single parabolic band, and we believe that this is a fair approximation near room temperature. One can see that the data fall on a straight line, except for the 0.5% Ni sample. Therefore, a clear doping effect is apparent and is consistent with Ni substituting for Co in the skutterudite lattice and donating its extra electron to the conduction band. However, at this temperature the efficiency of the ionization is only 30%. The most striking feature in the Hall data is the sharp peak at low temperatures. As the Ni concentration is increased, the peak position moves to lower temperatures. A similar effect was also seen in the magnetoresistance. Continued study is needed to clarify the origin of this unexpected behavior.

In Figure 7 we show the temperature dependence of the thermopower, S, for the Ni-doped series. Pure $CoSb_3$ is p-type and has positive room temperature thermopower. Ni-doped samples are n-type. At room temperature a Ni concentration of only 0.1% drastically affects the thermopower, changing the sign and showing the largest magnitude of approximately –500 µV/K. Further doping with Ni increases the carrier concentration and, therefore, decreases the magnitude of S at room temperature. The most puzzling aspect of these data is the dramatic low temperature anomaly. For all of the Ni-doped $CoSb_3$ samples, the temperature dependence of the peak positions in the Hall coefficient and magnetoresistance seem to follow a similar trend as the thermopower. In Figure 8, we plot the temperature dependence of the thermopower together with R_H and magnetoresistance for samples of Ni content of 0.1% and 0.5% for an illustration of this effect. For all samples of this series, the peak in R_H and magnetoresistance lies near the crossover from positive to negative thermopower. While the thermopower has distinctly positive values at low temperature, the Hall coefficient remains negative, indicating the importance of more than one carrier participating in the transport and with different weights to their partial thermopowers and mobilities.

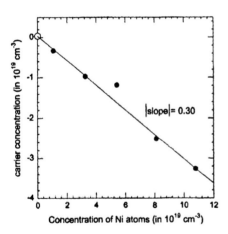

Figure 6. Room temperature carrier concentration for Ni-doped samples as a function of Ni concentration. The open circle represents data for the p-type pure $CoSb_3$ sample. The line is a guide for the eye.

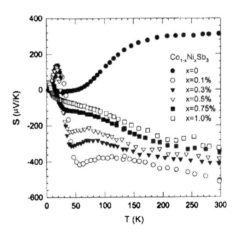

Figure 7. Temperature dependence of the thermopower of $Co_{1-x}Ni_xSb_3$.

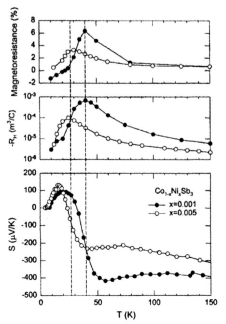

Figure 8. Temperature dependence of the thermopower, Hall coefficient and magnetoresistance for x=0.001 and 0.005 samples between 2 K and 150 K.

From the temperature dependence of the Hall mobility, $\mu_H = R_H/\rho$, we have determined that the dominant carrier scattering mechanism at room temperature for Ni-containing samples is acoustic lattice scattering. Then, assuming a single parabolic band, we calculated the reduced Fermi energy, η, from the experimental thermopower data using [25]:

$$S = -\frac{k_B}{e}[\frac{2F_1(\eta)}{F_0(\eta)} - \eta] \quad (1)$$

where k_B is Boltzmann's constant, e is the charge of the electron, and F_x is a Fermi integral of order x. Using the calculated η, the carrier concentration, n, can be expressed as [24]:

$$n = 4\pi(\frac{2m^*k_BT}{h^2})^{3/2}F_{1/2}(\eta) \quad (2)$$

where m^* is the effective mass, T is the temperature, and h is Planck's constant. The effective masses as a function of Ni concentration are calculated and have values between $3m_e$ and $4m_e$, where m_e is the free electron mass. These values seem to be higher than what is reported in the literature [19]. Upon analysis of the temperature dependence of the Hall mobility for the pure $CoSb_3$, it is clear that there is not a single dominant carrier scattering mechanism at room temperature. As a result, the equations above cannot be used to calculate the p-band effective mass. However, we can estimate that the hole effective mass should be between $0.18m_e$ and $0.05m_e$ using the above equations assuming only acoustic lattice scattering and only ionized impurity scattering, respectively.

Figure 9 displays the temperature dependence of the total thermal conductivity of all the samples; Ni doping clearly suppresses the thermal conductivity dramatically. Even 0.1 at% of Ni substituting Co decreases the peak value of the thermal conductivity by a factor of 5. As the Ni concentration increases, the peak position of the thermal conductivity shifts towards higher temperature. For all the samples, the electronic thermal conductivity is only a few percent of the total. In light of the heavy effective electron mass of the n-band, the suppression in the thermal conductivity is believed to be due to increasing electron-phonon interaction in addition to point defect scattering upon Ni doping [26]. A detailed study on the effect of Ni-doping on the lattice thermal conductivity will be published later.

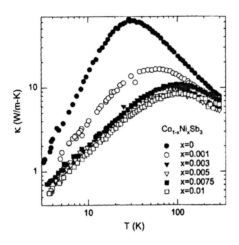

Figure 9. Temperature dependence of the total thermal conductivity for $Co_{1-x}Ni_xSb_3$.

The magnetic moment M varies linearly with the magnetic field H at all temperatures and for all six samples measured. Unlike the Fe-doped $CoSb_3$ samples [15], there is no ferromagnetic impurity detected. Susceptibilities were determined by $\chi = \frac{\partial M}{\partial H} = \frac{M}{H}$. We fit the magnetic susceptibilities of Ni-doped samples using the Curie-Weiss form:

$$\chi - \chi_{pure} = \chi_0 + \frac{C}{T + \theta_{cw}} \qquad (3)$$

where χ_{pure} is the magnetic susceptibility of the undoped sample, χ_0 is the temperature independent lattice susceptibility, C is the Curie constant and θ_{cw} is the Curie-Weiss temperature. Table II lists the fitting parameters for all Ni-doped samples.

Figure 10 displays $1/(\chi - \chi_{pure} - \chi_0)$ versus T and fits given by Eq. (3) for all Ni-doped samples. Paramagnetic effective Bohr magneton numbers are derived from the Curie constant C and plotted in Fig. 11 as a function of Ni concentration. Notice that all effective Bohr magneton numbers are close to 1.73 μ_B, which is the value expected for a single unpaired electron spin (low-spin s=1/2) configuration. Together with our Hall results, this suggests that the Ni impurity

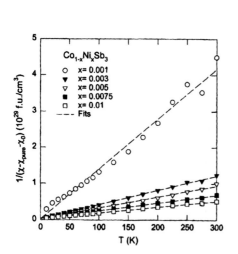
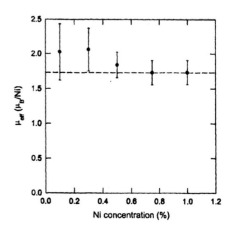

Figure 10. $1/(\chi - \chi_{pure} - \chi_0)$ versus T for all $Co_{1-x}Ni_xSb_3$ samples. Dashed lines are fits according to Eq. (3).

Figure 11. Effective Bohr magneton numbers for Ni-doped $CoSb_3$. The dashed line indicates the value of 1.73 μ_B/Ni expected for a single unpaired electron spin.

x	$\chi_0(10^{-30}$emu/f.u.)	$C(10^{-28}$emu-K/f.u.)	$\theta_{cw}(K)$	$P(\mu_B)$
0.001	7.766	8.526	15.58	2.027
0.003	-0.875	26.45	11.17	2.062
0.005	-2.388	35.23	15.99	1.843
0.0075	-1.978	46.70	17.32	1.733
0.01	-1.361	62.43	19.85	1.735

acts as a donor, giving an unpaired electron to the conduction band when ionized. Hence, we conclude that Ni takes the tetravalent state Ni^{4+} and assumes the zero-spin d^6 electronic configuration in the lower energy non-bonding orbitals.

SUMMARY

We performed transport and magnetic measurements on a series of polycrystalline samples of the form $Co_{1-x}Ni_xSb_3$ with x=0, 0.001, 0.003, 0.005, 0.0075, and 0.01. Over the entire temperature range (2 K to 300 K), all samples show an activated behavior. High temperature (T > 100 K) resistivity data suggest an intrinsic semiconducting gap of approximately 102 meV. For T < 20 K, an activation energy of 0.6 meV is attributed to impurity band conduction. Our Hall data indicate that room temperature carrier concentration increases with increasing Ni concentration. At low temperatures, very sharp peaks of Hall coefficient R_H are observed which correlate with the magnetoresistance peaks. The origin of this low temperature anomaly is not clear at this time. Ni doping dramatically changes the temperature dependence of the thermopower. The room temperature thermopower changes from positive to negative upon Ni doping. A very high thermopower (~ 500 µV/K) is achieved for the 0.1at% Ni-doped samples. The magnitude of the room temperature thermopower decreases with increasing Ni concentration. The low temperature anomalies (positive peaks) in our thermopower data are not fully understood but suggest multi-carrier thermal conduction at these temperatures. Ni doping dramatically suppresses the thermal conductivity of $CoSb_3$, which is believed to be a consequence of phonons scattered by lattice defects and carriers. Theoretical modeling is underway to shed some light on the phonon scattering mechanisms in these samples. Our magnetometry and Hall data indicate that Ni takes the tetravalent state Ni^{4+}, assumes the d^6 electronic configuration for the lower energy non-bonding orbitals, and gives one electron to the conduction band.

REFERENCES

1. C. Uher, **Semiconductors and Semimetals**, ed. T. M. Tritt, submitted.
2. G. S. Nolas, D. T. Morelli, and T. M. Tritt, *Ann. Rev. Mater. Sci.* **29** (1999) 89.
3. W. Jeitschko and D. J. Brown, *Acta Crystallogr.* **B33** (1977) 3401.
4. N. T. Stetson, S. M. Kauzlarich, and H. Hope, *J. Solid State Chem.* **91** (1994) 140.
5. B. C. Sales, B. C. Chakoumakos, and D. Mandrus, *Phys. Rev.* **B61** (2000) 2475.
6. G. A. Slack in *CRC* **Handbook of Thermoelectrics**, ed. D. M. Rowe, Boca Raton, FL, CRC Press, p. 407 (1995).
7. D. T. Morelli, T. Caillat, J.-P. Fleurial, A. Borshchevsky, J. Vandersande, B. Chen, and C. Uher, *Phys. Rev.* **B51** (1995) 9622.
8. J. W. Sharp, E. C. Jones, R. K. Williams, P. M. Martin, and B. C. Sales, *J. Appl. Phys.* **78** (1995) 1013.
9. T. Caillat, A. Borshchevsky, and J.-P. Fleurial, *J. Appl. Phys.* **80** (1996) 4442.
10. D. Mandrus, A. Migliori, T. W. Darling, M. F. Hundley, E. J. Peterson, and J. D. Thompson, *Phys. Rev.* **B52** (1995) 4926.
11. E. Arushanov, K. Fess, W. Kaefer, Ch. Kloc, and E. Bucher, *Phys. Rev.* **B56** (1997) 1911.
12. H. Anno, K. Hatada, H. Shimizu, K. Matsubara, Y. Notohara, T. Sakakibara, H. Tashiro, and K. Motoya, *J. Appl. Phys.* **83** (1998) 5270.
13. L. D. Dudkin and N. Kh. Abrikosov, *Zhurnal Neorganicheskoi Khimii*, **2** (1957) 212.
14. B. N. Zobrina and L. D. Dudkin, *Sov. Phys.—Solid State* **1** (1960) 1668.
15. J. Yang, G. P. Meisner, D. T. Morelli, and C. Uher, *Phys. Rev. B*—submitted.

16. J. Yang, D. T. Morelli, G. P. Meisner, and C. Uher, in **Thermal Conductivity 25/ Thermal Expansion 13**, ed. C. Uher and D. T. Morelli, Technomic Publishing, Lancaster PA, pp. 130 (2000).
17. S. Katsuyama, Y. Shichijo, M. Ito, K. Majima, and H. Nagai, *J. Appl. Phys.* **84** (1998) 6708.
18. K. L. Stokes, A. C. Ehrlich, and G. S. Nolas, *Mat. Res. Soc. Symp. Proc.* **545** (1999) 339.
19. H. Anno, K. Matsubara, Y. Notohara, T. Sakakibara, and H. Tashiro, *J. Appl. Phys.* **86** (1999) 3780.
20. C. Uher, *J. Appl. Phys.* **62** (1987) 4636.
21. D. J. Singh and W. E. Pickett, *Phys. Rev.* **B50** (1994) 11235.
22. J. O. Sofo and G. D. Mahan, *Mat. Res. Soc. Symp. Proc.* **545** (1999) 315.
23. L. D. Dudkin, *Sov. Phys.—Solid State* **2** (1960) 371.
24. B. Chen, J. Xu, C. Uher, D. T. Morelli, G. P. Meisner, J.-P. Fleurial, T. Caillat and A. Borshchevsky, *Phys. Rev.* **B55** (1997) 1476.
25. H. J. Goldsmid, **Electronic Refrigeration**, Pion Limited (1986).
26. J. M. Ziman, **Electrons and Phonons**, Oxford University Press (1960).

Structural Defects in a Partially-Filled Skutterudite

Jennifer S. Harper and Ronald Gronsky
Department of Materials Science and Engineering
University of California
Berkeley, California 94720-1760

ABSTRACT

The partially filled skutterudite structure is a candidate thermoelectric material with the capacity for phonon scattering by the decoupled rattling of filling ions. In this transmission electron microscopy investigation of a 1.6%Ce, 1.6%Ni, 4.9%Ge, 22.8%Co, and 69.1%Sb alloy, the structure is found to be that expected of a partially-filled skutterudite, but with a varied assortment of structural defects. These defects are characterized and their effect on thermoelectric properties is discussed.

INTRODUCTION

The generally accepted guidelines for selecting candidate thermoelectric materials include a large number of atoms per unit cell, a large average atomic mass, a high average coordination number, and a small band gap semiconductor with a complex electronic structure near the Fermi energy.[1,2] The skutterudites, based upon the original compound $CoAs_3$, meet these fundamental requirements in a body-centered unit cell with 32 atoms per unit cell and space group $Im\bar{3}$ (see Figure 1). Their thermoelectric properties are described by the dimensionless figure of merit, ZT, defined as

$$ZT = \frac{S^2\sigma}{\kappa}T \qquad (1)$$

where S is the Seebeck coefficient, σ the electrical conductivity, κ the thermal conductivity comprised of two components, κ_e the electronic thermal conductivity, and κ_g the lattice thermal conductivity, and T is the absolute temperature. In this context, the skutterudites have favorable electronic properties due to very high hole mobilities producing a high power factor ($S^2\sigma$), but they also have a relatively high thermal conductivity, resulting in a relatively low figure of merit $(ZT<<1)$[3].

Prospects for improving the performance of the skutterudites include the addition of rare earth atoms to fill the open interstices in the unit cell of Figure 1. These large weakly bound atoms are believed to undergo large anharmonic vibrations during thermal activation, in essence "rattling" within the "cages" imposed by the structure, thereby scattering phonons and impeding thermal conductivity with little or no effect on electrical conductivity[4].

A number of filled skutterudites have been synthesized, based upon the basic formula RM_4X_{12}, with R=Ce, Pr, Nd, Sm, Eu, La, U, Th, Tl, Sr, Ca, Yb, or Sn, M=Fe, Ru, Os, Ir, Co, or Ni, and X=P, Ge, or Sb[5,6,7,8,9,10,11]. Measurements of their properties show a decrease in thermal conductivity by a factor of 5 to 8 when sufficient charge compensation depresses the electronic contribution to thermal transport. This causes a corresponding improvement in their figure of merit, as expected[12]. More recently[13] it has been discovered that partial, rather than complete, filling of the interstices offers an even lower thermal conductivity, attributed to an increased entropy[14,15] from a random distribution of rattler atoms throughout the structure.

Higher complexity is also associated with defects in the crystalline lattice, since it can be expected that point, line, and planar imperfections will influence both electrical and thermal conductivities. Furthermore, these same defects could influence the distribution of filler constituents in the skutterudite compounds, making them the primary focus of the current investigation.

EXPERIMENTAL PROCEDURES

A filled skutterudite sample containing 1.6%Ce, 1.6%Ni, 4.9% Ge, 22.8%Co, and 69.1%Sb was provided by T. Caillait and J.P. Fleurial of the Jet Propulsion Laboratory. The sample was thinned to electron transparency using Ar^+ ion milling, followed by final plasma cleaning, before insertion into a JEOL JEM 200CX transmission electron microscope. Contrast analysis was impeded by the high density of diffraction spots under normal operating values of the camera constant, making it difficult to achieve proper "two-beam" conditions without appreciable parasitic scattering. Displaced-aperture dark field imaging was used extensively to localize the scattering conditions in images of defects. Experimental diffraction patterns were compared to those simulated on the basis of the skutterudite structures, both unfilled and filled, using the applications (Crystal Kit™ and MacTempas™) available at the National Center for Electron Microscopy, LBNL.

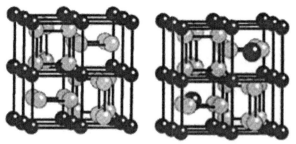

Figure 1. *Unit cells of unfilled skutterudite $CoSb_3$ at left, and filled skutterudite $CeCo_4Sb_{12}$ at right. Weakly-bound Ce atoms are shown occupying the large interstitial void in the structure.*

EXPERIMENTAL RESULTS

Electron diffraction confirms that the sample has the skutterudite structure with $Im\bar{3}$ space group, indicated by the excellent agreement between experimental and computed patterns shown in Figure 2. The sensitivity of selected area diffraction analysis to the filling of the large voids in the skutterudite unit cell is demonstrated in Figure 3. This direct comparison between the simulated electron diffraction patterns from unfilled $CoSb_3$ and filled $CeCo_4Sb_{12}$ reveals a subtle difference in intensities, one of which $(3\bar{2}\bar{1})$ is circled in the figure.

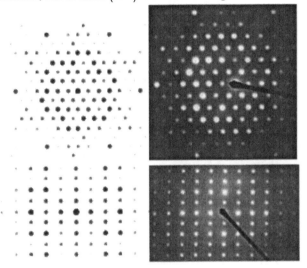

Figure 2. Comparison between simulated (left) and experimental (right) diffraction patterns in the same zone axis orientations [111] (top), [011] (bottom) demonstrating that the specimen has the expected skutterudite structure with space group $Im\bar{3}$.

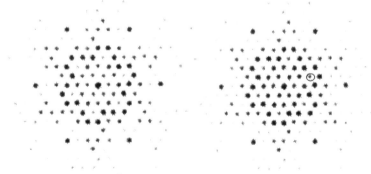

Figure 3. Simulated [111] electron diffraction patterns from $CoSb_3$ (left) and $CeCo_4Sb_{12}$ (right) showing intensity variations. Note circled $(3\bar{2}\bar{1})$ reflection.

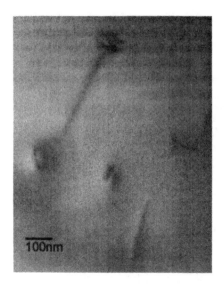
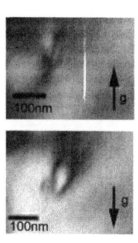

Figure 4. Bright Field TEM image displaying dislocations found in the filled skutterudite structure.

Figure 5. TEM image of same area showing inside-outiside contrast under ±g displaced aperture dark field imaging conditions suggesting evidence of dislocation dipoles.

The bright field image of Figure 4 shows contrast characteristic of dislocations. A standard slip dislocation in this body-centered structure has a burgers vector of a/2 <111> with unusually large magnitude (~8Å.). Consequently, it may be expected that these dislocations would interact or dissociate in order to minimize their strain energy imposed upon the lattice. One such interaction is shown in Figure 5. This +g /-g pair shows inside/outside contrast characteristic of a dislocation dipole.

Figure 6 shows an array of dipoles. Selected area diffraction shows no misorientation across the array, as expected for the compensation provided by adjacent dislocations with opposite burgers vectors.

A common method for determining the burgers vector of a dislocation is to perform **g • b** experiments and to establish an "effective invisibility" criterion. When the imaging condition satisfies $\mathbf{g} \cdot \mathbf{b} \times \mathbf{u} = 0$, the dislocation will exhibit no contrast. If this condition is achieved for more than two non-colinear **g**-vectors then the burgers vector may be inferred. This approach was not successful in the present run of imaging, possibly due to complications from elastic anisotropy or chemical segregation to the dislocation cores. It is unlikely that the closely-spaced dislocations are partials due to the high stacking fault energy in this material.

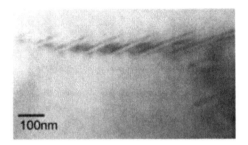
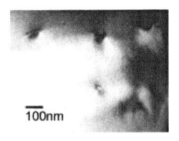

Figure 6. *Bright field TEM image of an array of dislocation dipoles.*

Figure 7. *TEM image displaying smaller defects with Ashby-Brown type contrast indicating small coherent particles.*

The last type of defect observed in this study is a much smaller one, sometimes displaying the "black-white" contrast of the Ashby-Brown type[16] shown in Fig. 7. These contrast effects may indicate the presence of small inclusions or second phase particles with a spherically-symmetric strain field. At larger dimensions, as might result from growth or coarsening of this initially small size distribution, strain field compensation could generate the dislocation dipoles observed above. If these clusters are Ce atoms that leave their interstitial cages under thermal activation, the implications for thermoelectric performance could be serious. It would therefore be important to assess the composition of these clusters and the evolution of the thermal conductivity of this material as a function of aging time.

CONCLUSIONS

This investigation of a 1.6%Ce, 1.6%Ni, 4.9%Ge, 22.8%Co, 69.1%Sb alloy shows that the material has the $Im\bar{3}$ space group expected for a skutterudite phase. Electron diffraction evidence that the large interstitial voids in the structure are filled by one or more of the alloying constituents is subtle and therefore not conclusive. Numerous structural defects exist in the alloy, including isolated dislocation dipoles, arrays of dislocation dipoles, and small coherent particles. There is some concern that these defects may result from exsolution of the additive elements, and their effect on thermoelectric performance has yet to be determined.

ACKNOWLEDGEMENTS

We would like to thank Dr. T. Caillat and Dr. J. P. Fleurial from JPL for supplying the filled skutterudite compound, Dr. Chris Caylor at U.C. Berkeley for assistance in the lab, and Dr. Eric Stach and Dr. Tamara Radetic of NCEM/LBNL for assistance with sample preparation and microscopy under Proposal #410. This research is funded by a California Legislative Grant and by the Department of Defense under a Multi-University Research Initiative, ONR-MURI 442444-25827.

REFERENCES

[1] G. Mahan, B. Sales, and J. Sharp, *Physics Today*, 42 (1997).
[2] M. G. Kanatzidis, and F. J. DiSalvo, *Naval Research Reviews* **48**, 14 (1996).
[3] T. Caillat, A. Borshchevsky, and J. P. Fleurial, *13th International Conference on Thermowlectrics*, 58 (1995).
[4] G. A. Slack and V. G. Tsoukala, *Journal of Applied Physics* **76**, 1665 (1994).
[5] G. P. Meisner, M. S. Torikachvili, K. N. Yang, M. B. Maple, and R. P. Guertin, *Journal of Applied Physics* **57**, 3073 (1985).
[6] D. J. Braun and W. Jeitschko, *Journal of the Less-Common Metals* **76**, 33 (1980)
[7] W. Jeitschko and D. Braun, *Acta Crystallographica* **B33**, 3401 (1977)
[8] G. S. Nolas, G. A. Slack, D. T. Morelli, T. M. Tritt, and A. C. Ehrlich, *Journal of Applied Physics* **79**. 4002 (1996).
[9] J. W. Kaiser and W. Jeitschko, *Journal Of Alloys and Compounds* **291**, 66 (1999).
[10] H. Takizawa, K. Miura, M. Ito, T. Suzuki, and T. Endo, *Journal of Alloys and Compounds* **282**, 79 (1999).
[11] B. C. Sales, B. C. Chakoumakos, and D. Mandrus, Physical Review B 61, 2475 (2000).
[12] J. P. Fleurial, A. Borshchevsky, T. Caillat, D. T. Morelli, and G. P. Meisner, 15th International Conference on Thermoelectrics, 91 (1996).
[13] D. T. Morelli, G. P. Meisner, B. Chen, S. Hu, and C. Uher, *Physical Review B* **56**, 7376 (1997).
[14] G. P. Meisner, D. T. Morelli, S. Hu, J. Yang, and C. Uher, *Physical Review Letters* **80**, 3551 (1998).
[15] G. S. Nolas, J. L. Cohn, and G. A. Slack, *Physical Review B* **58**, 164 (1998).
[16] M.F. Ashby and L.M. Brown, *Phil. Mag.* **8**, 1083 and 1649 (1963).

Devices, Measurements, and Applications

Optimization of Bismuth Nanowire Arrays by Electrochemical Deposition

J. H. Barkyoumb*, J. L. Price*, N. A. Guardala*, N. Lindsey*, D. L. Demske*, J. Sharma*, H. H. Kang[†] and L. Salamanca-Riba[†],
*Carderock Division, Naval Surface Warfare Center, West Bethesda, MD,
[†]Dept. of Materials and Nuclear Engineering U. of Maryland, College Park, MD 20742

ABSTRACT

Bismuth nanowires for thermoelectric applications have been made by electrochemical deposition from an aqueous Bi solution into nanoporous mica substrates, and Anopore (Al_2O_3) filters. The nanoporous substrates are created by acid etching damage tracks that are produced in the mica using heavy-ion irradiation. In this work, further improvements in the fabrication process are made by investigating the electrochemical growth at the 1-d to 2-d transition region at the nanochannel-substrate surface interface as a function of deposition time and electrochemistry. Issues related to doping of the Bi wires with Te along with the problem of electrochemical growth of contacts to the nanowires as compared to vacuum deposition of contacts will be discussed.

INTRODUCTION

Nanometer size wires of bismuth have been predicted to have a much higher thermoelectric figure of merit (ZT) than that of bulk bismuth.[1,2] This is due to bismuth's unique properties such as a long electron mean free path at low temperatures and a small carrier effective mass which allows quantum confinement effects to be observed for relatively large wire diameters. The low electron density and low thermal conductivity of bismuth also make it attractive as a high ZT material. Magnetoresistance results indicating quantum confinement and a semimetal-to-semiconductor transition for wire diameters less than 790 nm confirm these effects.[3]

Recently, Bi nanowires have been prepared by pressure injection[4] and by a vapor-phase technique into porous alumina hosts.[5] Also, stoichiometric Bi_2Te_3 alloys for thermoelectric applications have been grown successfully using electrochemical methods.[6] Electrochemical growth of Bi nanowires from an aqueous solution through the nanochannels is an alternative that may have advantages over these methods. A number of issues needs to be resolved to better quantify this technique in order to evaluate the utility of electrochemical growth. In this paper we examine the electrochemical growth process with the goal of ascertaining the quality of the nanowires during the growth process and to improve the quality control for the process. In anticipation of the need to eventually dope these Bi wires with tellurium for n-type operation, we also have investigated the conditions needed for tellurium incorporation into the electrochemically grown wires at the same time.

EXPERIMENT

The process by which nanochannels are created in a mica substrate has been described previously.[7] In brief, sheets of 10 μm thick muskovite mica are irradiated with high-energy (15 MeV) C^{4+} ions to fluences of between 10^{11} and 10^{13} ions/cm^2. At this ion-beam energy, the carbon ions traverse the thickness of the mica and are implanted behind the templates. Track densities of 10^9 channels per square cm are inferred from beam current flux measurements using in-line particle beam diagnostics and from AFM measurements.[7]

These structural damage tracks are much more chemically active than the undamaged material. A temperature-controlled (25 °C) 20% HF acid bath was used to preferentially etch the radiation-damaged mica with a very high selectivity compared to the undamaged material. After processing, the resulting etched pores have parallel sides with diameters as small as 50 to 80 Å. Exploiting the birefringence of the mica templates, the nanochannel etching process was continuously monitored through the use of a laser polarization discrimination technique with phase-sensitive detection to calibrate the etching time. After the etching, a metal film is evaporated on one side (the back side) of the template to serve as the working electrode in the electrochemical cell. The electrochemical cell is a EG&G model K0235 flat cell with a platinum mesh counter electrode and a 1 cm^2 aperture for the working electrode. Fig. 1 shows the cyclic voltammograms for both Bi and Te deposition. Bismuth ion concentrations in the solutions of 10^{-2} M and 10^{-3} M were investigated.

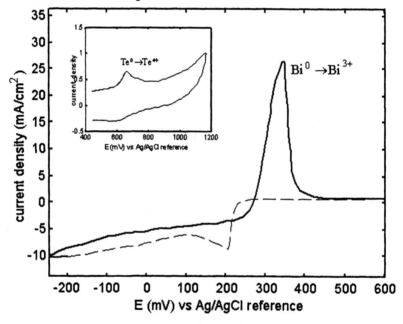

Fig. 1. Cyclic voltammogram of Bi in HNO_3 showing half of the cycle shown as a solid line and the return half as a dashed line. Positive current densities represent anodic processes (removal of Bi) and negative currents represent cathodic processes (deposition of Bi at the cathode). The inset shows a portion of the voltammogram for a 5×10^{-4} M Te solution. The Te curve was run to high voltages to ensure that no other processes were involved except Te^{4+} oxidation and reduction.

To make the electrolyte, Bi metal is dissolved in a small amount of concentrated perchloric acid. Nitric acid and water are then added to this solution to achieve the desired nitric acid molarity and Bi ion concentration. The electrochemical deposition process was carried out using potentiostatic control at -642 mV versus a Ag/AgCl reference electrode. Bi^{3+} ions are reduced to bismuth metal on the thin film electrode with the accompanying exchange of 3 electrons. Thus bismuth wires are grown up from the back to the front of the mica template. For producing tellurium-doped wires, high-purity Te shot is dissolved in nitric acid to produce solutions with concentrations of 10^{-4}, 10^{-5}, and 10^{-6} M of Te ions.

We used two methods to corroborate Te incorporation into the wires. The first was using electron microprobe energy dispersive X-ray spectroscopy to find the relative fractions of the Bi and Te in the samples. The spectrometer was a JEOL 8900R Superprobe operated at 30 kV. We also used nuclear reaction analysis to count the total amount of tellurium deposited.

The total number of Te atoms in a sample can be determined by producing and then following the characteristic decay radiations of the metastable nuclear state of ^{125}Te. This nuclear isomer has a 58-day halflife and decays predominantly by emission of Te atomic K-shell x rays. A 3 MeV tandem positive-ion accelerator was used to produce neutrons by bombarding a compound target of Be/Li_2O at the incoming proton energy of 5 MeV. Neutrons produced in this manner constitute a "white" spectrum, i.e. a continuous distribution from 140 to 4700 keV in energy. The Bi wires are essentially inert to the fast neutrons in terms of producing any radioactive nuclei which could contaminate the Te K x-ray spectrum. Comparison of the amount of Te K x rays produced in a standard and in the Bi wires gave a measure of the Te dopant levels of the wires. Initially, only Te-doped wires in Anopore (alumina) filters were examined to avoid problems with producing radioactive isotopes in the mica. One drawback to this method is that the ^{125}Te is only 7% abundant.

RESULTS AND DISCUSSION

Figure 2a shows the current vs. time for electrodeposition of Bi onto a stainless steel coupon (upper curve) and into an Anopore filter (lower curve). The electrochemical solution for both was 5×10^{-3} M Bi^{3+} ions in 0.5 M HNO_3. Since the geometry of the electrochemical cell, electrodes, and electrode separation are maintained for every run, the observed current should only depend upon the conductivity of the solution which is a function of the ion species and concentration of the solution. This can be expressed by

$$I = zFcf(c)\left(\sum_i \mu_i\right)\frac{A}{L}\Delta\Phi \quad (1)$$

where zF is the charge carried per Bi^{3+} ion, c is the number of Bi ions per unit volume in the solution, μ_i is the mobility of the ith ion species (including anions) and A, L, and $\Delta\Phi$ are the equivalent areas of the electrodes, length of the cell, and the potential difference across the cell respectively. Ion-ion interactions are represented by $f(c)$ and are generally not negligible except for extremely dilute solutions.[8] (The product of $f(c)$ and μ_i can be considered a concentration dependent mobility.) For this reason, one should not *a priori* assume that the mobility of each species of a mixture is the same as that which is measured independently. Also from Eq. 1, the deposition current should be directly proportional to the total active area of the working electrode exposed to the electrolyte. For a thin film or metallic coupon in the electrochemical cell, this is 1 cm^2, assuming the entire surface is electrochemically active. For an array of nanochannels in an insulating template with a metallic backing, the current is then proportional to the sum total areas of all the nanochannels.

This effect can be seen in Fig. 2, which shows the current vs. time for electrodeposition of Bi onto a stainless steel coupon (a) and of Bi into an Anopore filter (b) with 0.2 μm diameter channels. Since this is an electrochemical process, the total number of ions deposited (or electrons transferred) can be measured by integrating the deposition current (Faraday's law). The upper curve represents a deposition of a total of 1.3×10^{20} Bi atoms and the lower curve is 5.9×10^{19} Bi atoms. Also the mean current of deposition onto the bare coupon is 56 mA compared to a mean current in the Anopore sample of 27 mA. This corresponds to a ratio of active area of ~ 45%. The reported porosity of these Anopore filters is ~ 40% [9]. Fig. 2c is the current vs time curve for Bi wires grown into a 10 μm thick mica sample with channels etched to a nominal diameter of 80-100 Å. The total number of Bi atoms deposited in this sample is 1.2×10^{18}. From the mean current of 7 mA this typical sample has a total area of the nanochannels of 10% of the 1 cm^2 of the total surface area exposed.

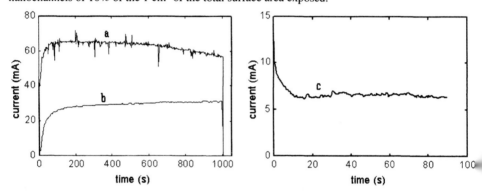

Fig. 2 Deposition current vs. time from a solution of 5×10^{-3} M Bi^{3+} in 0.5 M HNO_3 onto a stainless steel coupon (a), into an Anopore with 0.2 μm channels backed with a silver film (b), and into an etched mica sample (c).

Another geometrical effect observed is that the total deposition current increases after the Bi wires have grown through the mica to the front surface. It has previously been stated that this is due to the transition from a 1-d linear growth regime to a 2-d planar growth. We observe that the total current increases by a factor of 2-5 for nanochannels in mica for deposition times greater than 100-120 seconds. No such increase is observed for the Anopore samples. This can be explained by the observation that the electrochemically grown Bi does not form a 2-d plate growth over the insulator but does grow in long needle-like forms and into platelets for both mica and Anopore (Fig. 3). Our mica has much smaller pores and a lower total porosity by a factor of 4 and thus does show a marked change in growth rate upon wire emergence from the front surface even if the growth is dendritic.

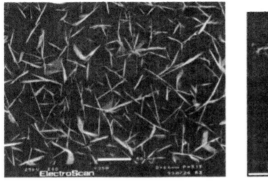
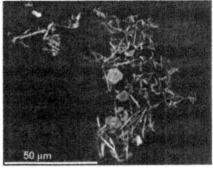

a b

Fig. 3a is an SEM image of Bi needles grown through an Anopore sample onto the free, front surface of the template. Fig 3b is the same phenomenon observed in a mica template using wavelength dispersive SEM. The width of 3a is about 320 μm and the width of 3b is 120 μm.

For sufficiently dilute concentrations of the electrolyte, the Bi^{3+} and Te^{4+} ions in solution should act independently of one another. To test this, we deposited wires from a solution with concentrations of 5×10^{-3} M Bi^{3+} and 5×10^{-5} M Te^{4+} (to make a 1% Te to Bi solution) and from separate solutions of each ion at these concentrations. All depositions were performed under the same conditions. For all the co-depositions, the final ratios of total number of Te to Bi ions deposited were between 8 and 15%. This indicates that a much greater amount of Te is being deposited than is stoichiometric in the solution. EDS measurements performed on a sample from the 1% Te solution also show that this solution incorporates Te into these wires to approximately 13%. The nuclear spectroscopy on the same wires from the 1% solution verified that the Te was present but we could not determine the concentrations accurately with this method.

CONCLUSIONS

One of the purposes of this work is to devise a protocol for determining semi-quantitatively the quality of the templates and nanowires as they are being created in the electrochemical process before making further and more difficult measurements. Using the relative current ratios during deposition we find the total porosity (integral cross-sectional area of the channels) of the samples. This is a general rule that is seen in all samples. This method alone does not tell us the quality of the individual channels or distribution of channel sizes without some independent tests. It does allow one to quickly survey processing parameters and their effects. The formation of a network of dendrites after growth through the channels is an interesting phenomenon that almost certainly needs to be controlled in order to produce known electrical contacts on the top side of the wire arrays.

It should not be surprising that the electrodeposition from binary solutions does not retain the stoichiometry of the solution. The dependence of the rate of deposition is highly dependent on the potential in a non-linear fashion. The mobilities and solvation (screening) of each ion species in solution will be different, especially when they possess differing charges. With the knowledge from these measurements, one should be able to scale the concentrations down to lower, more appropriate, doping levels.

ACKNOWLEDGMENT

The author(s) would like to acknowledge support by the Carderock Division of the Naval Surface Warfare Center's In-house Laboratory Independent Research Program sponsored by the Office of Naval Research and administered under Program Element 0601152N and Eric Wuchina at NSWC Carderock for the wavelength-dispersive SEM measurements.

REFERENCES

1. L. D. Hicks and M. S. Dresselhaus, Phys. Rev. B **47**, 16631 (1993).
2. D. A. Broido and T. L. Reinecke, Appl. Phys. Lett. **67**, 100 (1995).
3. Z. Zhang, X. Sun, M. S. Dresselhaus, J. Y. Ying and J. P. Heremans, Appl. Phys. Lett. **73**, 1589 (1998).
4. Z. Zhang, J. Y. Ying and M. S. Dresselhaus, J. Mater. Res. **13**, 1745 (1998).
5. J. Heremans, C. M. Thrush, Y. M. Lin, S. Cronin, Z. Zhang, M. S. Dresselhaus, and J. F. Mansfield, Phys. Rev. B **61**, 2921 (2000).
6. J.-P. Fleurial, A. Borshchevsky, M. A. Ryan, W. M. Phillips, J. G. Snyder, T. Caillat, E. A. Kolawa, J. A. Herman, P. Mueller, M. Nicolet, Mat. Res. Soc. Symp. Proc., **545**, 493 (1999).
7. D. L. Demske, J. L. Price, N. A. Guardala, N. Lindsey, J. H. Barkyoumb, J. Sharma, H. H. Kang and L. Salamanca-Riba, Mat. Res. Soc. Symp. Proc., **545**, 209 (1999).
8. $f(c)$ is given by $f(c) = 1 - Ac^{1/2}$ for dilute solutions where A is a temperature-dependent constant for a given ion species and can be derived from the Debye-Hückel-Onsager relation; J. O'M. Bockris and A. K. N. Reddy, *Modern Electrochemistry*, Vol. 1, Plenum, New York, NY, 1970.
9. G. P. Crawford, L. M. Steele, R. Ondris-Crawford, G. S. Iannacchione, C. J. Yeager, J. W. Doane, and D. Finotello, J. Chem. Phys., **96**, 7788 (1992).

Evaluation of a Thermoelectric Device Utilizing Porous Medium

Hideyuki Yasuda, Itsuo Ohnaka, Yoichi Inada, Kimitaka Nomura
Department of Adaptive Machine Systems,
Osaka University, Suita, Osaka 565-0871, Japan.
yasuda@ams.eng.osaka-u.ac.jp

ABASTRACT

A thermoelectric device consisting of the porous part and the bulk part was proposed. Increase of heat exchange area due to porous medium will improve the efficiency of heat exchange between heating/cooling sources and the device. Estimation based on physical properties of $FeSi_2$ indicated that generated power of the partially porous device per unit area can be several times higher than that of the bulk one in case of gas heating / cooling system. The partially porous thermoelectric devices were produced. In measurement of power, it has been confirmed that generated power per unit area of the partially porous $FeSi_2$ devices was roughly 10 times higher than that of the conventional device. The proposed porous device will exhibit its advantage in the case of low heat transfer coefficient between the device and the heating / cooling sources.

INTRODUCTION

Thermoelectric materials, which convert heat into electricity or electricity into heat, and system for the power generation, have been extensively studied. Development of microstructure that enhances the value of the figure-of-merit and discovery of new thermoelectric materials with a high figure-of-merit have been taken into attention in order to improve the efficiency of thermoelectric energy conversion. It is also required to design power generation or cooling / heating devices and systems in order to exhibit the best performance which is expected from the figure-of-merit.

From a view point of thermo-fluid mechanics, it is generally difficult to achieve good device performance because of low thermal conductivity of heating/cooling sources when heating / cooling sources are gas. Echigo et al have proposed a novel thermoelectric power generation system and a cooling/heating system using porous thermoelectric devices [1-3]. A flammable gas flows through the porous device and combustion occurs at the end of device. The flowing gas drives thermal energy from the cold side to the hot side like a heat pump, because heat exchange efficiently occurs at the large interface of the porous medium. As a result, large temperature difference between inlet and outlet of the porous device is caused without any additional cooling system. Materials processing for the porous thermoelectric device has been also reported [4,5].

Another thermoelectric device has been proposed, which consisted of the porous parts and the bulk part [6]. A characteristic feature of porous medium is large amount of surface area in comparison with its volume, causing efficient heat exchange between solid and fluid. This paper presents a model of the partially porous device to evaluate the generated power and experimental results of the power generation using the device. Usage of the partially porous thermoelectric device will be discussed by comparing with the bulk thermoelectric device.

MODEL OF THE PARTIALLY POROUS DEVICE

Generated power of the thermoelectric materials per unit area is generally given by eq. (1).

$$P = \frac{1}{4}\alpha^2 \sigma \frac{(T_1 - T_2)^2}{d}$$

Here, d: the thickness of the thermoelectric material, α: thermoelectric power, σ: electrical conductivity, T_1 and T_2: temperatures at the hot and the cold sides, respectively. Once the thermal resistance between the device and the fluid is determined, the thickness has an optimum value for maximum output [7]. Heat exchange efficiently occurs in porous medium, since it has large surface area with respect to its volume. Thus, thickness of the device can be reduced by utilizing the porous medium. Using porous medium has potential to improve the efficiency of heat exchange and increase the power generation per unit area, although the porous medium increases the electrical resistivity.

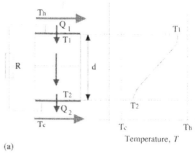

(a)

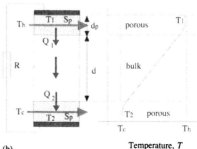

(b)

Figure 1. Configuration and temperature profiles of (a) the bulk device and (b) the partially porous device. T_h, T_c: temperatures of the hot and the cold fluids, T_1, T_2: temperatures of a thermoelectric material.

Figure 1(a) shows schematic chart of a conventional thermoelectric device and its temperature profile. Thickness and cross section area are d and S, respectively. Temperatures of heating and cooling sources are T_h and T_c, respectively. The h is heat transfer coefficient between the thermoelectric material and fluids for both hot and cold sides. Temperature difference in the thermoelectric material is express by eq. (2).

$$T_1 - T_2 = (T_h - T_c)\left[1 + \left(\frac{\lambda}{hd}\right)\left(2 + \frac{\alpha^2 \sigma}{\lambda}\frac{T_h + T_c}{2}\right)\right]^{-1} \quad (2)$$

The denominator reducing the temperature difference consisted of two factors. One is the heat resistance between a thermoelectric material and fluid express by $2\lambda/hd$. The other is the Peltier effect expressed by $(\lambda/hd)(\alpha^2\sigma/\lambda)(T_h+T_c)/2$. Generated power, P_{bulk}, per unit area is expressed by equation (3).

$$P_{bulk} = \frac{1}{4}\alpha^2 \sigma \frac{(T_h - T_c)^2}{d}\left[1 + \left(\frac{\lambda}{hd}\right)\left(2 + \frac{\alpha^2 \sigma}{\lambda}\frac{T_h + T_c}{2}\right)\right]^{-2} \quad (3)$$

Once physical properties of thermoelectric materials, the heat transfer coefficient and operating temperatures are determined, optimum value of the thickness is given by eq. (3). When gas heating / cooling sources are utilized, the optimum thickness becomes relatively large and the generated power becomes small, because low thermal conductivity of gas leads to low heat transfer coefficient.

Figure 1(b) shows structure of the partially porous thermoelectric device proposed in this paper and its temperature profile. In the device, not only thermoelectric materials but also electrodes are partially porous. Thickness of the porous part and the bulk part are d_p and d, respectively. Hot and cold fluids sufficiently flow through the porous parts. Temperature decrease of the fluids in the porous medium is ignored. Energy conservation equations are obtained in the same manner of the bulk device.

$$Q_1 = h_p(T_h - T_1)S_p = \lambda S \frac{T_1 - T_2}{d} + \alpha T_1 i_p - \frac{1}{2} r_p i_p^2 \tag{4}$$

$$Q_2 = h_p(T_2 - T_c)S_p = \lambda S \frac{T_1 - T_2}{d} + \alpha T_2 i_p + \frac{1}{2} r_p i_p^2 \tag{5}$$

Here, S_p: surface area for each porous part, h_p: heat transfer coefficient in the porous medium, i: is current intensity, r_p: internal electrical resistivity in the device. An assumption, that the external resistivity is equal to the internal resistivity and $T_h + T_c$ is equal to $T_1 + T_2$, gives the maximum power generation condition. Electrical resistivity of the device is expressed by equations (6).

$$r_p = \frac{1}{\overline{\sigma}_p} \frac{d}{S} \quad \text{here,} \quad \overline{\sigma}_p = \sigma \left[1 + 2\left(\frac{\sigma}{\sigma_p}\right)\left(\frac{d_p}{d}\right) \right]^{-1} \tag{6}$$

Here, σ_p is electrical conductivity of the porous thermoelectric medium. The external electrical resistivity is assumed to be equal to the internal electrical resistivity as well as the bulk device. The temperature difference between the hot and the cold sides of a thermoelectric material and the generated power per unit area for the porous device are given by eqs.(7) and (8), respectively.

$$T_1 - T_2 = (T_h - T_c)\left[1 + \left(\frac{S}{S_p}\right)\left(\frac{\lambda}{h_p d}\right)\left(2 + \frac{\alpha^2 \overline{\sigma}_p}{\lambda} \frac{T_h + T_c}{2}\right) \right]^{-1} \tag{7}$$

$$P_{porous} = \frac{1}{4} \alpha^2 \overline{\sigma}_p \frac{(T_h - T_c)^2}{d} \left[1 + \left(\frac{S}{S_p}\right)\left(\frac{\lambda}{h_p d}\right)\left(2 + \frac{\alpha^2 \overline{\sigma}_p}{\lambda} \frac{T_h + T_c}{2}\right) \right]^{-2} \tag{8}$$

The $1/4(\alpha^2 \sigma_p)(T_h-T_c)^2/d$ indicate the maximum generated power in the case of ideal thermal contact (no thermal resistance between a thermoelectric material and heating / cooling fluids). As shown in eqs. (7) and (8), the temperature difference is a function of (λ/hd) and $(S/S_p)(\lambda/h_p d)$. Here (λ/hd) is regarded as reciprocal of the Biot number of the device. The value of $(\alpha^2\sigma/\lambda)(T_h+T_c)/2$ corresponds to the dimensionless figure of merit at average operating temperature. According to experimental data [4], the (S/S_p) has the order of 10^{-1}-10^{-2} for the porous PbTe. The $(S/S_p)(\lambda/h_p d)$ for the partially porous device can be chosen to be smaller than (λ/hd) for the bulk device. Therefore, for the partially porous device, thinner device can be used and higher generated power is expected in comparison with the bulk device.

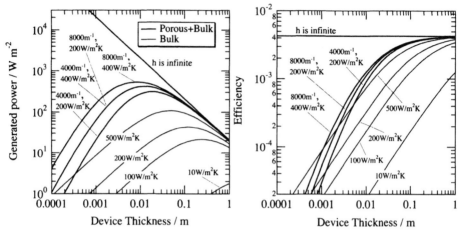

Figure 2 Estimated generated power of the partially porous device and the bulk device by using the physical properties of p-type FeSi$_2$.

Figure 2(a) shows estimation of generated power for the bulk device and the partially porous device. In the calculation, $\alpha=200\mu V/K$, $\sigma=1.3\times 10^4$ S m^{-1}, $\sigma_p=6.5\times 10^3$ S m^{-1}, $\lambda=12 W/mK$ are used, that are taken from those of p-type FeSi$_2$. The heat transfer coefficient is the order of 10^1-10^2 W/m^{2}K, since λ_{gas} is the order of 10^{-2} W/mK and R_p is 200-1000 μm [3]. The h used in the calculation is 10,100,200 and 500W/m^{2}K, while the h_p=200,400W/m^{2}K. Specific surface area is 4000, 8000m^{-1}. The porous media with the specific surface range of 4000-8000m^{-1} were experimentally produced for FeSi$_2$ and PbTe (200-1000μm particle diameter)[4]. T_h and T_c are 693 and 293K, respectively. Thickness of the porous part, d_p, is 5×10^{-3}m. Generated power for the partially porous device indicated by solid line tends to be higher than that for the bulk device. The partially porous device has smaller optimum thickness that gives maximum generated power than the bulk device. Figure 2(b) shows the efficiency of the two different devices, that is generated electricity / given thermal energy at the hot side. The efficiency of the proposed device is higher than that of the bulk device when the thickness is larger than about 0.001m.

GENERATED POWER OF FeSi$_2$ POROUS DEVICE

The FeSi$_2$ is one of the candidates for a material used for the porous device, since FeSi$_2$ is well known to have good oxidation resistance. Composition of the FeSi$_2$ used in this study was 32.25at%Fe-66.65at%Si-1.0at%Mn-0.1at%Cu. Addition of Cu drastically increases reaction rate in transformation from α-Fe$_2$Si$_5$ to β-FeSi$_2$ [8]. FeSi$_2$ particles to produce porous medium were produced by a rotating-water-atomization equipment [9]. Porous medium was produced by a sintering method. Sintering pressure and temperature are 8.5MPa and 1173K, respectively. Sintering time was 3.6ks. Particles with 590-710 μm diameter were used for the porous medium. Ni mesh and wire were used for electrodes. After producing the partially porous device, the device was annealed at 1123K for 96 hours in order that the device consisted of β- FeSi$_2$. Properties of the porous FeSi$_2$ have been reported [5]. The Seebeck coefficient of the annealed FeSi$_2$ was 175 μV/k. Dimension of the device was 15 mm in diameter.

In measurement of the power generation, gas was used for heating and cooling sources. Temperature of the hot gas was 593K, while that of the cold gas was 283K. Quantity of the gas that heated or cooled the device was 28 l/min. Voltage of the device tended to saturate at 28l/min. Generated power in the external resistivity was measured as a function of the external resistivity. Figure 3 shows the measured power for the partially porous devices and the bulk device. In the porous device, length of

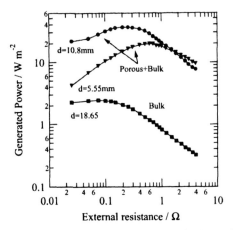

Figure 3 Generated power of the bulk device and the partially porous device using Mn-doped p-type FeSi$_2$.

the porous parts was 4.65mm. In the case of the bulk device with 18.65 mm thickness, maximum generated power was about 2 W/m^2 when the external resistivity was 0.1 Ω. The generated power decreased with decreasing the device thickness. On the other hand, generated power of the partially porous devices with 10.8 and 5.55 mm thickness (bulk part) 10-20 times larger than that of the bulk device (conventional one). The external resistivity at which generated power exhibited a maximum for the partially porous devices was slightly larger than that for the bulk device. It means that internal resistivity of the porous device is larger than that of the bulk one due to utilizing porous medium. Large interface area in the porous medium enhanced heat exchange between the gas and the device and large temperature difference was obtained in thinner device. Consequently, larger thermoelectric power was obtained in the porous medium, although using the porous medium in the thermoelectric device increased the internal resistivity. The experimental results coincide with the calculated results from the model mentioned in the previous section.

CONCLUSIONS

A thermoelectric device consisting of the porous part and the bulk part was examined by considering the model to evaluate the power generation and the experiments to measure the generated power.

In the model, enhancement of heat exchange around electrodes by introducing porous medium will improve the efficiency of heat exchange between heating / cooling sources and the device. The enhancement leads to increase the temperature difference in the thermoelectric materials even in thinner device. The model suggested that enhancement of the generated power was expected, although introducing porous medium increases the internal resistivity. Especially, the partially porous medium will be effective when gas that has rather low thermal conductivity is used for heating / cooling sources. Estimation based on physical properties of $FeSi_2$ indicated that generated power of the partially porous device per unit area can be higher than that of the bulk one in case of gas heating / cooling system.

A procedure has been developed in order to produce the partially porous thermoelectric devices. By applying the procedure, the partially porous thermoelectric devices were produced using $FeSi_2$. In measurement of the power generation, it has been confirmed that generated power per unit area of the partially porous $FeSi_2$ device can be more than 10 times higher than that of the bulk device (conventional one). The proposed porous device exhibits its advantage in the case of low heat transfer coefficient between the device and the heating / cooling sources, i.e. exhaust gas is used for heating source.

ACKNOWLEDGMENTS

This work is supported by a Grant-in-Aid for Scientific by the Ministry of Education, Science, Sports and Culture, Japan.

REFERENCES

[1] R. Echigo; J. Jpn Mech. Soc., **96**, 204 (1993).
[2] K. Hanamura, R. Echigo, S.A. Zhdanok; Int. J. Heat Mass Trans. **36**, 3201 (1993).
[3] S. Tada, R. Echigo, H. Yoshida; Proc. 15th Int. Conf. on Thermoelectrics, IEEE, (1996), pp.264
[4] H. Yasuda, I. Ohnaka, H. Kaziura, T. Yano, Proc. 17th Int. Conf. on Thermoelectrics, IEEE, (1998), pp.502.
[5] H. Yasuda, I. Ohnaka, H. Kaziura, T. Yano, Materials Science Forum **308-311**, 736 (1999).
[6] H. Yasuda, I. Ohnaka, Y. Inada, Mater. Trans. JIM, **40**, 447 (1999).
[7] G. Min and D. M. Lowe; CRC Handbook of THERMOELECTRICS, ed. by D.M. Lowe, CRC Press, (1995), pp.479.
[8] I. Yamauchi, A. Suganuma, T. Okamoto, I. Ohnaka, J. Mater. Sci., **32**,4603,(1997).
[9] I. Ohnaka, T. Fukusako, T. Ohmichi, T. Masumoto, A. Inoue, M. Hagiwara, Proc. 4th Int. Conf. on Rapidly Quenched Metals, Sendai, (1981), pp.31.

Electrochemical Deposition of (Bi,Sb)$_2$Te$_3$ for Thermoelectric Microdevices

Jean-Pierre Fleurial, Jennifer A. Herman, G. Jeffrey Snyder, Margaret A. Ryan,
Alexander Borshchevsky, and Chen-Kuo Huang
Jet Propulsion Laboratory/California Institute of Technology
4800 Oak Grove Drive
Pasadena, CA 91109, USA

ABSTRACT

New experimental methods have been developed to electrochemically deposit p-type Sb-rich Bi$_{2-x}$Sb$_x$Te$_3$, Pb-doped and Bi-doped Bi$_2$Te$_3$, and PbTe thick films. Some of the deposited films were dense and had a smooth surface morphology. These films were deposited potentiostatically at room temperature in an acidic aqueous electrolyte. Experimental deposition of Bi$_2$Te$_3$ alloys into various thick nanoporous templates made out of anodized alumina has also been achieved. Miniaturized thermoelements for microdevices (25 µm tall, 60 µm diameter) were grown by plating through thick photoresist templates. The experimental techniques developed, as well as the transport properties of some of the films and filled templates, will be presented.

INTRODUCTION

Electrochemical deposition (ECD) is a promising alternative technique to bulk synthesis for fabrication of thermoelectric microdevices. It is a simple, inexpensive, fast and safe technique, in which compounds are deposited on a substrate using an electrolyte and controlled current or potential. Deposition parameters determine the film composition, growth rate and morphology. Electrochemical deposition is a well-known technique for growth of metals, but is a relatively new method for growth of semiconductors [1]; nevertheless, it holds great promise for the fabrication of miniaturized thermoelectric elements and devices [2].

Our interests focus on n-type and p-type Bi$_{2-x}$Sb$_x$Te$_3$ materials as well as other known compounds such as PbTe, Zn$_4$Sb$_3$ and CoSb$_3$. We have previously reported the successful development of ECD techniques for the growth of n-type Bi$_2$Te$_3$ and p-type Bi$_{2-x}$Sb$_x$Te$_3$ [3]. However, p-type films with thicknesses greater than 5 µm typically possess a rough morphology unsuitable for application to devices. This paper briefly reports progress in optimizing ECD experimental conditions to grow smooth p-type Bi$_{2-x}$Sb$_x$Te$_3$ films, the growth of microelements and nanostructures using templates, and the fabrication of complete thermoelectric devices using a combination of electrochemical deposition and integrated circuit techniques.

ELECTRODEPOSITION OF $Bi_{2-x}Sb_xTe_3$ THERMOELECTRIC FILMS

Experimental Method

All experiments were run in a three electrode configuration (Pt counter electrode, Pt or Au working electrode and a Standard Calomel Electrode reference) using an EG&G Princeton Potentiostat/Galvanostat 273A. The standard electrolytes contained millimolar amounts of oxidized high purity elements, salts or chelating agents in aqueous 1 M HNO_3 (pH = 0). Film thickness was measured using a Dektak profilometer. The atomic composition was obtained using a Siemens D-500 x-ray diffractometer or with a Perkin Elmer 3300 DV inductively coupled plasma/optical emission spectrometer. The crystallographic orientation of the films were studied with a JEOL JXA-733 electron superprobe. The Seebeck coefficient was measured in a cross-plane direction using a simple differential thermocouple setup.

N-type Bi_2Te_3 Films

We have previously reported our capability of producing good quality n-type Bi_2Te_3 films and thermoelectric elements (legs) [3, 4]. Since that time, we have run a matrix of experiments to optimize our experimental parameters for the n-type material. This matrix followed the Robust Design "Taguchi" approach, a mathematical model that allows one to evaluate a matrix of experimental parameters without having to perform every possible experimental combination [5]. This method allowed us to perform only 18 experiments (twice) instead of 4374 of them. Experimental parameters included in the study were deposition potential, temperature, metal concentrations, solution pH, stirring rate, Ar de-aeration and substrate quality. Results to date indicate that a combination of slightly elevated temperature (45°C), slow solution stirring rate and low voltage (-2 mV vs. SCE) worked best for depositing smooth, uniform stoichiometric films. De-aerating the electrolyte with Argon or small changes in solution pH (0 to 1.0) did not have any significant impact on film morphology or properties. Pulse deposition techniques, alternating potential between deposition and short etch-back sequences can decrease even further the surface roughness from about 1 µm to less than 0.2 µm.

However, post deposition heat treatments are still needed to anneal lattice defects, reduce electron concentration and optimize electrical properties. Figure 1(a) displays the morphology of n-type Bi_2Te_3 thermoelements electrodeposited under these conditions and then heat-treated at 250°C for 1 hour.

Excess Bi and Pb-doped Bi_2Te_3 Films

It is well known that for bulk Bi_2Te_3, p-type conductivity can be obtained by doping with excess Bi (stoichiometric deviation) or Pb [6]. We were able to successfully grow films with excess Bi or Pb doping, up to a few atomic percent. However, these films were still n-type, even after high temperature annealing. It appears that defects in the film as a result of electrochemical deposition overrides the effect of the doping. A detailed study of the type and concentration of defects in electrodeposited Bi_2Te_3 films is needed to understand the source of the high electron concentration in as-deposited films. Studies on thick films deposited by the hot-wall epitaxy method suggested that Te vacancies and not antisite defects (i.e. Te atom on Bi site) could be responsible [7].

As a result of these studies, we have focused on finding the optimal experimental conditions for electrodepositing p-type Sb-rich $Bi_{2-x}Sb_xTe_3$ films.

P-type BiSbTe₃ Films

Sb_2Te_3-rich $Bi_{2-x}Sb_xTe_3$ compositions are best for p-type thermoelements. Such alloys can be obtained in electrodeposited films by controlling the [Sb]/[Bi] and [Sb]/[Te] ratios in the electrolyte. However, the solubility of Sb into the pH = 0 acidic electrolyte is very low (up to 0.0008 M) and relatively high deposition voltages (-120 mV and more negative values) must typically be used [3], resulting in slow deposition rates (1 μm/hr vs. the 15 μm/hr growth rate of Bi_2Te_3) and columnar, dendritic growth and highly porous films. At less negative voltages, Bi and especially Te deposition tend to dominate the process and off-stoichiometric films are obtained. Much higher Sb concentrations can be obtained by using chelating agents such as EDTA or tartaric acid (up to 0.03 M). However the reaction kinetics are also changed (slow release of chelated Sb for deposition). Using again a Robust Design "Taguchi" approach to cover a wide experimental space, conditions to work around these limitations and produce p-type legs with suitable composition and morphology have been identified. The p-type legs produced as a result of these changes are shown in Figure 1(b).

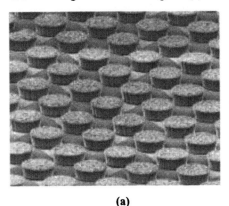
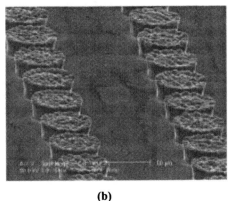

(a)   (b)

Figure 1. (a) *Micrograph of electrodeposited n-type Bi_2Te_3 25 μm thick thermoelements subsequently heat-treated at 250°C for one hour in inert atmosphere;* (b) *Micrograph of electrodeposited p-type $Bi_{0.5}Sb_{1.5}Te_3$ 25 μm thick thermoelements; notice the increased surface roughness compared to n-type deposits.*

Further improvements in surface morphology can be obtained by using a pulse deposition technique (similarly to that for n-type deposition), albeit resulting in a slower deposition rate. In addition to completing optimization of deposition conditions in the pH = 0 acidic electrolytes, recent studies on near neutral pH solutions (pH = 6-7) with very high elemental solubility (0.01

to 0.1 M) could offer an attractive alternative [8]. Experimental details and deposition results will be reported later.

ELECTRODEPOSITION OF NEW MATERIALS AND STRUCTURES

PbTe and $Bi_xPb_yTe_z$ Films

There is interest in investigating the electrochemical deposition of novel thermoelectric materials such as Zn_4Sb_3 and $CoSb_3$-based skutterudites, materials that are already being studied at JPL in bulk form. Our recent work on Pb-doped Bi_2Te_3 films has led us to consider the deposition of a more traditional thermoelectric material, PbTe, as well as layered compounds based on both Bi_2Te_3 and PbTe, such as $PbBi_4Te_7$.

The setup for the PbTe experiments was very similar to that of Bi_2Te_3: three electrode configuration, aqueous HNO_3-based (pH = 1) electrolyte. In order to achieve a 50:50 atomic ratio of lead and tellurium in the film, it was found that close concentrations of Pb and Te were required, in a departure from earlier results [9]. Current experimental conditions are a 2:1 [Pb]:[Te] concentration ratio and deposition voltages in the –150 to –200 mV range (vs. SCE). Figure 2(a) shows the voltammogram of [Pb]+[Te] in 0.1 M [HNO_3] aqueous solution. The electrolyte with a high Pb concentration (25:1) shows an anodic Pb^0 oxidation peak near –0.4 V (vs. SCE). Both solutions show the anodic oxidation of Te near +0.5 V. Pb and Te codeposition into a near-stoichiometry PbTe polycrystalline film (Figure 2(b) micrograph) occur in the –0.1 to –0.3V potential range. Films with excess Pb or excess Te can be prepared, with the x-ray diffraction pattern of a stoichiometric film shown in Figure 2(c). To date, deposited PbTe films were found to be n-type (Seebeck on average of $-76\,\mu V/K$) single phase, face-centered cubic crystals (Figure 2). This result matches the bulk material well, however, the film's surface quality is not yet appropriate for use in a thermoelectric device. Future experiments will focus on improving the film morphology by slowing down the growth rate and varying the electrolyte, potentially using a chelating agent. Finally, initial experiments on depositing $Pb_xBi_yTe_z$ films have shown that a number of crystalline compositions could be obtained with $PbBi_4Te_7$ and possibly $Pb_4Bi_2Te_7$ identified. Further study of this system with attractive bulk properties [10] through electrochemical deposition techniques is in progress.

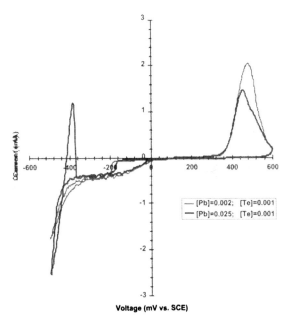

Figure 2. (a) *Voltammogram of [Pb]+[Te] electrolytes in 0.1 M [HNO$_3$] aqueous solution. Only the 25:1 electrolyte shows an anodic Pb^0 oxidation peak near −0.4 V (vs. SCE).*

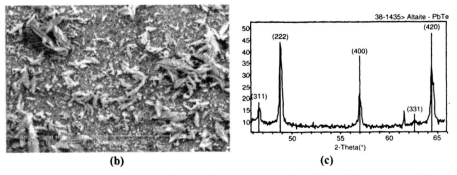

Figure 2. (b) *Micrograph of an as-deposited stoichiometric PbTe film and* (c) *corresponding x-ray diffraction pattern.. The small response at 61.5 degrees is believed to be from either an intermetallic associated with the substrate or a small amount of film oxide.*

Nanowires

Recent theoretical and experimental results [11,12] have shown the potential of low dimensional thermoelectric structures, such as nanowires, for altering electrical and thermal transport properties, potentially leading to significant improvements in the figure of merit. Using anodized alumina nanotube templates (Figure 3(a)) metallized on one side, Bi_2Te_3 nanowires with extremely high length over cross-section ratio can be grown, as shown on Figure 3(b). Nanowires as small as 8 nm in diameter were recently grown using custom-made templates. Initial Seebeck measurements on the 200nm diameter nanowires did not show any significant departure from values typical of as-deposited films, around –50 µV/K at 300 K.

This technique can be applied to other thermoelectric materials of interest, including Bi, and more experimental data will be reported later.

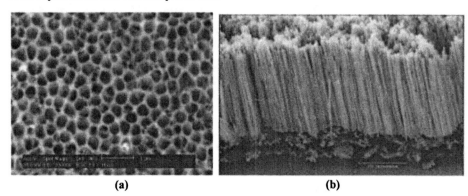

(a)   **(b)**

Figure 3. *Micrographs of an empty nanotube alumina template* **(a)** *and side view of n-type Bi_2Te_3 nanowires* **(b)**, *200nm in diameter and 20 µm long, grown using such templates (the template has been removed by etching and the Pt bottom metallization can be seen).*

ELECTRODEPOSITION OF COMPLETE THERMOELECTRIC MICRODEVICES

There is a growing interest in highly miniaturized multifunctional electronic and optoelectronic devices [4,13]. The high degree of integration sought for such devices requires novel approaches to thermal management, power generation, storage, management and distribution issues. Thermoelectrics can address some of these issues provided that high performance, miniaturized devices can be embedded into such multifunctional structures.

A key aspect of our research effort into thermoelectric microdevices is to develop techniques that are compatible with micromachining, integrated circuit and semiconductor technology. We have reported previously on our approach that combines electrochemical deposition of thermoelectric thick films, advanced UV photolithography and standard vapor deposition of thin metal films [4]. This combination offers a high degree of flexibility in designing and fabricating thermoelectric microdevices. A single photolithography mask can usually combine all of the necessary patterns to completely fabricate one microdevice configuration. Several fully interconnected microdevice structures have recently been successfully synthesized, as shown in Figures 4(a) and 4(b). Even though various fabrication steps still need to be optimized, this picture demonstrates that complete thermoelectric microdevices can be successfully built on top of a silicon substrate without any soldering step. Efforts are now focusing on fabricating several operational devices and testing their electrical and thermal performance.

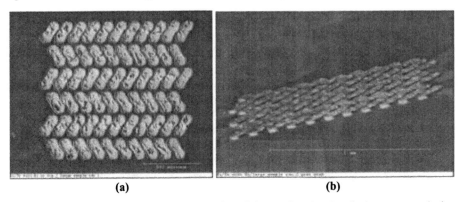

(a) (b)

Figure 4. *Micrographs of a complete Bi_2Te_3-based thermoelectric microdevice structure (only n-type material was used here). Each leg is 25 µm thick and 60 µm in diameter. Bottom interconnects are sputtered Au films and top interconnects are electroplated Ni films.* **(a)** *after Ni deposition and* **(b)** *after photoresist removal.*

CONCLUSION

We have discussed here several areas of electrochemical deposition research being carried out by the Thermoelectrics Group at the Jet Propulsion Laboratory. Recent advances in electrochemical deposition of p-type $Bi_{2-x}Sb_xTe_3$ and PbTe films and nanostructures are promising. Complete microdevice fabrication is under development; a working device will soon be ready for testing of its thermoelectric properties. We will investigate novel thermoelectric materials such as Zn_4Sb_3 and $CoSb_3$, as well as new methods to deposit $Bi_{2-x}Sb_xTe_3$ alloys in the next year. We expect that electrochemical deposition of this group of semiconductor alloys will become a standard manufacturing process, allowing further miniaturization of thermoelectric microdevices.

ACKNOWLEDGEMENTS

The authors would like to thank S. Cronin of the Massachusetts Institute of Technology for providing some of the alumina templates. They also thank J. Kulleck, J. Blosiu, G. Plett and A. Zoltan for technical contributions and R. Bugga and T. Olson for helpful discussions. The work described in this paper was performed at the Jet Propulsion Laboratory/California Institute of

Technology under contract with the National Aeronautics and Space Administration. Part of this work was supported by the U.S. Office of Naval Research, award No. N00014-96-F-0043 and the U.S. Defense Advanced Research Projects Agency, award No. 99-G557.

REFERENCES

1. R.K. Pandey, S.N. Sahu and S. Chandra in Handbook of Semiconductor Deposition, Ed. M. Dekker, New York (1996).
2. M. Muraki and D.M. Rowe, *Proc. X^{th} Int. Conf. on Thermoelectrics*, Cardiff, Wales, UK, 174 (1991).
3. J.-P. Fleurial et al., in: Thermoelectric Materials 1998 eds. T.M. Tritt, M.G. Kanatzidis, H.B. Lyon, and G.D. Mahan, MRS Volume 545, *MRS 1998 Fall Meeting Symp. Proc.*, (1998).
4. J.-P. Fleurial et al., *Proceedings, 18th International Conference on Thermoelectrics*, ed. A.C. Ehrlich (IEEE Catalog 99TH8407), p. 294 (1999)
5. W.Y. Fowlkes, C.M. Creveling, in Engineering Methods for Robust Design Using Taguchi Methods in Technology and Product Development, Addision-Wesley Publishing Company (1995).
6. M.K. Zhitinskaya et al., *Proc. XIV Intl. Conf. Thermoelectrics*, St. Petersburg, Russia, June 27-30, 231 (1995).
7. B.B. Anisev et al., *Inorg. Materials*, 8, 1380 (1991).
8. R. Williams, private communication.
9. E.A. Streltsov et al., *Electrochimica Acta*, 407-413 (1998).
10. Y. Oosawa et al., *Proceedings, 18th International Conference on Thermoelectrics*, ed. A.C. Ehrlich (IEEE Catalog 99TH8407), p. 550 (1999).
11. X. Sun, Z. Zhang and M.S. Dresselhaus, *Appl. Phys. Lett.*, 74, 4005 (1999).
12. S.B. Cronin et al., *Proceedings, 18th International Conference on Thermoelectrics*, ed. A.C. Ehrlich (IEEE Catalog 99TH8407), p. 554 (1999).
13. J.-P. Fleurial, A. Borshchevsky, T. Caillat and R. Ewell, *Proc. 32nd IECEC*, July 27-August 1, Honolulu, Hawaii (2), 1080 (1997).

TRANSIENT THERMOELECTRIC COOLING OF THIN FILM DEVICES

A. RAVI KUMAR[1], R.G. YANG[1], G. CHEN[1] and J.-P. FLEURIAL[2]
[1]Mechanical and Aerospace Engineering Department,
University of California at Los Angeles
Los Angeles, CA 90095-1597
[2]Jet Propulsion Laboratory/California Institute of Technology,
4800 Oak Grove Drive, MS 277-207,
Pasadena, California 91109

Abstract

We report theoretical analysis for the transient thermal response of thermoelectric (TE) element and the integrated thin-film devices. It is predicted that the TE element geometry and applied current pulse shape influences the transient response of the system. Analysis for the integrated systems shows that the transient response is affected by the effusivity of the attached mass. This analysis provides a means to examine the effectiveness of thermal management of the thin-film devices, particularly semiconductor lasers, using the transient mode operation of thermoelectric coolers, and also suggests geometry constraints and optimum pulse shapes for an integrated system.

Introduction

For a thermoelectric element maintained at a steady temperature, the Peltier cooling being a local effect is confined to the cold end of the element and the Joule heating is volumetric in nature. At steady-state optimum conditions these two effects combined with the heat conduction from hot end to the cold end will allow the cold side of the element to be maintained at a steady low temperature. If a current pulse with magnitude several times higher than the steady state optimum current (I_{opt}) is applied to the element, intense cooling is achieved at the cold end. This coupled with the delay of the diffusion of Joule heat will lead to an instantaneous lower temperature than that reachable at the steady state. This phenomenon is referred to as transient thermoelectric effects.

Stilbans and Fedorovich [1] first reported the transient effects in TE elements. They analyzed response of a TE cooler consisted of two elements soldered together, which is equivalent to zero cold junction mass. The experiments indicated a drop of 12 K compared to the steady state optimum temperature for an applied current of twice the value of I_{opt}. The analysis for semi-infinite TE element shows that the temperature decrease is fifty percent more compared to the steady state operation [2]. Experiments by Landecker [3,4] show that sub 100 K at cold junction is possible by applying a time dependent current pulse. However, these measurements are subjected to question because of lack of good temperature sensing instrumentation. Further there are no experiments cited in the literature to corroborate those results. Yamamoto [5] designed an integrated structure in which a semiconductor light-emitting diode (LED) is sandwiched between the p and n junctions of two TE elements. The transient mode operation of this device resulted in doubling the output power from the LED indicating that the lower temperature due to the transient effect in the TE element assisted the improved performance of the LED.

The cross sectional area of the hot end (A_h) and cold end (A_c) determines the thermal resistance for Joule heat. By increasing the ratio of A_h to A_c, Hoyos et al [6] experimentally showed that it is possible obtain better transient performance (that include lower temperature and recovery time) compared to the TE elements having equal A_h and A_c. Recently proposed MEMS based thermal switch utilizes intermittent contact of a mechanical element synchronized with transient effect in TE element to achieve lower temperatures in the mechanical element [7].

Recent developments in the fabrication of thermoelectric microcoolers allow placing the TE coolers near the high heat flux producing active regions of thin film devices that need to be cooled [8]. This will enable a compact and more effective design of thermal system for package level cooling. The integration of micro TE coolers into the semiconductor laser and detectors packaging is being actively pursued.

The transient effects in TE element will allow us to further improve the performance of these devices. We are interested in employing the transient effect of TE element for pulsed mode operation of mid-IR lasers. The mid-IR lasers, which operate continuously at cryogenic temperatures, can also be operated at higher temperatures in pulsed mode operation [9]. The output power tends to decrease and the duty cycle increases as the operating temperature increases. Integrated laser and TE element can take advantage of lower transient temperatures of TE element thus improving output power and duty cycle of the pulsed mode operation of the mid-IR lasers.

Previously mentioned transient analyses of TE system are exclusively for free standing TE elements. The transient response of integrated laser and TE system would require to address the effect of the attached laser mass and its thermal properties. In this work the transient response of an integrated laser and TE element is discussed as a function of the thermal properties of the laser and the TE element length. First, a problem is formulated assuming a semi-infinite mass is attached to a semi-infinite TE element and an analytical solution is provided for the interface temperature as a function of the thermal properties of the attached mass. Later this result is modified to study the transient response of a finite TE element and that of the finite TE element with a finite mass attached to it. The effect of the current pulse shape and the length of the TE are studied for the lowest possible temperature and the holding period. Results provided in this work will help in understanding transient effects and design of laser package integrated with TE elements.

Transient Response of Semi-infinite TE Element Attached to A Semi-infinite Mass

In this section we will examine the temperature response of the cold end of a TE element with a mass attached to it. The hot end of the TE element is assumed to be at room temperature and the side surfaces are assumed to be adiabatic. This will lead to a one dimensional distribution of the temperature. When the transient process occurs at very short time scales, thermal diffusion lengths may be much shorter compared to the actual leg length. Under this assumption it is valid to assume that both the attached mass and the TE element are semi-infinite and the hot end of the TE element does not affects the temperature at the cold end. The coordinate system chosen for this situation is described in Fig.1. The energy balance equations for the TE element

$$\frac{\partial^2 T}{\partial^2 x^2} + \frac{j^2 \rho}{k} = \frac{1}{\alpha}\frac{\partial T}{\partial t} \qquad (1)$$

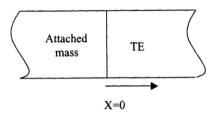

Figure 1 Coordinate system for semi infinite mass attached to a semi infinite TE element

For the attached mass

$$\frac{\partial^2 T_L}{\partial^2 x^2} = \frac{1}{\alpha_L}\frac{\partial T_L}{\partial t} \qquad (2)$$

With the initial condition

$$T_L(x,0) = T(x,0) = T_0 \qquad (3)$$

The boundary conditions at the interface are

$$-k_L \frac{\partial T_L}{\partial x} = -k\frac{\partial T}{\partial x} + SjT \qquad (4)$$

$$T_L(0,t) = T(0,t) \qquad (5)$$

and $\qquad T_L(\infty,t) = T(\infty,0) = T_0 \qquad (6)$

Here, k is thermal conductivity and α is thermal diffusivity of the TE material and the subscript L used to denote these properties of the attached mass, S is seebeck coefficient, ρ is resistivity of the TE material, and j is the applied current density. Solving the above set of equations using the Laplace transform technique and omitting the intermediate steps, the final solution for the temperature at the interface is

$\Delta T = T_0 - T(0,t) =$

$$T_0\left[(1-\exp A^2 \; \text{erfc} A)\left\{\frac{1+ZT_0}{ZT_0}\right\} - \frac{2}{\sqrt{\pi}}\frac{A}{ZT_0}\right] - \frac{k_L P}{ZSj\sqrt{\alpha_L}}\left[1-\exp A^2 \; \text{erfc} A - \frac{2A}{\sqrt{\pi}}\right] \qquad (7)$$

In the above expression the modified figure of merit ZT is

$$ZT = \frac{S^2 k}{\left(\frac{k}{\sqrt{\alpha}}+\frac{k_L}{\sqrt{\alpha_L}}\right)^2 \alpha \rho} T \qquad (8)$$

and $\qquad A = P\sqrt{t} \; \text{with} \; P = \dfrac{Sj}{\left(\dfrac{k}{\sqrt{\alpha}}+\dfrac{k_L}{\sqrt{\alpha_L}}\right)} \qquad (9)$

Equation (7) suggests that the temperature response of the TE element is affected by the effusivity $(k\rho c_p)_L$ of the attached mass. For $(k\rho c_p)_L = 0$, Eq. (7) reduces to the transient response of a freestanding TE element [5]. Figure 2 shows the ΔT given by Eq. (7) for different values of $(k\rho c_p)_L$. The value of ρ equal to 5500 kg.m^{-3} and c_p equal to 250 J kg^{-1} k^{-1} are corresponding to that of a typical mid-IR laser substrate material.

The plot shown for $k_L = 0$ corresponds to freestanding semi-infinite TE element, this achieves the lowest possible temperature during the transient mode operation [2]. With the mass attached at the cold end, the cooling power produced diffuses into the attached mass thus reduces the temperature of the cold junction. Equation (7) suggests that the decrease in the value of

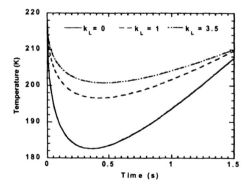

Figure 2. The transient temperature behavior of the cold junction of a semi infinite TE elemen attached to semi-infinite mass

effusivity helps in decreasing the thermal diffusion into the attached mass hence a higher temperature drop at the cold junction.

The above discussion suggests guideline for analysis of an integrated system. In practice both the TE element and the attached mass are finite in size, their length affects the transient performance of the system. In addition, the time dependence of the applied current pulse also influences the evolution of Peltier and Joule effects. This can be observed from Eq. (9) which suggests optimal pulse shape $j\sqrt{t}=\text{const}$. In the following sections the effect of the TE element length and the applied pulse shape are discussed. First, performance of a freestanding TE element with finite length is analyzed; later, integrated laser and TE system is studied.

Transient Response of Freestanding TE Element of Finite Length

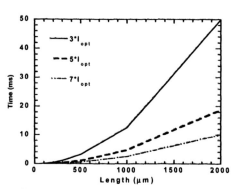

Figure 3 Effect of the TE element length on the holding period

The length of TE element determines its thermal inertia, the shorter the length the lower will be the thermal inertia and vice versa. The steady state analysis suggests that the optimum current is inversely proportionate to the length of the TE element. These two conditions indicate that its length affects the holding period as well as the recovery period. Transient response of a thermoelectric element of finite length is solved. We define the holding periodic as the time period over which it is possible to maintain the temperature within the range of 2 K from the lowest temperature possible. Figure 3 shows the holding period as a function of length for different pulse magnitudes. The boundary conditions assumed here are: the cold end is at adiabatic conditions and the hot end is maintained at room temperature.

As shown in Fig.3 the holding period is longer as the length of he element increases. This is because the applied current flux is much larger for the shorter elements and thus high heat dissipation density close to the cold junction, for example the applied current pulse ($5*I_{opt}$) for the 50 μm element is ten times larger than that of the 500 μm element. However, the recovery

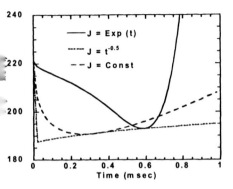

Figure 4 Effect of the pulse shape on the transient response

period, the time required for cold end to reach its steady state temperature after it has reached its lowest temperature (and removing the applied current pulse) will be longer for longer element and vice versa. The current pulse shape also affects the transient response; Fig. 4 shows the temperature response for three different pulse shapes for a 0.5 x 0.5 x 0.5 mm³ TE element. We found numerically that the lowest temperature that can be obtained is the same for any pulse shape, but the holding period deviates for different pulse shapes. In order to take advantage of the spatial difference between the Peltier and Joule effects, the better approach would be that the applied current pulse should be higher at beginning and subsequently it should be reduced similar to pulse $j = t^{-0.5}$ [2,3,10]. This will enable a longer holding period compared to the ramp pulse or the square pulse also described in the figure. In general for longer holding period the applied current pulse shape should be of the form $I \sqrt[n]{t}$ = const, where I is the current and t is the time with n >1 [4].

Transient Response of Integrated Laser and TE Element

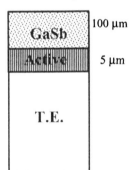

Figure 5. Integrated laser and TE element

Currently we are pursuing improving the performance of the mid-IR laser through their integration with TE coolers operated in transient mode. The transient effects in TE element can improve the performance of pulsed mode operated lasers by lowering the temperature of the active region before it is fired. The longer holding periods help to offset the temperature rise of the laser during its firing. For repetitive operation, the recovery time must also be taken into consideration. In previous section, it was shown that the longer holding period requires longer TE element. In order to study the transient response of TE element integrated with laser, finite element simulations were carried out. The device structure considered for the simulation is shown in Fig. 5.

Figure 6 shows the effect of the TE length on the temperature obtained at the interface. Since the active region is placed near the cold end it is valid to assume that the active region temperature is same as the interface temperature, particularly when the time to reach minimum is relatively long. The simulations were carried for 0.5, 2 and 5 mm length TE elements and the length of the active region and the laser substrate are kept constant at 5 and 100 μm respectively. A large decrease in temperature can be observed from Fig.6 as the TE element length increases. For 5 mm length element, sub 200 K can be achieved at the active junction.

The thickness of the attached laser and its substrate also affects the temperature response. Figure 7 shows the transient response of a 0.5 mm thick TE element for different thickness of the

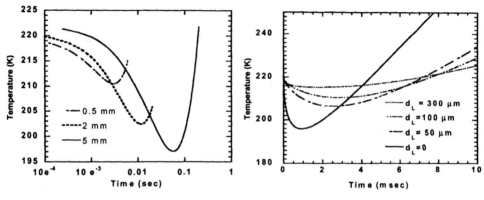

Figure 6. Transient response of the integrated systems for different lengths of TE element

Figure 7. Temperature response for different thick nesses of laser substrate

attached substrate, and also the response of a freestanding element is compared. The increase in the substrate length affects the lowest temperature possible as well as the inertia of the integrated system. Better performance can be obtained if the substrate thickness is reduced.

Conclusions

This work explores the potential application of transient Peltier effect for cooling thin film devices, particularly midIR lasers, mounted junction down to the cold end of the TE element. It was shown that the length of the TE element and the applied pulse shape affects the holding period. The effusivity of the attached device determines the cooling effect and the transient response of the integrated system.

Acknowledgments

This work is supported by DARPA HERETIC program.

References

[1] L.S. Stilbans and N.A. Fedorovich, Soviet Phys-Tech. Phys., 3, p. 460-463 (1958)
[2] V.P. Babin and E.K. Iordanishvili, Soviet PhysTech. Phys., 14, p. 293-298 (1969)
[3] K. Landecker and A.W. Findlay, Solid State Electronics, 2, p. 239-260 (1961)
[4] M. Idnurm and K. Landecker, Journal of Appl. Phys., 34, p 1806-1810 (1963)
[5] T. Yamamoto, Proc. of IEEE, pp. 230-232 (1968)
[6] G.E. Hoyos, K.R.Rao and D. Jerger, Energy Conversion, 17, pp.45-54 (1977)
[7] A. Miner, A. Majumdar, U. Ghoshal Appl. Phys. Lett. 75, 1176 (1999)
[8] J.-P. Fleurial, A. Borshchevsky, M.A. Ryan, W. Phillips, E. Kolawa, T. Kacisch, and R. Ewell, p. 641-645, 16th ICT (1997).
[9] R.Q. Yang, Microelectronics Journal, 30, p. 1043-1056 (1999).
[10] K. Woodbridge and M.E. Ertl, Phys. Stat. Sol. (a), 44, p. k123-k126 (1977)

P-type SiGe/Si Superlattice Cooler

Xiaofeng Fan, Gehong Zeng, Edward Croke[1], Gerry Robinson, Chris LaBounty, Channing C. Ahn[2], Ali Shakouri[3], and John E. Bowers
Department of Electrical and Computer Engineering, University of California, Santa Barbara, CA 93106, U.S.A.
[1]HRL Laboratories, LLC, Malibu, CA 90265, U.S.A.
[2]California Institute of Technology, Pasadena, CA 91125, U.S.A.
[3]Baskin School of Engineering, University of California, Santa Cruz, CA 95064, U.S.A.

ABSTRACT

The fabrication and characterization of single element p-type SiGe/Si superlattice coolers are described. Superlattice structures were used to enhance the device performance by reducing the thermal conductivity between the hot and the cold junctions, and by providing selective emission of hot carriers through thermionic emission. The structure of the samples consisted of a 3 μm thick symmetrically strained $Si_{0.7}Ge_{0.3}$/Si superlattice grown on a buffer layer designed so that the in-plane lattice constant is approximately that of relaxed $Si_{0.9}Ge_{0.1}$. Cooling up to 2.7 K at 25 °C and 7.2 K at 150 °C were measured. These p-type coolers can be combined with n-type devices that were demonstrated in our previous work. This is similar to conventional multi element thermoelectric devices, and it will enable us to achieve large cooling capacities with relatively small currents.

INTRODUCTION

Effective cooling is essential for many high power or low noise electronic and optoelectronic devices. Thermoelectric (TE) refrigeration is a solid-state active cooling method with high reliability. Unlike conventional air-cooling, it can spot cool discrete or localized devices and reduce the temperature of the device below ambient. For a material to be a good thermoelectric cooler, it must have a high value of the dimensionless figure of merit ZT [1] which is given by $ZT=S^2\sigma T/\kappa$, where S is the Seebeck coefficient, σ is the electrical conductivity, T is the temperature, and κ is the thermal conductivity. The use of quantum-well structures to increase ZT was proposed by Hicks and Dresselhaus [2]. Since then much work has been done in the study of superlattice thermoelectric properties, mostly for the in-plane direction [3-6]. The physical origin of the increase in ZT comes mainly from the enhanced density of electron states due to the reduced dimensionality. Recent study shows that superlattice thermal conductivity of cross-plane direction is even lower than that of in-plane direction [7], which can further increase ZT. In addition, Shakouri and Bowers proposed that heterostructure could be used for thermionic emission to enhance the cooling [8]. Large ZT improvement is possible for the cross-plane transport [9, 10].

SiGe is a good thermoelectric material especially for high temperature applications [11]. Superlattice structures can enhance the cooler performance by reducing the thermal conductivity between the hot and the cold junctions [7, 12], and by selective emission of hot carriers above the barrier layers in the thermionic emission process [8, 9]. N-type SiGe/Si cooler was reported in our previous work [13]. In this paper, single element p-type SiGe/Si superlattice coolers with electrical transport in the cross-plane direction is demonstrated. This paves the road to make n-type and p-type superlattice coolers electrically in series and thermally in parallel, similar to conventional TE coolers, and thus achieve large cooling capacities with relatively small currents.

MATERIAL AND DEVICE FABRICATION

The p-type SiGe/Si superlattice cooler sample was grown with molecular beam epitaxy (MBE) on a five-inch diameter (001)-oriented Boron doped Si substrate with resistivity less than 0.006 Ω-cm. The cooler's main part is a 3 μm thick 200 × (5 nm $Si_{0.7}Ge_{0.3}$/10 nm Si) superlattice grown symmetrically strained on a buffer layer designed so that the in-plane lattice constant was approximately that of relaxed $Si_{0.9}Ge_{0.1}$. The buffer layer consisted of 1 μm 5 × (150 nm $Si_{0.9}Ge_{0.1}$/50 nm $Si_{0.845}Ge_{0.150}C_{0.005}$) and 1 μm $Si_{0.9}Ge_{0.1}$. Both the superlattice and the buffer layer are doped to 5×10^{19} cm^{-3} with Boron. 0.5 μm thick $Si_{0.9}Ge_{0.1}$ cap layer was grown on the superlattice with the top 0.25 μm doped to 2×10^{20} cm^{-3} to achieve good ohmic contact. The $Si_{0.7}Ge_{0.3}$/Si superlattice has a valence band offset of about 0.2 eV [14], and hot holes over this barrier produce thermionic cooling. In addition, superlattice structure has many interfaces that increase phonon scattering, and therefore gets lower thermal conductivity. The material growth procedure is similar to that of n-type SiGe/Si superlattice, described in reference [13]. A transmission electron microscopy (TEM) image of the grown p-type SiGe/Si superlattice cooler sample is shown in figure 1.

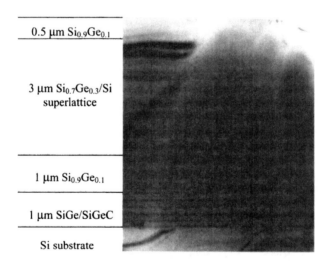

Figure 1. *TEM image of the p-type SiGe/Si superlattice cooler sample*

For the cooler device fabrication, mesas with an area of 50×50 μm^2 were etched down to the $Si_{0.9}Ge_{0.1}$ buffer layer using reactive ion etching. Metallization was made on the mesa and $Si_{0.9}Ge_{0.1}$ buffer layer for top and bottom contact respectively. Electrical current goes from the top contact to bottom contact for cooling. This is a cross-plane transport in the superlattice. To reduce contact resistance and facilitate wire bonding, Ti/Al/Ti/Au was used for contact metallization. Annealing was accomplished at 450 °C, and specific contact resistance of 3.6×10^{-7} Ω-cm^2 was measured.

TEST RESULTS AND DISCUSSIONS

The p-type SiGe/Si superlattice coolers were tested on a temperature controlled copper plate that worked as the heat sink. The cooling area of the single element device is 50×50 μm^2. The device cooling temperatures were measured with micro thermocouples, and they are relative to the device temperature at zero current. Figure 2 shows the measured cooling temperature versus current with the heat sink at 25 °C. Cooling up to 2.7 K with respect to the heat sink was obtained, corresponding to maximum cooling power densities on the order of 100 W/cm^2 at zero delta T.

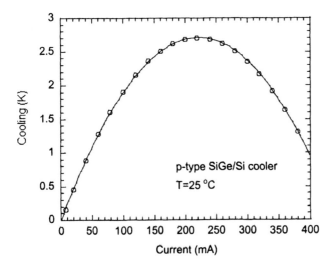

Figure 2. *Measured cooling of p-type SiGe/Si superlattice cooler at 25 °C (heat sink). The dots are measured data and the line is their quadratic fitting curve.*

The device cools better at higher temperatures. The measured cooling of the 50×50 μm^2 p-type SiGe/Si device at 150 °C (heat sink temperature) is shown in figure 3. The maximum cooling increased from 2.7 K at 25 °C to 7.2 K at 150 °C. The reason for the improved performance with the increase in temperature is two fold. First, in the temperature range of our measurements, the figure of merit ZT of SiGe alloy increases with temperature due to smaller thermal conductivity and larger Seebeck coefficient [15], and second, the thermionic emission cooling power increases due to the larger thermal spread of carriers near the Fermi energy.

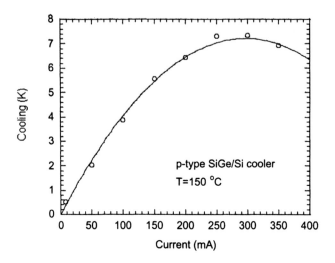

Figure 3. *Measured cooling of p-type SiGe/Si superlattice cooler at 150 °C (heat sink). The dots are measured data and the line is their quadratic fitting curve.*

Since the devices here are single element superlattice coolers, heat conduction to the cooling side from the bonding wires or probes are unavoidable. This reduces the maximum cooling. To solve this problem, n-type and p-type SiGe/Si superlattice coolers can be made in an array format electrically in series and thermally in parallel, similar to conventional thermoelectric coolers. In this way, both electrical terminals can be made at the heat sink side, and large cooling capacities can be achieved with relatively small currents. With optimized superlattice material and device design and packaging, cooling up to tens of degrees is possible. More important, the processing of SiGe/Si superlattice coolers is compatible with that of very-large-scale-integration (VLSI) technology, thus it is possible to integrate these coolers monolithically with Si and SiGe devices to achieve compact and efficient cooling.

CONCLUSIONS

P-type SiGe/Si superlattice cooler was demonstrated. Cooling up to 2.7 K at 25 °C and 7.2 K at 150 °C was obtained, corresponding to maximum cooling power densities of hundreds of watts per square centimeter at zero delta T.

ACKNOWLEDGMENTS

The authors would like to acknowledge many stimulating discussions with Professor Venky Narayanamurti. This work was supported by the DARPA HERETIC program and the Army Research Office.

REFERENCES

1. H. J. Goldsmid, *Thermoelectric Refrigeration* (Plenum, New York, 1964).
2. L. K. Hicks and M.S. Dresselhaus, *Phys. Rev. B*, **47**, 12727 (1993).
3. P. J. Lin_Chung and T. L. Reinecke, *Phys. Rev. B*, **51**, 13244 (1995).
4. R. Venkatasubramanian, E. Siivola, and T. S. Colpitts, *Proceedings of the 17th International Conference on Thermoelectrics*, 191 (1998).
5. T. Koga, T. C. Harman, S. B. Cornin and M. S. Dresselhaus, *Phys. Rev. B*, **60**, 14286 (1999).
6. T. Koga, X. Sun, S. B. Cronin and M. S. Dresselhaus, *Apple. Phys. Lett.*, **75**, 2438 (1999).
7. G. Chen, S. Q. Zhou, D.-Y. Yao, C. J. Kim, X. Y. Zheng, Z. L. Liu and K. L. Wang, *Proceedings of the 17th International Conference on Thermoelectrics*, 202 (1998).
8. A. Shakouri and J. E. Bowers, *Apple. Phys. Lett.*, **71**, 1234 (1997).
9. A. Shakouri, C. Labounty, P. Abraham, J. Piprek, and J. E. Bowers, *Material Research Society Symposium Proceedings*, **545**, 449 (1999).
10. L. W. Whitlow and T. Hirano, *J. Apple. Phys.*, **78**, 5460 (1995).
11. C. B. Veining, *J. Apple. Phys.*, **69**, 331 (1991).
12. S.-M Lee, D. G. Cahill and R. Venkatasubramanian, *Appl. Phys. Lett.*, **70**, 2957 (1997).
13. G. Zeng, A. Shakouri, C. LaBounty, G. Robinson, E. Croke, P. Abraham, X. Fan, H. Reese and J. E. Bowers, *Electronics Letters*, **35**, 2146 (1999).
14. R. People and J. C. Bean, *Apple. Phys. Lett.*, **48**, 538 (1986).
15. J. P. Dismukes, L. Ekstrom, E. F. Steigmeier, I. Kudman and D. S. Beers, *J. Apple. Phys.*, **35**, 2899 (1964).

Progress in the development of segmented thermoelectric unicouples at the Jet Propulsion Laboratory

T. Caillat, J.- P. Fleurial, G. J. Snyder, and A. Borshchevsky
Jet Propulsion Laboratory, California Institute of Technology, MS 277/207, 4800 Oak Grove Drive, Pasadena, CA 91109 USA

ABSTRACT

A new version of a segmented thermoelectric unicouple incorporating advanced thermoelectric materials with superior thermoelectric figures of merit has been recently proposed and is currently under development at the Jet Propulsion Laboratory (JPL). This advanced segmented thermoelectric unicouple includes a combination of state-of-the-art thermoelectric materials based on Bi_2Te_3 and novel materials developed at JPL. The segmented unicouple currently being developed is expected to operate between 300 and about 975K with a projected thermal to electrical efficiency of up to 15%. The segmentation can be adjusted to accommodate various hot-side temperatures depending on the specific application envisioned. Techniques and materials have been developed to bond the different thermoelectric segments together for the n- and p-legs and low contact resistance bonds have been achieved. In order to experimentally determine the thermal to electrical efficiency of the unicouple, metallic interconnects must be developed for the hot side of the thermocouple to connect the n- and p-legs electrically. The latest results in the development of these interconnects are described in this paper. Efforts are also focusing on the fabrication of a unicouple specifically designed for thermal and electrical testing.

INTRODUCTION

The segmented unicouple under development incorporates a combination of state-of-the-art thermoelectric materials and novel p-type Zn_4Sb_3, p-type $CeFe_4Sb_{12}$-based alloys and n-type $CoSb_3$-based alloys developed at JPL. In a segmented unicouple as depicted in Figure 1, each section has the same current and heat flow as the other segments in the same leg. Thus in order to maintain the desired temperature profile (i.e. keeping the interface temperatures at their desired level) the geometry of the legs must be optimized. Specifically, the relative lengths of each segment in a leg must be adjusted, primarily due to differences in thermal conductivity, to achieve the desired temperature gradient across each material. The ratio of the cross sectional area between the n-type and p-type legs must also be optimized to account for any difference in electrical and thermal conductivity of the two legs. A semi-analytical approach that includes smaller effects such as the Peltier and Thompson contributions and contact resistance in order to optimize and calculate the expected properties of the device has been used to solve the problem [1]. For each segment, the thermoelectric properties are averaged for the temperature range it is used. At each junction (cold, hot, or interface between two segments), the relative lengths of the segments are adjusted to ensure heat energy balance at the interface. Without any contact resistance between segments, the efficiency is not affected by the overall length of the device; only the relative length of each segment needs to be optimized. The total resistance and power output, however, does depend on the overall length and cross sectional area of the device. The calculated optimized thermoelectric efficiency is about 15% with the hot junction at 975K and the cold junction near room temperature. The optimal geometry is illustrated in Figure 1.

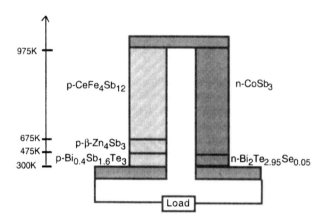

Figure 1. Illustration of the advanced unicouple incorporating new high performance thermoelectric materials. The relative lengths of each segment and the cross-sectional areas for the p- and n-legs are drawn to scale. The calculated thermoelectric efficiency is about 15%.

High contact resistance between the thermoelectric segments and interconnects at the cold and hot sides can dramatically reduce the efficiency of a generator. Calculations show that a low contact resistance, less than about 20 $\mu\Omega cm^2$, is required to keep the efficiency from being significantly degraded by the contact resistance. Techniques and materials have been developed to bond the different thermoelectric segments of the unicouple together [2]. Electrical contact resistance lower than 5 $\mu\Omega cm^2$ have been obtained for each of the junctions at its projected operating temperature. The n- and p-legs have also been successfully connected to a "cold shoe" for heat transfer to the heat sink using a Bi-Sn solder and Ni as a diffusion barrier [2]. While some success has been achieved in brazing Nb metal to the top skutterudite segments [2] using a $Cu_{28}Ag_{72}$ alloy, the relatively high brazing temperature required (780°C) may impact the mechanical integrity of the legs after brazing. Some alternative ways of creating metallic interconnects for the hot-side of the legs have been explored and the results are reported in this paper.

EXPERIMENTAL PROCEDURES

Since both entire n- and p-segmented legs can be fabricated by uniaxial hot-pressing the various thermoelectric materials separated by thin metallic foils [2], an alternative approach to create a metallic layer on the top of the legs is to add some metallic powder on the top of the legs during the hot-pressing. Two type of metals were investigated : Ni and Nb. In some cases entire legs were fabricated with the top metallic portion while in some other cases only the top skutterudite materials were hot-pressed with the metallic powders. After pressing, a small strip of the samples was polished along the pressing axis to reveal the microstructure of the junction which was investigated by both optical microscopy and electron microprobe analysis. In addition, the electrical contact resistance was measured by a four probe technique up to the predicted optimum temperature of operation. One voltage probe is located at one end of the sample while the second probe can move along the sample. The variations of the electrical contact resistance is therefore recorded as a function of the distance of the moving probe to the fixed probe.

RESULTS AND DISCUSSION

Figures 2 and 3 show samples of cut p- and n-entire, hot-pressed legs with Ni as the top metallic segment. The density of the hot-pressed Ni was determined to be 91% of the theoretical density. Microprobe analysis of the Ni/skutterudite interfaces showed a good quality bond. However, in both cases, lateral cracks are visible within the skutterudite materials in the region adjacent to Ni. This is probably due to the relative large difference between the thermal expansion of Ni, 13.3×10^{-6}/K, compared to those for p-$CeFe_4Sb_{12}$ and n-$CoSb_3$, 7.5 and 6.36×10^{-6}/K, respectively.

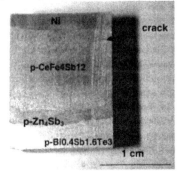 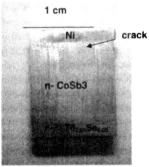

Figure 2. Photograph of a three segments p-leg cut in half and with Ni as the top metallic interconnect. Cracks are visible in the top thermoelectric material next to the interface with Ni.

Figure 3. Photograph of a two segments n-leg cut in half and with Ni as the top metallic interconnect. Cracks are visible in the top thermoelectric material next to the interface with Ni.

Figures 4 and 5 show p-CeFe$_4$Sb$_{12}$/Nb and n-CoSb$_3$/Nb junctions after hot-pressing. Microprobe analysis of the Nb/skutterudites interface regions showed a good quality bond. No cracks were observed in the skutterudite materials which is presumably due to the reasonable match between the thermal expansion coefficient of the skutterudite materials and that of Nb (7.1 x 10^{-6}/K)

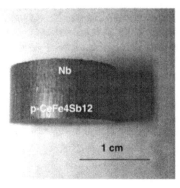
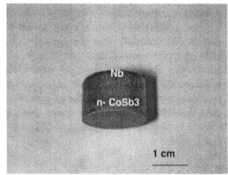

Figure 4. Photograph of a Nb/p-CeFe$_4$Sb$_{12}$ junction fabricated by hot-pressing. No cracks are visible in the skutterudite material.

Figure 5. Photograph of a Nb/n-CoSb$_3$ junction fabricated by hot-pressing. No cracks are visible in the skutterudite material.

The electrical contact resistance between the skutterudite materials and hot-pressed Nb powder was measured and the results are reported in Figures 6 and 7. The results show that electrical resistance lower than 5 μΩcm^2 were achieved at temperatures close to the projected operation temperature. No Nb diffusion into the skutterudite materials was observed by microprobe analysis. The density of the hot-pressed Nb powder was determined to be about 80% of the theoretical value which results in an increased electrical resistivity and could potentially increase the total resistance of the unicouple. The added resistance due to the addition of 1.5 mm of hot-pressed Nb on the top of 1.5 cm long segmented legs would however result in a negligible 0.41% of the total resistance of the unicouple. Once both n- and p-legs can be fabricated with a Nb layer of hot-pressed Nb on the top they would have to be joined by an electrical connector. For electrical and thermal performance testing of the unicouple, this connector could also be used as a heater. A 12 mm diameter heater was fabricated out of Nb. It is composed of a heating element introduced into a Nb shell filled with an electrically insulated refractory cement. Brazing of Nb metallic samples to hot-pressed Nb samples was successfully achieved at temperatures as low as 600°C using a Cu$_{22}$Ag$_{56.2}$Zn$_{17}$Sn$_5$ brazing alloy. The resulting bond was found to be of good mechanical quality and no electrical contact resistance was measured at the interface. Fabrication of an entire unicouple with Nb heater brazed on the top of the legs is currently underway.

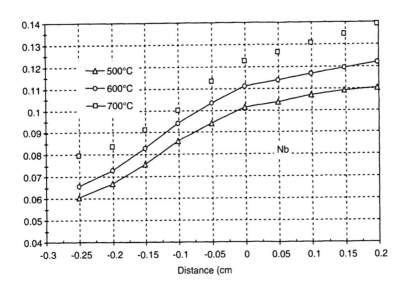

Figure 6. Electrical contact resistance (in mΩcm^2) as a function of distance for a hot-pressed n-CoSb$_3$/Nb junction. The origin corresponds to the interface position.

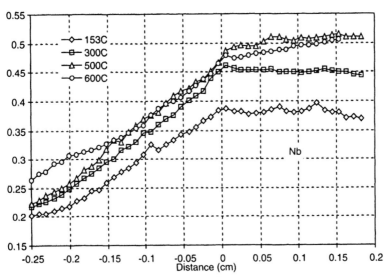

Figure 7. Electrical contact resistance (in mΩcm^2) as a function of distance for a hot-pressed p-CeFe$_4$Sb$_{12}$/Nb junction. The origin corresponds to the interface position.

CONCLUSION

Some alternative ways of creating metallic interconnects for the hot-side of segmented legs of an advanced thermoelectric unicouple have been explored. Hot-pressing Nb on the top of the segmented thermoelectric legs seem to provide a good electrical and mechanical bond. In addition, hot-pressed Nb was successfully brazed to a custom made Nb heater. The most critical technical issues associated with the fabrication of an advanced unicouple designed for thermal and electrical testing have been addressed. Efforts are currently focusing on the fabrication of a unicouple to experimentally determine its efficiency.

ACKNOWLEDGMENTS

The work described in this paper was carried out at the Jet Propulsion Laboratory/California Institute of Technology, under contract with the National Aeronautics and Space Administration. This work is supported by the U. S. Defense Advanced Research Projects Agency, Grant No. E748.

REFERENCES

1. B. W. Swanson, E. V. Somers, and R. R. Heikes, *Journal of Heat Transfer*, 77 (1961).
2. T. Caillat, J. –P. Fleurial, G. J. Snyder, A. Zoltan, D. Zoltan, and A. Borshchevsky, in *Proceeding of 18th International Conference on Thermoelectrics*, edited by A. Ehrlich, IEEE Catalog Number 99TH8407, pp. 473-476 (1999).

Clathrates

THERMAL CONDUCTIVIY OF TYPE I AND II CLATHRATE COMPOUNDS

G.S. Nolas[1], J.L. Cohn[2], M. Kaeser[3] and T.M. Tritt[3]
[1] R&D Division, Marlow Industries, Inc., 10451 Vista Park Road, Dallas, Texas 75238
[2] Department of Physics, University of Miami, Coral Gables, Florida 33124
[3] Department of Physics, Clemson University, Clemson, South Carolina 29634

ABSTRACT

Compounds with clathrate-hydrate type crystal lattice structures are currently of interest in thermoelectric materials research. This is due to the fact that semiconducting compounds can be synthesized with varying doping levels while possessing low, even 'glass-like', thermal conductivity. Up to now most of the work has focused on type I Si and Ge clathrates. Sn-clathrates however are viewed as having the greatest potential for thermoelectric cooling applications due to the larger mass of Sn and the expected small band-gap, as compared to Si and Ge clathrates. Transport properties on type I Sn-clathrates has only recently been reported [1-3]. In this report we present ongoing experimental research on both type I and II clathrates with an emphasis on the thermal transport of these novel materials. We present thermal conductivity data Si-Ge and Ge-Sn alloys as well as on a type II Ge clathrate for the first time, and compare these data to that of other clathrate compounds.

INTRODUCTION

The recent interest in thermoelectrics is fueled by new and novel materials and synthesis techniques presently under development in an effort to improve the efficiency of thermoelectric devices. One approach that has received enormous attention in the past few years is the 'phonon-glass-electron-crystal' (PGEC) approach introduced by Slack [4]. In such an approach the best thermoelectric material would be 'engineered' to have thermal properties similar to those of an amorphous material, 'a phonon-glass', and electronic properties similar to that of an ordered, highly covalent single crystal, 'an electron single crystal'. The importance of this approach emerges very clearly from the definition of the dimensionless thermoelectric figure of merit, ZT, with $ZT=TS^2\sigma/\kappa$. In this equation S is the Seebeck coefficient, σ the electrical conductivity, κ the thermal conductivity and T the absolute temperature. In practice, however, the development of such a material system is not straightforward.

Slack[4] has identified a series of crystal structures that can potentially be 'engineered' for PGEC properties. These are crystal systems with an 'open structure', i.e. with high coordination number, large voids in their lattice wherein loosely-bound 'guest' atoms may be inserted to create phonon scattering centers that substantially suppress κ. Much of the work on this approach has been dominated by the skutterudite family of compounds [5] however most recently Nolas and co-workers have presented evidence that certain compounds with the clathrate-hydrate crystal structure possess PGEC properties.[1-3, 6-12]

The semiconductor clathrate compounds have crystal structures that are formed by fullerene-like polyhedra made up of group IV elements (i.e. Si, Ge or Sn). The polyhedra are covalently bonded to each other by shared faces. These crystal structures are similar to those found in clathrate-hydrates. In clathrate hydrates, or ice-clathrates, each (water) oxygen is four-coordinated with neighboring oxygen atoms. Ice-clathrates form several different crystal structures, in each of which the water molecules form a matrix encapsulating guest atoms or molecules [13]. It is fortuitus to us in thermoelectrics material research that this bonding scheme also nicely applies to the group IV elements. In addition, depending on the guest atom, these compounds, with their 'guests' in place, can be more thermodynamically stable than their diamond structured counterparts (i.e. Si, Ge and gray-Sn) [14]. Silicon clathrate compounds with crystal structures similar to the ice-clathrates were first synthesized by Cros et al [15]. These compounds formed Si_{46} and Si_{136} host matrices with alkali metal atoms in their 'cages'. Since that time there have been other similar compounds synthesized with these two crystal structures based on Ge and Sn. In this report we describe thermal transport and structural properties of semiconducting compounds with the type I and II clathrate crystal structure. To our knowledge no published κ data on type II clathrates exists.

RESULTS AND DISCUSSION

The type I structure has space group $Pm\overline{3}n$ comprised of two dodecahedra, M_{20}, and six tetrakaidecahedra, M_{24}, per cubic unit cell. The type II structure has space group $Fd\overline{3}m$ comprised of 16 dodecahedra, M_{20}, and 8 hexakaidecahedra, M_{28}, per cubic unit cell. Until recently very little has been know of the transport properties and atomic displacement parameters of type-I compounds. An in depth investigation of the type II compounds has not yet begun. The semiconductor clathrate compounds presented in this report are representative of the types of compounds under investigation as potential thermoelectric materials. The clathrate compounds presented in this report were prepared as described previously [1, 7]. The structural and chemical properties of these phase-pure specimens were analyzed by x-ray diffraction and Rietveld refinement, and by optical, electron-beam and transmission electron microscopies.

Figure 1 shows the lattice thermal conductivity, κ_g, as a function of temperature of five polycrystalline compounds with the type I crystal structure. For Cs_8Sn_{44} κ_g exhibits a sharp temperature maximum at T_{max}~10 K and an approximately 1/T dependence for T>T_{max}, both typical of crystalline insulators dominated by phonon-phonon scattering. κ_g for both $Cs_8Zn_4Sn_{42}$ and $Rb_8Zn_4Sn_{42}$ also increases with decreasing temperature, however not as strongly, and T_{max} is shifted to higher T, presumably due in part to mass-fluctuation disorder scattering of the phonons.

Structural information and the hypothesis that resonant interaction of guest-host atom vibrations are a mechanism for κ_g suppression, suggest a simplified interpretation of the differences in κ_g for these compounds. First consider Cs_8Sn_{44} and $Cs_8Zn_4Sn_{42}$. Both have Cs atoms in cages, but the former compound has two Sn vancies per formula unit. The additional bonding induced between the Cs and Sn atoms neighboring the vacancies in Cs_8Sn_{44} apparently constrains the displacement of Cs atoms in the tetrakaidecahedra, Cs(2). This hypothesis is supported by recent temperature-dependent neutron scattering data [2] that reveal a larger thermal motion for Cs(2) in the $Cs_8Zn_4Sn_{42}$ compound but not in the Cs_8Sn_{44} compound. The additional bonding induced between the Cs and Sn atoms neighboring the vacancies in Cs_8Sn_{44} apparently constrain the Cs(2) displacement [2]. We therefore conclude that the additional atomic displacements are the source of κ_g suppression in $Cs_8Zn_4Sn_{42}$ relative to Cs_8Sn_{44}. Rb atoms are smaller than Cs and thus it is expected that Rb 'rattles' more freely inside the polyhedra formed by the (Zn,Sn) atoms. These localized vibrations may provide more prominent phonon scattering that is reflected in the still lower κ_g of $Rb_8Zn_4Sn_{42}$. While the enhanced thermal motion of the Cs(2) and Rb(2) atoms in these Sn clathrates appear to diminish κ_g, the effect is not as great as that caused by the Sr(2) motion in $Sr_8Ga_{16}Ge_{30}$ [8, 11] where the Sr vibration frequencies lie in the range of the acoustic phonons resulting in a 'glass-like' temperature

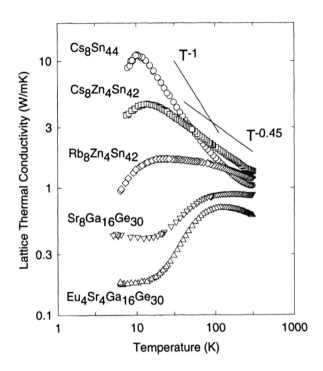

Figure 1. Lattice thermal conductivity of five representative type I semiconducting clathrates and fits to T^{-1} and $T^{-0.45}$ indicating the temperature dependence of Cs_8Sn_{44} and $Cs_8Zn_4Sn_{42}$, respectively.

dependent κ_g [6, 7, 11]. In the case of $Sr_4Eu_4Ga_{16}Ge_{30}$ there are two different atoms in the voids of the crystal structure which introduces six different resonant scattering frequencies (three for each ion). This compound exhibits the lowest κ_g values in the temperature range shown and is very similar to the temperature dependence of amorphous Ge (not shown in the figure).

Figure 2 shows κ_g as a function of temperature for polycrystalline semiconductor Si, Ge, a Si-Ge alloy and a Ge-Sn alloy with the type I crystal structure. The $Ba_8Ga_{16}Si_{30}$ sample has a relatively low κ_g, however the temperature dependence is not similar to that of $Sr_8Ga_{16}Ge_{30}$. Although Ba is much more massive than the elements that make up the host matrix (i.e. Ga and Si), a prerequisite for glass-like κ_g [4], the temperature dependence of κ_g does contain a somewhat crystalline character. This is due to the fact that Ba^{2+} is similar in size to the Si_{20} and Si_{24} cages whereas Sr^{2+} (or Eu^{2+}) are smaller than the Ge_{20} and Ge_{24} cages. From x-ray diffraction data, it does not appear that Ga changes the average size of these cages very much. In the case of the Si-Ge alloy clathrate, κ_g exhibits characteristics similar to that of $Sr_8Ga_{16}Ge_{30}$, as expected, due to the fact that Si does not substantially change the size of the dodecahedra. In the case of

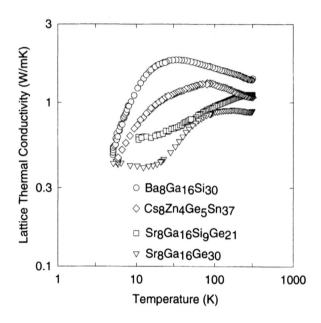

Figure 2. Lattice thermal conductivity of Si, Ge, Si-Ge alloy and Ge-Sn alloy with the type I clathrate crystal structure.

$Cs_8Zn_4Ge_5Sn_{37}$, alloying Ge with Sn reduced κ_g as compared to $Cs_8Zn_4Sn_{42}$ (see Figure 1).

Figure 3 shows κ for polycrystalline $Cs_8Na_{16}Ge_{136}$, polycrystalline $Sr_8Ga_{16}Ge_{30}$, single crystal diamond structured Ge, amorphous Ge (a-Ge) and vitreous silica (a-SiO_2). The polycrystalline $Sr_8Ga_{16}Ge_{30}$ has a temperature dependence that is similar to that of a-SiO_2, while being less than a factor of 2 larger than that of a-Ge in the temperature range shown in Figure 3. In addition κ_g of $Sr_8Ga_{16}Ge_{30}$ is a factor of 5 lower than that of $Cs_8Na_{16}Ge_{136}$. Seebeck and resistivity measurements on $Cs_8Na_{16}Ge_{136}$ indicate this material to be metallic,[16] unlike what was reported by Bobev and Sevov [17]. This may be the reason for the relatively high thermal conductivity measured for this specimen, although it does not appear to be a 'typical' metal.

CONCLUSIONS

The thermal conductivity of type I clathrates correlates with structural features. This is very important for thermoelectric materials research as it helps to elucidate the mechanism(s) that produce the low thermal conductivity in these semiconductors. The thermal conductivity of the type II Ge-clathrate $Cs_8Na_{16}Ge_{136}$ as well as alloys of Si-Ge and Ge-Sn is reported for the first

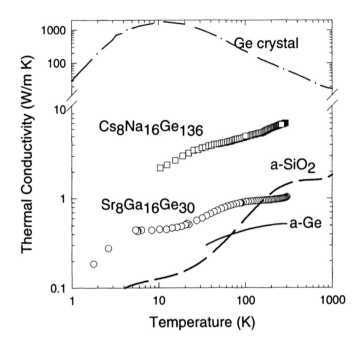

Figure 3. Thermal conductivity of four germanium compounds (type I clathrate $Sr_8Ga_{16}Ge_{30}$, type II clathrate $Cs_8Na_{16}Ge_{136}$, diamond structured Ge and amorphous Ge, a-Ge) and vitreous silica, a-SiO_2.

time. Although the thermal conductivity of $Cs_8Na_{16}Ge_{136}$ is much lower than that of germanium, it is a metal and therefore not of interest for thermoelectric applications. The synthesis of type II semiconducting materials is currently underway by the authors in an effort to investigate the transport properties of these interesting materials. In the semiconductor $Sr_8Ga_{16}Ge_{30}$ both static as well as dynamic disorder results in a 'glass-like' thermal conductivity.

ACKNOWLEDGMENT
    The authors are grateful for support from the U.S. Army Research Laboratory under contract number DAAD17-99-C-0006.

REFERENCES
1. G.S. Nolas, T.J.R. Weakley and J. L. Cohn, Chem. Mater. **11**, 2470 (1999).
2. G.S. Nolas, B.C. Chakoumakos, B. Mahieu, G.J. Long and T.J.R. Weakley, Chem. Mat., in press.
3. G.S. Nolas, J.L. Cohn and E. Nelson, Proceedings of the Nineteenth International Conference on Thermoelectrics, in press.
4. G.A. Slack, in *Thermoelectric Materials -- New Directions and Approaches*, edited by T.M. Tritt, M.G. Kanatzidis, H.B. Lyon, Jr., and G.D. Mahan (Mat. Res. Soc. Symp. Proc. Vol. 478, Pittsburgh, PA, 1997) p. 47.
5. G.S. Nolas, D.T. Morelli and T.M. Tritt, Annu. Rev. Mater. Sci. **29**, 89 (1999).
6. G. S. Nolas, J. L. Cohn, G. A. Slack and S. B. Schujman, Appl. Phys. Lett. *73*, 178 (1998).
7. J.L. Cohn, G.S. Nolas, V. Fessatidis, T.H. Metcalf and G.A. Slack, Phys. Rev. Lett. **82**, 779 (1999).
8. B. C. Chakoumakos, B. C. Sales, D. G. Mandrus and G. S. Nolas, J. Alloys and Comp. **296**, 801 (1999).
9. G.S. Nolas, Mater. Res. Soc. Symp. Proc. Vol. 545, Pittsburgh, PA 1999) p 435.
10. B.B. Iverson, A.E.C. Palmqvist, D.E. Cox, G.S. Nolas, G.D. Stucky, N.P. Blake and H. Metiu, J. Solid State Chem **149**, 455 (1999).
11. G.S. Nolas, T.J.R. Weakley, J.L. Cohn and R. Sharma, Phys. Rev. B **61**, 3845 (2000).
12. S.B. Schujman, G. S. Nolas, R. A. Young, C. Lind, A. P. Wilkinson, G. A. Slack, R. Patschke, M. G. Kanatzidis, M. Ulutagay, S. -J. Hwu, J. of App. Phys. **87**, 1529 (2000) .
13. See for example, F. Franks, *Water, A Comprehensive Treatise* (Plenum Press, New York, 1973).
14. A.A. Demkov, W. Windl and O.F. Sankey, Phys. Rev. B **53**, 11288 (1996) and references therein.
15. C. Cross, M. Pouchard and P. Hagenmuller, C.R. Acad. Sc. Paris **260**, 4764 (1965).
16. G. S. Nolas and J.L. Cohn, unpublished.
17. S. Bobev and S.C. Sevov, J. Am. Chem. Soc. **121**, 3795 (1999).

FRAMEWORK STOICHIOMETRY AND ELECTRICAL CONDUCTIVITY OF Si-Ge BASED STRUCTURE-I CLATHRATES

Ganesh K. Ramachandran[1], Paul F. McMillan[1,2], Jianjun Dong[3], Jan Gryko[4] & Otto F. Sankey[3]

[1]*Department of Chemistry & Biochemistry, and Materials Research Center, Arizona State University, AZ 85287-1604;* [2]*Center for Solid State Science, Arizona State University, AZ 85287-1504;* [3]*Department of Physics & Astronomy, and Materials Research Center, Arizona State University, AZ 85287-1704;* [4]*Department of Earth & Physical Sciences, Jacksonville State University, AL 36265*

ABSTRACT

We report the synthesis and structural characterization of two Structure I clathrates in the K-Si and Rb-Si systems. The alkali-Si clathrates are fully stoichiometric at the framework sites, i.e., devoid of framework vacancies. This is in sharp contrast to the analogous K-Ge, Rb-Ge and Rb-Sn, Cs-Sn systems, where vacancies are formed at one-third of the crystallographic 6c tetrahedral sites. This is rationalized in terms of Zintl-Klemm rules to remove the tetrahedral atom of its hypervalency. The contrasting behavior is understood in terms of weaker Tt-Tt (Tt – tetrelide, Si, Ge, Sn) bonding as one descends the periodic table, and results in poorly metallic conductivities for vacancy-free K_7Si_{46} and Rb_6Si_{46}, but semiconducting behavior of K_8Ge_{44}. The observation suggests tuning of the electronic properties of Tt clathrates by substitution of (Si,Ge,Sn) on framework sites, for thermoelectric applications. We describe preliminary results designed to synthesize "mixed" Si-Ge clathrate structures. Thermal decomposition of K_2SiGe results in formation of a Structure I clathrate with mixing of Si and Ge on framework sites. The lattice constant a_o = 10.523(6) Å, is intermediate between those of K_8Si_{46} and K_8Ge_{44}.

INTRODUCTION

Group IV analogs of the Structure I and II clathrate hydrates [1,2] have received increasing attention over the past several years, particularly following observation of promising electrical and thermal conductivities for thermopower applications [3-6]. The recent focus on clathrates as potential thermoelectric materials stems from Slack's concept of the "phonon glass-electron crystal" (PGEC). The figure of merit ($ZT = \alpha^2\sigma T/\kappa$) requires that materials for thermoelectric applications be semiconducting with a large Seebeck coefficient and low thermal conductivity. The clathrates are an ideal testing ground for the PGEC concept, as there are many opportunities for varying tuning parameters to optimize the electrical conductivity, the Seebeck coefficient, and the thermal conductivity. Slack suggested that "rattling" of guest atoms would be key role in reducing the thermal conductivity of these materials [3]. Clathrate structures built from Ge and Ga frameworks with alkaline-earth atoms (and Eu) as guests have now been studied, and show promisingly low values of κ [4-6]. The thermal conductivity of disordered Ga,Ge clathrates is lowered further, due to alloy-scattering of phonons.

Although several clathrate structure types with similar energies have been predicted, only two (types I and II) have been identified experimentally [1,2]. Among the group IV clathrates with the type I structure ($M_{8-x}T_{46-y}$), there has been controversy surrounding the framework stoichiometry. Si and Ge members have been reported to be either metallic or semiconducting, with or without vacancies on framework sites [7]. The vacancy model stems from work on semiconducting compounds M_8Sn_{44}, where framework vacancies are created to satisfy Zintl-Klemm rules for electroneutrality [8]. Studies on Si clathrate (($Na,Ba)_8Si_{46}$) showed instead that the framework is fully stoichiometric, and the compounds are metallic [9]. Careful studies are required to correlate framework stoichiometry with electrical properties of group IV (Tt) clathrates, to establish "design rules" for tuning of thermoelectric properties in mixed Si-Ge clathrates, for example. We describe the synthesis and structure refinement of stoichiometric K_8Si_{46}, Rb_6Si_{46} and the defect stabilized $K_8Ge_{44\ 2}$, and their electrical conductivities as a function of temperature. We also report preliminary results on Si-Ge based clathrates.

PART I: VACANCY STABILIZATION IN GROUP IV CLATHRATES

Discussion of the stoichiometry of the type I clathrate A_8Tt_{46} structure (A: Na, K, Rb, Cs) has centered around the conflict between the ideal structural stoichiometry and the observed semiconducting behavior of many of the compounds [7-8]. According to electron counting, fully stoichiometric A_8Tt_{46} structures would be electron-rich, because of additional electrons donated from the electropositive metals. The total valence electron count indicates that the Tt atoms are hypervalent (i.e., 192/46 = 4.17 electrons per Tt atom).

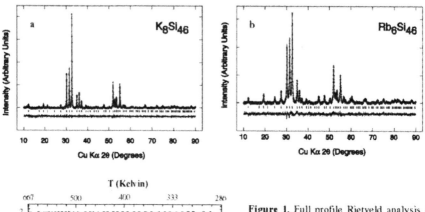

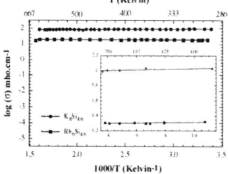

Figure 1. Full profile Rietveld analysis of the powder x-ray diffraction pattern of (a) K_8Si_{46} and (b) Rb_6Si_{46}. In either structure, there were no vacancies at the framework sites. The "extra" electrons from the guest atoms simply populate the conduction bands of pristine Si_{46}, resulting in poorly metallic conductivities as is observed for both members (left panel).

Following the Zintl-Klemm concept [7], the formation of two framework vacancies, i.e. $A_8Tt_{44\ 2}$, would allow accommodation of the 8 extra electrons from the alkali metals (A) onto the framework, while maintaining electroneutrality. The phenomenon was demonstrated for phases in the Rb-Sn and (K,Cs)-Sn systems. From x-ray studies, stoichiometries were found to be Rb_8Sn_{44} and $(K,Cs)_8Sn_{44}$ [$Rb_8Sn_{44\ 2}$ and $(K,Cs)_8Sn_{44\ 2}$] respectively [5, 8]. The Zintl-Klemm model offers a simple explanation for the formation of many Structure I clathrates with partial substitutions on framework and guest atom sites, i.e., $A_8Tr_8Tt_{38}$ (Tr = Al, Ga, In) and $B_8Tr_{16}Tt_{30}$ (B = Ba, Sr, Eu), which have "fully occupied" framework stoichiometries [6, 10]. Inclusion of pnictides (P, As) is permitted in the framework when halogens (Br, I) are incorporated as guests [12]. The case of Si and Ge clathrates is more problematic [7, 9]. We undertook to synthesize samples of alkali (K, Rb)-containing Si and Ge clathrates, and determine their framework stoichiometry correlated that with their electrical conductivity.

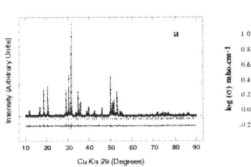 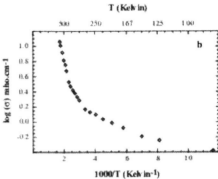

Figure 2. (a) Full profile Rietveld analysis of the clathrate K_8Ge_{44} revealed that two atoms are vacant at the crystallographic 6c sites. The formation of vacancies results in the ionic compound $8(K^+)(Ge_{44})^{8-}$. The semiconducting behavior as shown in panel (b) agrees well with this formulation. From the high temperature limiting slope, an activation energy of $E_a \sim 0.15$ eV was determined.

Samples were synthesized by thermal degradation of KSi, RbSi or KGe. Framework site occupancies were determined from full profile Rietveld analyses of x-ray powder data sets. Fig. 1 shows the x-ray pattern along with the theoretical and difference profiles of K_8Si_{46} and Rb_6Si_{46}, and conductivity vs. temperature for these materials.

TABLE 1
Atomic Positions, Site Occupancies and Atomic Displacement Parameters for K_8Si_{46}

	[10167 data points; a = 10.27518(5), χ^2 = 1.491, R_p = 0.041, R_{wp} = 0.052]					
Atom	Site	x	y	z	Occupancy	100 U_{iso} (Å^2)
Si1	6c	1/4	0	1/2	1.006(9)	3.64(6)
Si2	16i	0.1848(2)	0.1848(2)	0.1848(2)	1.010(5)	3.64(6)
Si3	24k	0	0.3062(2)	0.1183(5)	1.018(2)	3.64(6)
K1	2a	0	0	0	0.885(4)	3.3(2)
K2	6d	1/4	1/2	0	0.975(3)	3.6(2)[a]

[a]U_{11} = 4.1(3)Å^2, U_{22} = 3.4(2)Å^2 U_{33} = 3.4(2)Å^2, U_{ij} = 0 Å^2 ($i \ne j$)

TABLE 2
Atomic Positions, Site Occupancies and Atomic Displacement Parameters for Rb_6Si_{46}

	[10167 data points; a = 10.27188(6), χ^2 = 1.769, R_p = 0.029, R_{wp} = 0.037]					
Atom	Site	x	y	z	Occupancy	100 U_{iso} (Å^2)
Si1	6c	1/4	0	1/2	0.995(1)	1.42(3)
Si2	16i	0.1845(2)	0.1845(2)	0.1845(2)	1.013(5)	1.42(3)

Si3	24k	0	0.3037(4)	0.1187(8)	1.004(2)	1.42(3)
Rb1	2a	0	0	0	0.218(5)	4.1(1)
Rb2	6d	1/4	1/2	0	0.952(7)	2.54(9)[a]

[a] $U_{11} = 3.3(2)$ Å^2, $U_{22} = 2.1(1)$ Å^2 $U_{33} = 2.1(1)$ Å^2, $U_{ij} = 0$ Å^2 ($i \neq j$)

TABLE 3
Atomic Positions, Site Occupancies and Atomic Displacement Parameters for K_8Ge_{46}
[10403 data points; $a = 10.66771(1)$, $\chi^2 = 1.949$, $R_p = 0.035$, $R_{wp} = 0.045$]

Atom	Site	x	y	z	Occupancy	100 U_{iso} (Å^2)
Ge1	6c	1/4	0	1/2	0.691(3)	3.99(2)
Ge2	16i	0.1832(3)	0.1832(3)	0.1832(3)	0.997(2)	3.99(2)
Ge3	24k	0	0.3151(2)	0.1198(7)	0.998(5)	3.99(2)
K1	2a	0	0	0	0.854(6)	3.5(3)
K2	6d	1/4	1/2	0	0.969(4)	4.5(1)[a]

[a] $U_{11} = 3.8(3)$ Å^2, $U_{22} = 5.0(3)$ Å^2 $U_{33} = 5.0(3)$ Å^2, $U_{ij} = 0$ Å^2 ($i \neq j$)

Fig. 2 depicts the x-ray diffraction pattern of K_8Ge_{44}, along with the variation of conductivity with T. For K_8Si_{46}, residuals obtained were $R_p = 4.1\%$ and $R_{wp} = 5.2\%$ and $\chi^2 = 1.491$. Similar residuals for Rb_6Si_{46} were $R_p = 2.9\%$ and $R_{wp} = 3.7\%$, $\chi^2 = 1.782$. We conclude that both silicon clathrates are devoid of framework vacancies, as found earlier for Na_8Si_{46} [9]. The Ge clathrate bears vacancies on one-third of the 6c sites (Table 1). Refined residuals were $R_p = 3.5\%$ and $R_{wp} = 4.5\%$; $\chi^2 = 1.949$. For Si clathrates, the temperature dependence of conductivity indicates metallic behavior (Fig. 1) while the Ge clathrate is semiconducting (Fig. 2). The limiting activation energy is $E_a \sim 0.15$ eV.

The introduction of vacancies is rationalized in "chemical" terms by maintaining electroneutrality and removing the Tt atom of its hypervalency. The formation of vacancies involves breaking bonds around the 6c framework sites, leading to formation of defect states in the band-gap where the "extra" electrons from the guest are localized [7]. Electron pairing and localization energy then outweighs the energy required to break Tt-Tt covalent bonding to form the defect sites. The process is obviously favorable for Ge and Sn clathrates, where the bonding is weaker, resulting in "vacancy-stabilized clathrates" that are semiconducting. In the case of Si clathrates, the Si-Si bonds are stronger, and "extra" electrons from the guest atoms are distributed in the conduction bands of the Si framework resulting in the observed metallic behavior.

PART II: Si-Ge BASED CLATHRATES

We next attempted to form "mixed" Si_xGe_{1-x} clathrates. Possible synthetic routes include "direct" synthesis from the elements (or from a Si_xGe_{1-x} solid solution) at high temperatures. Some Ga-Ge/Si clathrates have been obtained directly from a melt in single crystalline form [4-6, 12]. Others are not thermally stable, and are obtained metastably *via* thermal degradation of an appropriate Zintl phase [1]. We attempted "direct" syntheses by heating mixtures targeted for $M_8Si_{23}Ge_{23}$ (M = K, Rb, Cs) at 1250 K. The resulting x-ray patterns were dominated by lines from Si_xGe_{1-x} solid solutions and occasional additional lines that could be attributed to clathrate. The result shows that mixed Si-Ge clathrates are metastable with respect to the diamond structured solid solution at high T, and must be obtained by a metastable route. A "Zintl-phase" route was attempted from Na_2SiGe and K_2SiGe, formed by reacting elements or pure Zintl phases in a Ta crucible at 650 °C for 24 h. The diffraction pattern of the product consists of lines intermediate between the parent phases (Fig. 3), indicating the formation of "K_2SiGe". The "mixed" Zintl phases are new, and it is not known yet if discrete Si_4^{4-} and Ge_4^{4-} tetrahedra exist within the structure, or if mixed $(Si, Ge)_4^{4-}$ tetrahedra are present. NMR and Raman investigations are under way. The x-ray pattern was indexed within $P\bar{4}3n$ with a_o=12.677(1) Å, intermediate between KSi and KGe [13]. Na_2SiGe was also prepared by this route.

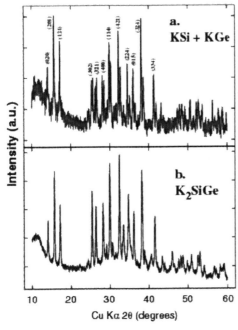

Figure 3. (a) A physical mixture of a sample of silicide KSi and germanide KGe. Both KSi and KGe crystallize in the cubic space group $P\bar{4}3n$. They are characterized by the presence of isolated tetrahedra of Si or Ge with a formal charge of -1 [the compounds are formulated as $4K^+(T_4)^{4-}$]. KSi has a lattice constant of 12.62 Å and that of KGe is 12.78 Å. Sample (a) was heated at 700 °C for 48h, the resulting x-ray diffraction pattern is shown in the lower panel (b). All lines which were doublets in the physical mixture have collapsed into sharp single lines in the heat treated sample (b). The resulting compound in panel (b) had an unit cell of 12.677(1) Å, intermediate between that of KSi and KGe, suggesting that it may consist of tetrahedra formed by both Si and Ge and can be regarded as a solid solution with formula "K_2SiGe".

K_2SiGe was then placed in a Ta boat and "degassed" at 400 °C for 72 h at $\sim 10^{-6}$ torr. The resulting x-ray pattern showed mainly peaks indicative of a (Si_xGe_{1-x}) clathrate structure (Fig. 4). Preliminary Rietveld analysis gives a_o=10.523(6) Å, intermediate between K_8Si_{46} and K_8Ge_{44}. Mixing was observed at the three framework sites, with ~50% occupancy of Si and Ge at $24k$ sites; 25% Si-50% Ge at $16i$ sites, and 50% Si-35% Ge at the $6c$ sites.

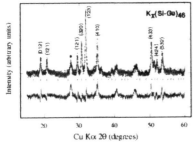

Figure 2. X-Ray diffraction pattern and Rietveld profile analysis of a sample that was obtained by thermal degradation of "K_2SiGe" at 400 °C for 72h under vacuum. The sample had poorly crystallized, however, it could be indexed to the Strcutre I clathrate structure with a lattice constant of 10.523(6) Å, intermediate between that of K_8Si_{46} [10.27518(5) Å] and K_8Ge_{44} [10.66771(1)Å]. Mixing of Si and Ge at each of the three crystallographic sites was also observed.

CONCLUSIONS

Preliminary attempts to synthesize "mixed" clathrate structures containing Si and Ge atoms on framework sites show promising results. Structure I silicon clathrates K_8Si_{46} and Rb_6Si_{46} are shown to have fully stoichiometric frameworks, and are metallic. The germanium clathrate $K_8Ge_{44\ \square 2}$, however bears vacancies at one-thirds of the $6c$ sites, and is semiconducting, with an effective bangap calculated from the activation energy of $E_a \sim 0.15$ eV. The absence of vacancies in the silicon compounds, correlated with the different electrical behavior, is rationalized on the relative strength of Si-Si bonds compared with Ge-Ge bonds. This provides the basis for designing clathrates with "tunable" electronic properties, by forming solid solutions in the (Si-Ge) series.

ACKNOWLEDGEMENTS

This study was supported by the NSF MRSEC program.

REFERENCES

1. C. Cros, M.Pouchard and P. Hagenmuller, *J. Solid State Chem.* **2**, 570 (1970); J. S. Kasper, P. Hagenmuller, M. Pouchard and C. Cros, *Science* **150**, 1713 (1965)
2. G. B. Adams, M. O'Keeffe, A. A. Demkov, O. F. Sankey and Y. Huang, *Phys. Rev. B* **49**, 8048 (1994); M. O'Keeffe, G. B. Adams and O. F. Sankey, *Phil. Mag. Letts.* **78**, 21 (1998)
3. G. A. Slack, *Mat. Res. Soc. Symp. Proc.* **478**, 47 (1997)
4. G. Nolas, J. L. Cohn, G. A. Slack and S. B. Schujman, *Appl. Phys. Lett.* **73**, 178 (1998)
5. J. L. Cohn, G. S. Nolas, V. Fessatidis, T. H. Metcalf and G. A. Slack, *Phys Rev Lett.* **82**, 779 (1999)
6. S. B. Shujman, G. S. Nolas, R. A. Young, C. Lind, A. P. Wilkinson, G. A. Slack, R. Patschke, M. G. Kanatzidis, M. Ulutagay and S-J. Hwu, *J. Appl. Phys.* **87**, 1529 (2000); B. B. Iversen, A. E. C. Palmqvist, D. E. Cox, G. S. Nolas, G. D. Stucky, N. P. Blake and H. Metiu, *J. Solid State Chem.* **149**, 455 (2000); B. C. Chakoumakos, B. C. Sales, D. G. Mandrus and G. S. Nolas, *J. Alloys Comp.* **296**, 80 (2000)
7. *Chemistry, Structure and Bonding of Zintl Phases and Ions*, ed. Susan M. Kauzlarich, (VCH Publishers, NY, 1996
8. H. G. von Schnering, *Nova Acta Leopoldina* **59**, 168 (1985); J-T. Zhao and J. D. Corbett, *Inorg. Chem.* **33**, 5721 (1994)
9. G. K. Ramachandran, P. F. McMillan, J. Diefenbacher, J. Gryko, J. Dong and O. F. Sankey, *Phys. Rev. B* **60**, 12294 (1999) ; Shimizu, Y. Maniwa, K. Kume, H. Kawaji, S. Yamanaka and M. Ishikawa, *Phys. Rev. B* **54**,13242 (1996); S. Yamanaka, E. Enishi, H. Fukuoka and M. Yasukawa, *Inorg. Chem.* **39**, 56 (2000)
10. B. Eisenmann, H. Schafer, and R. Zagler, *J. Less-Comm. Metals* **118**, 43 (1986); G. Cordier, and P. Woll, *J. Less-Comm. Metals* **169**, 291 (1991)
11. T. L. Chu, S. S. Chu, R. L. Ray, *J. Appl. Phys.* **53**, 7102 (1982); M. M. Shatruk, K. A. Kovnir, A. V. Shevelkov, I. A. Presniakov, B. A. Popovkin, *Inorg. Chem.* **38**, 3455 (1999)
12. S. Bobev, and S. C. Sevov, *J. Am. Chem. Soc.* **121**, 3795 (1999)
13. J. Witte and H. G. Schnering, *Z. Anorg. Chem.* **327**, 260 (1964); E. Busmann, *Z. Anorg. Chem.* **313**, 90 (1961).

Ultrasound Studies of Clathrate Thermoelectrics

Veerle Keppens*, Brian. C. Sales**, David Mandrus**, Bryan C. Chakoumakos**, and Christiane Laermans***
*National Center for Physical Acoustics, University of Mississippi, Oxford, MS 38677, U.S.A.
**Solid State Division, Oak Ridge National Laboratory, Oak Ridge, TN 37831-6056, U.S.A.
***Dept. of Physics, Katholieke Universiteit Leuven, B-3001 Leuven, Belgium.

ABSTRACT

Resonant Ultrasound Spectroscopy and low-temperature ultrasonic attenuation measurements are reported for filled and unfilled skutterudites and for Ge-clathrates. These data reveal that an unusual elastic behavior complements the thermal properties of the filled skutterudites and indicate the presence of low-energy vibrational modes. The attenuation at low-temperatures in the single-crystalline Ge-clathrate is glasslike and can be described by the same phenomenological Tunneling Model that has been developed to describe the low-temperature properties of amorphous solids.

INTRODUCTION

Motivated by the search for improved thermoelectric materials, several compounds have attracted attention that combine the high electron mobilities found in crystals with a low thermal conductivity κ, approaching κ-values typical for glasses. The common structural feature of these "electron-crystal phonon-glass" (ECPG) materials is that they contain loosely bound atoms that reside in a large crystalline "cage"; these materials are thus "inclusion compounds" or "crypto-clathrates". A particular class of clathrates is formed by the filled skutterudite antimonides [1]. More recently, the Ge-clathrate $Sr_8Ga_{16}Ge_{30}$ was found to be an ECPG material, having a truly glasslike thermal conductivity while maintaining crystalline electronic properties [2,3].

In this paper, we present Resonant Ultrasound Spectroscopy (RUS) measurements on $CoSb_3$ and $La_{0.75}Fe_3CoSb_{12}$ skutterudites, and ultrasonic attenuation measurements on single crystals of $Sr_8Ga_{16}Ge_{30}$. We find that we can describe the elastic response of filled skutterudites in terms of two distinct two-level systems. In $Sr_8Ga_{16}Ge_{30}$, ultrasonic attenuation measurements clearly indicate the presence of tunneling states. The elastic properties of both filled skutterudites and Ge-clathrates are strikingly different from the behavior expected for normal crystalline solids.

EXPERIMENTAL DETAILS

The skutterudite and Ge-clathrate specimens were synthesized at Oak Ridge National Laboratory by melting stoichiometric quantitities of high-purity elements in carbon-coated, evacuated, and sealed silica tubes [1,4]. High-quality samples of $La_{0.75}Fe_3CoSb_{12}$ with a density that is 98% of the theoretical value, and $CoSb_3$ (95% dense) were cut into 2 x 2.5 x 3 mm^3 rectangular parallelepipeds and used for Resonant Ultrasound Spectroscopy (RUS) measurements. RUS is a novel ultrasonic technique, developed by Migliori et al. [5] for determining the complete set of elastic moduli of a solid by measuring the free-body resonances of the sample. This method is unique in that all moduli can be determined simultaneously, avoiding remounts of transducers and multiple temperature sweeps. For the attenuation measurements, a single crystal of $Sr_8Ga_{16}Ge_{30}$ was cut into a bar with dimensions 8x2x1 mm^3. Pulse-echo measurements of the ultrasonic attenuation were performed as a function of temperature (0.3-10 K) and at frequencies of 155 and 250 MHz.

RESULTS AND DISCUSSION

Measurements of the elastic moduli were performed as a function of temperature for both filled ($La_{0.75}Fe_3CoSb_{12}$) and unfilled ($CoSb_3$) skutterudites. A polycrystal has two elastic moduli: a shear modulus c_{44} governing transverse waves and a longitudinal modulus c_{11} governing longitudinal waves.

Figure 1 shows both the shear and longitudinal modulus for the unfilled skutterudite

CoSb$_3$. The solid line through the CoSb$_3$ data is a model calculation using the so-called Varshni function [6]:

$$c_{ij}(T) = c_{ij}^0 - s/(e^{t/T} - 1)$$

with T the temperature, c_{ij}^0 the elastic constant at 0 K and s and t fitting parameters. This function, which has some theoretical justification, was shown by Varshni to describe the temperature-dependence of the elastic constants of many simple substances and characterizes to some extent "normal" elastic behavior. It is immediately apparent form Figure 1 that the elastic response of CoSb$_3$ can be well described by this Varshni model. However, the moduli of the filled skutterudite La$_{0.75}$Fe$_3$CoSb$_{12}$ behave anomalously at low temperatures. As shown in Figure 2, these moduli display a much stronger temperature dependence, which is suggestive of unusual low-energy vibrational modes in addition to the normal acoustic modes.

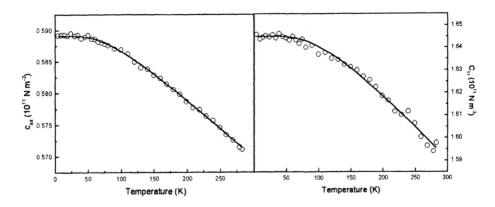

Figure 1. Shear modulus c$_{44}$ and longitudinal modulus c$_{11}$ as a function of temperature for CoSb$_3$ obtained using RUS. The solid lines through the data are fits to the Varshni-function. The fitting parameters used were (for c$_{44}$): c$_0$=0.589x10^{11}N/m^2, s=0.022x10^{11}N/m and t=229 K, and (for c$_{11}$): c$_0$=1.645x10^{11}N/m^2, s=0.1x10^{11}N/m and t=312 K.

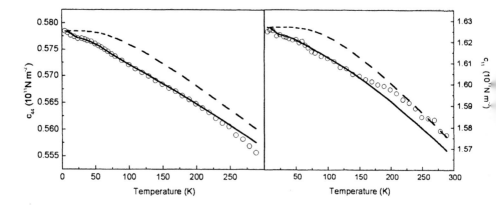

Figure 2. Shear modulus c_{44} and longitudinal modulus c_{11} as a function of temperature for $La_{0.75}Fe_3CoSb_{12}$. The dashed line is the estimated background, obtained from the Varshni-fit for $CoSb_3$, with $c_0=0.5785 \times 10^{11} N/m^2$ for c_{44} and $c_0=1.625 \times 10^{11} N/m^2$ for c_{11}. The solid lines represent model calculations using 2 two-level systems as explained in the text.

A first attempt to model this unusual behavior, taking into account the presence of harmonic Einstein oscillators, failed to describe the "dip" in the data at low temperatures. The elastic data can, however, be described by considering the elastic response of a two-level system (TLS) with level-spacing Δ. We can calculate the TLS-contribution to the elastic response, $c = \partial^2 F / \partial \varepsilon^2$, with $F = -N_A k_B \ln(1+e^{-\Delta/T})$ the Helmholtz free energy of a TLS, and assuming that the level spacing Δ depends linearly on the strain ε, that is $\Delta = \Delta_0 + A\varepsilon$ with A a coupling constant. As illustrated in Figure 2, both the c_{11} and c_{44} modulus can be modeled reasonably well by adding two TLS's with level spacings of 50 K and 200 K to a background contribution, which is estimated from the Varshni-fit for the unfilled skutterudite $CoSb_3$. The local modes found in the La-filled skutterudite are not observed in the unfilled skutterudite $CoSb_3$, suggesting that they have to be ascribed to the "rattling" rare-earth. Inelastic neutron scattering results provide strong evidence that both local modes are indeed associated with the presence of the La ion [7]. As explained in Ref. 7, the vibrational spectrum associated with the La-atom shows two distinct features: one at 7 meV (80 K) and one at 15 meV (175 K). Although these features in the phonon DOS have been reproduced by some recent lattice dynamical calculations by Feldman, et al. [8], no simple picture of the origin of these features has yet emerged.

$Sr_8Ga_{16}Ge_{30}$

The temperature-dependence of the ultrasonic attenuation, $\alpha(T)$, was measured on a single crystal of $Sr_8Ga_{16}Ge_{30}$ at two frequencies. The results appear in Figure 3. Clearly, the behavior of $\alpha(T)$ at the lowest temperatures is much different than that expected for an ordinary crystalline material, for which $\alpha(T)$ is small and practically independent of temperature below a few degrees K. Instead, the attenuation from 0.3 K to 1 K shows a T^3 dependence and strongly resembles the $\alpha(T)$ of a structural glass such as GeO_2 [9]. The low-temperature thermal and acoustic properties of glasses were given a phenomenological explanation by the Tunneling Model (TM) [10]. The TM postulates the existence of tunneling states that couple to strain fields and that have a broad distribution in energy splittings. The TM successfully accounts for the thermal and acoustic behavior of glasses below 1 K, and can explain both the T^2 behavior of the thermal conductivity and the T^3 behavior of the ultrasonic attenuation in terms of the interaction between phonons and tunneling states. The observation of the T^3- dependence of the ultrasonic attenuation in $Sr_8Ga_{16}Ge_{30}$ is strong evidence that tunneling states exist in this crystalline material.

As shown in Ref. 3, the thermal conductivity of $Sr_8Ga_{16}Ge_{30}$ is also glasslike at low temperatures with a temperature-dependence consistent with both the TM and the attenuation data. Recent measurements on $Ba_8Ga_{16}Ge_{30}$ show that the thermal conductivity of the Ba-filled clathrate, although small, has a crystalline temperature-dependence [11]. In both clathrates, the alkaline earth clearly "rattles" in its cage, as observed from the large atomic displacement parameter. However, nuclear density plots, obtained from single-crystal neutron diffraction data, reveal that in the Sr case, the density at the alkaline earth site shows 4 lobes and is much more smeared out than in the Ba case [11]. These results, together with the observation of tunneling states in $Sr_8Ga_{16}Ge_{30}$ suggest that the 4 lobes represent tunneling sites, and imply that the origin of the glasslike properties in $Sr_8Ga_{16}Ge_{30}$ involves the tunneling of Sr ions between the 4 sites displaced from the cage center. This makes $Sr_8Ga_{16}Ge_{30}$ one of the few materials in which the

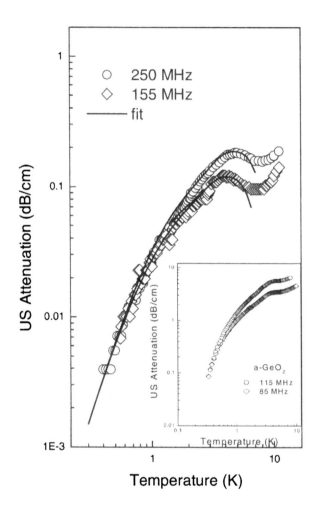

Figure 3. Ultrasonic attenuation, $\alpha(T)$, obtained on a single crystal of $Sr_8Ga_{16}Ge_{30}$ measured at 250 MHz and 155 MHz. The attenuation strongly resembles the attenuation of a structural glass like GeO_2 [9] (shown in the inset). The solid line through the $Sr_8Ga_{16}Ge_{30}$ data is a fit using the tunneling model with a restricted energy-distribution.

microscopic origin of the tunneling states is identified. A detailed analysis of the acoustic attenuation data reveals that the energy-distribution of the tunneling states in $Sr_8Ga_{16}Ge_{30}$ is much narrower than in a structural glass such as GeO_2. This is not surprising, because although the environment of the Sr-atom can be different from site to site, it is not as random as it would be in an amorphous solid.

CONCLUSION

We have shown that both the La-filled skutterudite and the $Sr_8Ga_{16}Ge_{30}$ have a very unusual elastic behavior that is characterized by local modes. In the La-filled skutterudite, the temperature dependence of the elastic moduli can be modeled with 2 distinct two-level systems with splittings that correlate well with peaks in the vibrational DOS associated with the La ion as determined from inelastic neutron scattering. The attenuation at low-temperatures in single crystals of $Sr_8Ga_{16}Ge_{30}$ is glasslike and can be explained with the tunneling model used to describe the low-temperature properties of amorphous solids. Our results suggest that in order to produce true glasslike behavior, both rattling and tunneling states are necessary.

ACKNOWLEDGEMENTS

This work was supported by the Division of Materials Sciences, U.S. Department of Energy. Work at the K. U. Leuven was sponsored by the F.W.O.-Vlaanderen.

REFERENCES

1. See, for example, B. C. Sales et al., *Phys. Rev. B* **56**, 15081 (1997) and references therein.
2. G. S. Nolas, J. L. Cohn, G. A. Slack, and S. B. Schujman, *Appl. Phys. Lett.* **73**, 178 (1998).
3. J. L. Cohn et al., *Phys. Rev. Lett.* **82**, 779-782 (1999).

4. B. C. Chakoumakos, B. C. Sales, D. Mandrus, and G. S. Nolas, *J. Alloys and Compounds*, in press.
5. A. Migliori et al., *Physica B* **183**, 1 (1993).
6. P. Varshni, *Phys. Rev. B* **2**, 3952 (1973).
7. V. Keppens et al., *Nature* **395**, 876 (1998).
8. J. Feldman, unpublished results.
9. C. Laermans, V. Keppens, and R. Weeks, *Phys. Rev. B*, **55**, 2701 (1997).
10. P. W. Anderson, B. I. Halperin, and C. M. Varma, *Phil. Mag.* **25**, 1 (1972); W. A. Phillips, *J. Low Temp. Phys.* **7**, 351 (1972).
11. V. Keppens et al., *Phil. Mag. Lett*, accepted for publication.

Synthesis and Characterization of Large Single Crystals of Silicon and Germanium Clathrate-II Compounds and a New Tin Compound with Clathrate Layers

Svilen Bobev and Slavi C. Sevov[*]
Department of Chemistry and Biochemistry
University of Notre Dame
Notre Dame, IN 46556-5670
E-mail: ssevov@nd.edu

ABSTRACT

We have synthesized large single crystals of clathrate-II compounds with frameworks of silicon and germanium by employing mixed alkali metal countercations. The combinations of alkali metals are rationally selected in order to fit the different cages of the clathrate-II structure. This approach leads to the following stoichiometric and fully "stuffed" compounds: $Cs_8Na_{16}Si_{136}$, $Cs_8Na_{16}Ge_{136}$, $Rb_8Na_{16}Si_{136}$ and $Rb_8Na_{16}Ge_{136}$. The structures and the corresponding Si–Si and Ge–Ge distances are elucidated and established with high accuracy from extensive single crystal X-ray diffraction work. The compounds are stoichiometric, metallic, and are very stable at a variety of extreme conditions such as heat, concentrated acids, hydrothermal treatment etc. No evidence was found for vacancies in the silicon and germanium networks or partial occupancies of the alkali metal sites. The stoichiometry of these fully "stuffed" clathrates is consistent with the measured temperature independent Pauli paramagnetism, supported also by the conductivity measurements on single crystals and thermopower measurements on pellets. A new compound with novel clathrate-like structure forms when small and large cations are combined with tin. The new materials, $A_6Na_{18}Sn_{46}$ (A = K, Rb, Cs), are made of clathrate layers and the interlayer space filled with Sn_4-tetrahedra and alkali-metal cations. Its formula can be rationalized as $A_6Na_6Sn_{34} + 3 \cdot Na_4Sn_4$ (one clathrate layer and three tin tetrahedra). The compound is stable in air and is being currently tested at other conditions. Detailed measurements of its transport properties are under way.

INTRODUCTION

The clathrates are inclusion compounds with open three-dimensional frameworks, made of tetrahedrally coordinated group 14 element with voids that are occupied by alkali metals. Their structures are isotypical with the structures of the gas hydrates, reported as early as 1811. Until now, two types of such compounds are known, clathrates of types I and II, with prototypes the gas hydrates $(G_8(H_2O)_{46})$, where G = Kr, Xe, Cl_2, CH_4, etc., and liquid or double hydrates $(G_{24}(H_2O)_{136}$ or $G_{16}G_8(H_2O)_{136})$, where G = H_2S, $CHCl_3$, CH_3NO_2, C_3H_8, etc., respectively.

An analogy can be also drawn between the zeolites and the clathrates - the latter being often referred to as "zeolites without oxygen" or "oxygen-free zeolites". Due to the many ways to link $SiO_{4/2}$ tetrahedra in the zeolites via the rather flexible Si–O–Si angle, there are a great variety of zeolite structures, both naturally occurring and experimentally synthesized. Formal removal of the oxygen, i.e. replacement of the X–O–X bonding (X = Al, Si, Ge) with direct X–X bonds while preserving the tetrahedral coordination, and appropriate "rescaling" of the unit cell, would result in the formation of isomorphous open frameworks. The bond energy of Si–Si or Ge–Ge is such that stretching the bonds and bending the angles is no longer favorable. As a result, there are only two types of silicon or germanium clathrates are known thus far, clathrate-I and clathrate-II, isomorphous with the zeolites melanophlogite and zeolite ZSM-39 (also known as Dodecasil 3C or MTN), respectively.

The discovery of the silicon clathrates in 1965 [1] stimulated more interest towards the compounds of C, Ge, Sn and Pb with the alkali metals. The efforts made in this direction led to the discovery (1969) of the first germanium and tin clathrates of type-I, K_8Ge_{46} and K_8Sn_{46} [2]. Since then, much experimental and theoretical work has been devoted to the clathrates of group 14 with alkali/alkaline earth metals. Moreover, a hypothesis was made that variable physical properties of these phases, such as the thermopower (S), the electrical (σ) and thermal (λ) conductivities can be tuned by altering the guest atoms in the cages. This brought new motivation for clathrate studies as potential thermoelectric materials. The stability of the clathrates, the fact that they conduct electricity relatively well, and the possibility of "rattling" guest cations in the cavities bring the clathrates closer to the ultimate thermoelectric PGEC ("Phonon-Glass and an Electron Crystal") material [3]. In order to increase the figure of merit ZT, an empirical coefficient that measures the thermoelectric performance (T = temperature and $Z = \sigma S^2/\lambda$), one needs a material with high thermopower and low thermal conductivity. Clearly, since the electrical conductivity and the electronic component of the thermal conductivity can not be varied independently, higher ZT values can be expected for material with low lattice thermal conductivity. Group 14 clathrates are considered as promising candidates for thermoelectric applications since their relatively large unit cells with "voids and rattlers inside them" are important prerequisite for scattering of the heat carrying phonons and thereby for low phonon-contributed thermal lattice conductivity. All this interest has led to the search for reliable and reproducible ways for their synthesis in high yields and with defined stoichiometry.

EXPERIMENTAL DETAILS

All manipulations were performed inside a nitrogen-filled glove box. The starting materials were Na (Alfa, 99.9%), K (Alfa, 99.95%), Rb (Alfa, 99.8%), Cs (Acros, 99.95 %), Si (Alfa, 99.9999%), Ge (Acros, 99.999%), Sn (Alfa, 99.999%). The reaction mixtures in appropriate stoichiometric ratios were loaded in arc-welded niobium containers (under argon) that were subsequently sealed in evacuated fused silica jackets. The syntheses involving Si and Ge were carried out at 650°C for 3-4 weeks, because at higher temperatures the stable phases $NbSi_2$ and $NbGe_2$ are being formed and this causes leaks of alkali metals. The products of these reactions were well shaped crystals with triangular faces and had distinctive bluish metallic luster. The crystals were then easily washed and separated under microscope from the traces of unreacted dust-like Si and Ge. Reactions involving tin were carried out at 400°C for 1 week. It is important to note that better quality crystals were obtained from reactions, loaded with excess of tin, most likely due to the more facile conditions for crystal growth in a presence of a tin-flux.

X-ray powder diffraction patterns were taken on an Enraf-Nonius Guinier camera with

Cu Kα_1 radiation. They were also used to verify the unit cell parameters by a least-squares refinement of the positions of the lines, calibrated by NIST silicon as an internal standard.

Single crystal data were collected on an Enraf-Nonius CAD-4 diffractometer with graphite monochromated Mo Kα radiation at room temperature (ω-2θ scans). The structures were solved by direct methods and refined on F^2 with the aid of the SHELXTL-V 5.1 software package. Further comprehensive details from the refinements can be found elsewhere [4].

Magnetic measurements were carried out on a Quantum Design MPMS SQUID magnetometer at a field of 3 T over a temperature range of 10-270 K. The samples were prepared by selecting crystals under microscope and then checking their quality by XRD.

Four-probe resistivity measurements were done on single crystals and pellets using a home made set-up (Van der Pauw's method). The pellets for the measurements were prepared from carefully picked single crystals that were then ground and pressed under pressure of 100 tons in a 1/2" die at room temperature and at 200°C. Independent measurements were also performed on monodisperse (particle size within the range of 250-425 µm) selected single crystals of $Cs_8Na_{16}Ge_{136}$ using a standard, high frequency, contact-free Q-technique.

The thermopower of a hot-pressed pellet of $Cs_8Na_{16}Si_{136}$ was measured by using steady-state technique over the temperature range 70-300 K.

RESULTS AND DISCUSSION

The stoichiometries of the two clathrate types do not differ much. When normalized, their ideal chemical formulas are $AX_{5.75}$ and $AX_{5.67}$, for clathrates-I and -II, respectively. Structurally, nevertheless, they are noticeably different.

Clathrate-I (A_8X_{46}) crystallizes in the primitive cubic space group *Pm-3n* (No. 223). The framework building atoms are tetrahedrally coordinated and their arrangement in the unit cell is such as to define two polyhedra of different sizes as shown in Figure 1(a). The smaller is a 20-atom pentagonal dodecahedron $\{5^{12}\}$, and the larger is a 24-atom tetrakaidecahedron $\{5^{12}6^2\}$ (the symbol $\{5^{12}6^2\}$ denotes polyhedron with 12 pentagonal and 2 hexagonal faces). The guest atoms or molecules can be encapsulated into these cages if their sizes are comparable with the size of

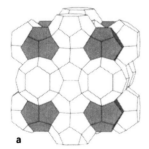 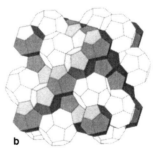

Figure 1. General view of the primitive cubic structure of clathrate-I (a) and the face-centered cubic structure of clathrate-II (b). Shaded are the smaller pentagonal dodecahedra $\{5^{12}\}$

the available empty space (Table I). There are eight cages per unit cell, two smaller and six bigger, and therefore the formula of the clathrate-I can be rewritten as $A'_2A''_6X_{46}$, where X denotes the clathrand element, and A' and A" stand for the atoms/molecules occupying the $\{5^{12}\}$ and $\{5^{12}6^2\}$ polyhedra, respectively.

Clathrate-II ($A_{24}X_{136}$) crystallizes in the face centered cubic space group $Fd\text{-}3m$ (No. 227). The framework-building atoms are again four-bonded in nearly ideal tetrahedral environment and also form two types of polyhedra. The key difference is that here, in addition to the pentagonal dodecahedra, the second type polyhedra are built of 28 atoms, the hexakaidecahedra $\{5^{12}6^4\}$, that have twelve pentagonal and four hexagonal faces. They share their hexagonal faces, forming a diamond-like structure. The smaller pentagonal dodecahedra are not isolated as in the clathrate-I but are rather linked through common faces to form layers, as shown in Figure 1(b). The ratio of smaller to larger cavities in clathrate-II is 16 : 8, and therefore the formula can be written as $A'_{16}A''_8X_{136}$.

Since there are two types of cages in each clathrate, two alkali metals will be preferred for their formation and stabilization (Table I). Therefore, the choice of cations with appropriate sizes and in appropriate ratio will be the product-directing factors. Thus, cations of only one size as in binary A–Tt systems stabilize the structure with cavities of similar sizes, i.e., that of clathrate-I, even when the reactions are designed to produce clathrate-II. Also, different cations but with similar sizes will stabilize clathrate-I as well. The smaller sodium and potassium form clathrate-I with the smaller silicon and germanium, while the larger rubidium and cesium match with tin. Since the two types of cavities in clathrate-II are quite different in size, on the other hand, its formation will be facilitated by combination of cations with very different sizes. The radius of the smaller cavity in Si and Ge phases is comparable only with the radii of Na^+ and K^+. Both Rb^+ and Cs^+ are too big to be incorporated into these cages but they can fit well into the larger $\{5^{12}6^4\}$. No single size cation can fit efficiently enough in both cavities of clathrate-II.

Following the approach described above, we synthesized the following new clathrates-II: $Cs_8Na_{16}Si_{136}$, $Cs_8Na_{16}Ge_{136}$, $Rb_8Na_{16}Si_{136}$ and $Rb_8Na_{16}Ge_{136}$ [4]. Prior to this work, the synthesis of phases with the clathrate-II structure such as Na_xSi_{136} was done by the thermal decomposition of the Zintl compounds NaSi at high temperatures (300-400°C) and dynamic vacuum. The resulting products are usually inhomegeneous powders with undefined stoichiometry, contain both clathrates and unreacted elements. Furthermore, small changes in the experimental conditions result in quite different overall compositions, and therefore poor reproducibility.

Table I. Approximate van der Waals radii of the available empty space in clathrate-I and -II. The ionic radii of the alkali metals are as follows: Na^+ 1.02; K^+ 1.38; Rb^+ 1.49; Cs^+ 1.70 Å [5]

	CLATHRATE-I		CLATHRATE-II	
	5^{12}-cage	$5^{12}6^2$-cage	5^{12}-cage	$5^{12}6^4$-cage
Si	1.10 Å	1.35 Å	1.10 Å	1.85 Å
Ge	1.25 Å	1.55 Å	1.25 Å	2.00 Å
Sn	1.65 Å	1.85 Å	1.65 Å	>2.2 (est.) Å

Powder XRD in this case is not a very useful tool for stoichiometry confirmation since the composition could vary substantially from particle to particle without noticeable changes in the lattice parameters due to the rigidity and covalency of the framework.

Our comprehensive work on the clathrates-II revealed that the structure is very stable at different extreme conditions (heat, acids, heat and vacuum, etc.) and is invariant with respect to the size of the cation. The structure refinements confirm the ideal stoichiometry for fully "stuffed" structure $A_{24}X_{136}$. No evidences for any partially occupied positions were found. Varying the occupancies of any site, while keeping the remaining ones fixed, does not lead to deviations of more than 2σ for the network and 5σ for the alkali metal sites, respectively [4]. Therefore, there are 24 "extra" electrons per 136 clathrand atoms that could be either delocalized over the four-bonded network or localized on the alkali metals. Theoretical calculations for A_xSi_{136} predict significant overlap between the A and Si orbitals at x >10 [6]. Thus, it is anticipated that all four compounds should be metallic conductors. Indeed, our four-probe measurements showed resitivities ranging from 30 to 160 $\mu\Omega\cdot cm$ at 293 K with temperature dependence that clearly indicates a metallic state. These results are also consistent with the temperature independent Pauli paramagnetism for all samples (χ_m in the order of 10^{-4} emu/mol). Also, no superconductivity at low field (0.1 T) and temperatures down to 2 K were found.

As already discussed, the clathrates are considered to be promising thermoelectric materials due to the believe that "the rattling of the atoms within the voids of the structure" would improve their thermoelectric performance [3]. In the context of this idea, the larger equivalent isotropic displacement parameter for the alkali metals residing in the clathrate-II cages indicate larger mean-square displacement amplitude. This is in agreement with the fact that the sodium thermal parameters in the Ge-clathrates are larger compared to those in the Si-clathrates. Na has ionic radius of 1.02 Å and therefore fits better the smaller cage for the silicon clathrates (1.10 Å) than that of the germanium clathrates (1.25 Å). Similarly, the rubidium thermal parameters are larger than those of cesium, because Cs^+ fits better in the larger cage than Rb^+ for both silicon and germanium clathrates (Table I). These oversized cages, with respect to the cations, are believed to be responsible for the good thermoelectric efficiency as demonstrated for some clathrate-I phases (e.g., $Sr_8Ga_{16}Ge_{30}$) and some filled skutterudites (e.g., $LaFe_4Sb_{12}$) [7]. However, the results from the conductivity measurements discussed above and the thermopower value at room temperature (-29 $\mu V/K$) imply that clathrate-II will not have high performance at ambient conditions. Indeed, simple calculations show that in order to have $ZT \geq 1$ at 293 K, $Cs_8Na_{16}Si_{136}$ should have total thermal conductivity of less than 0.8 W/m·K.

In case of the Sn-clathrates, both cavities are with a suitable size to be "stuffed" with K^+, Rb^+ and Cs^+ but too large for Na^+ and therefore no Na-containing clathrates of type I or II shall be expected. Moreover, the clathrate-II structure will not be thermodynamically favorable, since the size of the larger void in the latter is too big (r > 2.2 Å). Therefore, the most reoccurring compounds in the binary and pseudo-binary systems A-Sn (A = K, Rb, Cs) are clathrate-I and the clathrate-alike phase $A_{8\pm x}Sn_{25}$ with chiral chains of Sn-dodecahedra [8]. Nevertheless, employing appropriate mixtures of Na and A it is also possible to stabilize novel structure types that are otherwise inaccessible in the A-Sn systems, such as the new compounds $A_6Na_{18}Sn_{46}$, containing clathrate layers, Sn_4^{4-}-tetrahedra and alkali metal cations as shown in Figure 2. No evidence for Sn-vacancies within the clathrate layers was found. The formula of these "more reduced" clathrate derivatives suggests electronically balanced compounds. All of them seem to be surprisingly stable at ambient conditions although good single crystals have not been found

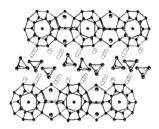

Figure 2. General view of part of the rhombohedral cell of $A_6Na_{18}Sn_{46}$ (A= K, Rb, Cs). Shown are the clathrate layers Sn_{34}^{12-} (stuffed with A^+) and the layers of Sn_4^{4-}, separated by layers of Na^+ (open circles). The formula can be rationalized as $A_6Na_{18}Sn_{46} = A_6Na_6Sn_{34} + 3Na_4Sn_4$

after being treated with solvents. The most likely reason for that is the partial and slow oxidation of the Sn_4^{4-} tetrahedra in the interlayer space that causes disorder of the structure. Preliminary magnetic and conductivity measurements confirm the proposed structure and stoichiometry [9].

CONCLUSIONS

For the first time, large single crystals of four clathrate-II type compounds have been synthesized from the pure elements. The rational synthesis is reproducible and quantitative. The structures of these four compounds were elucidated with high accuracy from single crystal X-ray diffraction data and the refinements unambiguously confirmed the first "defect-free and fully-stuffed" clathrate-II type structures. Transport and magnetic properties support the suggested "ideal" clathrate-II stoichiometry. Using the same approach, three new compounds with novel clathrate-like structure type have been synthesized and structurally characterized.

REFERENCES

1. J. S. Kasper, P. Hagenmuller, M. Pouchard and C. Cros, *Science* **150**, 1713 (1965).
2. J. Gallmeier, H. Schäfer and A. Weiss, *Z. Naturforsch.* **24b**, 665 (1969).
3. G. A. Slack, in *Thermoelectric Materials - New Directions and Approaches*, T. M. Tritt, M. G. Kanatzidis, H. B. Lyon and G. D. Mahan, Eds.; MRS Symposium Proceedings Vol. 478; Pittsburgh, PA (1997) p. 47.
4. S. Bobev and S. C. Sevov, *J. Am. Chem. Soc.* **121**, 3795 (1999); S. Bobev and S. C. Sevov, *J. Solid State Chem.*, accepted.
5. R. D. Shannon and C. T. Prewitt, *Acta Cryst.* **B25**, 925 (1969)
6. V. I. Smelyanski and J. S. Tse, *Chem. Phys. Lett.* **264**, 459 (1997).
7. G. S. Nolas, J. L. Cohn, G. A. Slack and S. B. Schbujman, *Appl. Phys. Lett.* **73**, 178 (1998); B. C. Sales, D. Mandrus and R. K. Williams, *Science* **272**, 1325 (1996).
8. J. T. Zhao and J. D. Corbett, *Inorg. Chem.* **33**, 5721 (1994).
9. S. Bobev and S. C. Sevov, to be published.

Thin Films TE

Electrodeposition of Bi_2Te_3 Nanowire Composites

Amy L. Prieto[1], Melissa S. Sander[1], Angelica M. Stacy[1], Ronald Gronsky[2], Timothy Sands[2]
[1]Department of Chemistry, University of California, Berkeley
[2]Department of Materials Science, University of California, Berkeley
Berkeley, CA, 94720

ABSTRACT

Widespread applications of thermoelectric materials are limited due to low efficiency. Currently, the most widely used thermoelectric devices consist of alloys based on Bi_2Te_3. In such devices, the thermoelectric figure-of-merit (ZT) of bulk Bi_2Te_3 has been increased through doping. It is postulated that further enhancements in ZT may be attained by engineering the microstructure of the material to enhance carrier mobility while suppressing the phonon component of the thermal conductivity. This may be achieved by fabricating Bi_2Te_3 in the form of one-dimensional (1D) nanowires. We have deposited nanowires of Bi_2Te_3 with two different diameters (200 nm and 40 nm) by electrodeposition into porous anodic alumina. Characterization of the Bi_2Te_3/porous Al_2O_3 composite materials has been accomplished using X-ray diffraction (XRD), scanning electron microscopy (SEM) and transmission electron microscopy (TEM). Energy dispersive X-ray spectroscopy (EDS) has been used to determine the stoichiometry of the wires.

INTRODUCTION

There has been much excitement recently over theoretical predictions that quantum confinement of bulk thermoelectric materials could result in a composite material with a ZT significantly above that of bulk values.[1-3] An increase in the Seebeck coefficient due to quantum confinement of carriers, and a decrease in the thermal conductivity due to the scattering of phonons off interfaces is predicted to be the result. $PbTe/Pb_{1-x}Eu_xTe$ superlattices have confirmed these predictions for 2D systems.[4-9] The increase in ZT is expected to be more dramatic in 1D (nanowires) as opposed to 2D systems (superlattices). Our goal is to prepare and characterize Bi_2Te_3 nanowire arrays in order study to the effects of quantum confinement on 1D thermoelectric materials.

Arrays of nanowires have been fabricated in a wide variety of template materials.[10,11] We have chosen templates of anodic alumina due to the easily tunable pore diameters (from 9-300 nm), the high pore densities (to $7 \times 10^{10}/cm^2$), and the high aspect ratio pores (~ 100 μm long pores) these templates provide.[12] The control of these variables in porous anodic alumina templates is a well-established process. In addition, alumina is an electrical and thermal insulator and is thermally stable, making it a good matrix material for the composite arrays.

To fabricate wires within the templates, several different deposition techniques have been developed. We have chosen to employ electrochemical deposition because it offers a high degree of control over the synthesis conditions. The wire stoichiometry and microstructure can be controlled through careful manipulation of a broad range of deposition variables, including concentrations of species in solution, potential, current, electrode material, and temperature. Also, electrochemical deposition is fast, easily adaptable to larger scales and assures continuous wires.

Bi_2Te_3 is one of several thermoelectric materials, including $Bi_{0.5}Sb_{1.5}Te_3$, that have been electrodeposited and studied in bulk and thin films. Fleurial, *et al.* have combined lithography technology and electrodeposition to fabricate Bi_2Te_3 and $Bi_{0.5}Sb_{1.5}Te_3$ legs for integration into micron scale thermoelectric devices.[13] Martin, *et al.* have deposited Bi_2Te_3 into commercially available porous alumina membranes, with average pore diameters of 200 nm.[14] Since the electrodeposition of bulk Bi_2Te_3 is well understood[15], and since Bi_2Te_3 is currently one of the most efficient and well-studied thermoelectric materials, our goal is to fabricate nanowire arrays of Bi_2Te_3 in order to see if quantum confinement of a 1-dimensional thermoelectric will result in enhanced thermoelectric properties. Our immediate goal is to develop a methodology for the synthesis and characterization of Bi_2Te_3 nanowire arrays in order to understand the correlation between structure, composition, and properties.

EXPERIMENTAL

Template Synthesis

Two different template sizes were used to fabricate nanowire arrays: 200 nm and 40 nm. The fabrication of 40 nm anodic alumina templates was adapted[16,17] Aluminum foil (Alfa Aesar, $4cm^2$, 0.13mm thick, 99.9995%) was mounted on a glass slide with crystal bond. The foil was polished with 1-3 micron diamond paste until smooth. The foil was then heated on a hot plate for several hours to grow a thin oxide layer on the surface. A non-conducting epoxy was used to coat the edges of the Al, and allowed to dry thoroughly. The resulting foil was electrochemically polished in a solution of 95 vol % H_3PO_4, 5 vol % H_2SO_4, and 20 g/L CrO_3 at 85 °C and 20 V several times for a few seconds at a time. The polished foil had a mirror-bright finish. Immediately before anodization, the foil was rinsed in a 3.5 vol % H_3PO_4, 45g/L CrO_3 solution at 90°C for two minutes. The film was rinsed with distilled, deionized water, and then anodized.

To make 40 nm porous alumina, the Al foil was anodized in 4 wt % oxalic acid (2 °C) at 30 V for roughly 50 hours. This resulted in an alumina film approximately 100 microns thick. After anodization, the glass slide and foil were placed on a hot plate on low heat to melt the crystal bond. Once removed, the back of the foil was rinsed with acetone. The foil/alumina composite was soaked in saturated $HgCl_2$ to remove the remaining Al. The template was rinsed well with distilled, deionized water, and allowed to air dry. Once dry, the film was floated on the surface of solution of ethylene glycol saturated with NaOH with the barrier oxide layer down. The barrier oxide dissolved after approximately 90 minutes. The porous membrane was rinsed several times and allowed to air dry. Approximately 1 micron of Pt was sputter-deposited on one side.

The 200 nm porous alumina was purchased from Whatman. Approximately 1 micron of Pt was sputter deposited on one side.

Nanowire Electrodeposition

A piece of Cu wire was attached to the Pt side of the porous alumina with Ag paint. The back and edges of the alumina were coated with a non-conducting lacquer. Once dry, the electrode was used for deposition. Electrodeposition was conducted with a three-electrode system. The working electrode (porous alumina) and counter electrode (Pt gauze) were submerged in a solution of 1 M HNO_3, 7.5 x 10^{-3} M BiO^+, and 1.0 x 10^{-2} M $HTeO_2^+$. The reference electrode (Hg/Hg_2SO_4) was in a separate beaker, in a 1 M KNO_3 solution. The two beakers were connected via a 1 M KNO_3/agar salt bridge, and submerged in an ice bath. Depositions were conducted under potentiostatic control, at –740 mV versus the Hg/Hg_2SO_4 reference electrode.

RESULTS

We have successfully fabricated nanowire arrays of 200 nm and 40 nm diameter wires. We confirmed that crystalline Bi_2Te_3 was deposited by X-ray diffraction (Figure 1).

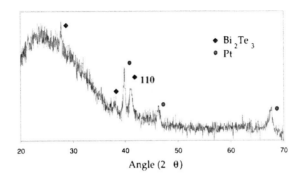

Figure 1: Representative X-Ray diffraction pattern of a filled 200 nm template. The pattern indexes to Bi_2Te_3, and shows strong texturing along the wire axis as indicated by the dominant (110) peak. The large, broad low angle peak is due to the porous Al_2O_3.

Further analysis using scanning electron microscopy (SEM) and energy dispersive spectroscopy (EDS) confirmed that many of the pores had been filled to the top, and that the stoichiometry of the resulting wires was roughly 40 atomic % Bi and 60 atomic % Te (Figure 2).

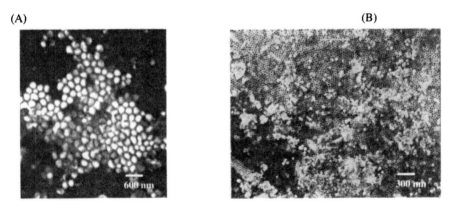

Figure 2: SEM pictures of Bi_2Te_3 nanowire array composites. The bright regions are the filled pores. Both samples were mechanically polished to remove excess Bi_2Te_3 from the surface. (A) 200 nm diameter wires, deposited into commercially available filter. (B) Approximately 40 nm diameter wires, deposited into porous alumina template fabricated in-house. EDS of both samples shows 40 atomic % Bi; 60 atomic % Te.

In order to probe the morphology and grain size of the Bi_2Te_3 wires, transmission electron microscopy was used to view individual wires. The porous template was selectively etched using a CrO_3/H_3PO_4 bath. The TEM images show that the wires are polycrystalline. The grains have aspect ratios of several times the wire width (see Figure 3).

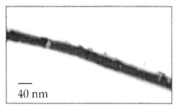 

bright field TEM　　　　　　　　　　　　　　dark field TEM

Figure 3: Bright field and dark field images of a single Bi_2Te_3 nanowires. The grains are the size of the diameter of the wire, but do not run the length of the wire. EDS of the individual wires indicates that the wire is Te-rich at the base, and over the distance of a few microns approaches the 2:3 Bi:Te stoichiometry expected.

CONCLUSIONS

We are able to control the fabrication of the porous alumina to generate desired pore diameters, which are well-ordered and have a narrow size distribution. The porous alumina membranes that we have synthesized are up to 100 microns thick. Using these templates we have successfully fabricated dense Bi_2Te_3 nanowire arrays of two different sizes (200 nm and 40 nm) and characterized the structure and composition of the arrays and the individual wires. The wires are polycrystalline Bi_2Te_3 with strong texturing along the wire axis. The stoichiometry of the wires is 40% Bi, 60 % Te as shown by EDS.

ACKNOWLEDGEMENTS

The authors would like to thank Ron Wilson (Department of Materials Science, UCB) for help with the SEM/EDS data, and Dr. J.-P. Fleurial, Dr. G.J. Snyder, and Dr. A. Borchshevsky (JPL) for helpful discussions. A.L. Prieto would like to thank Lucent Technologies, Bell Labs for funding. This work was made possible through funding provided by the Department of Defense ONR-MURI on Thermoelectrics, #N00014-97-1-0516.

REFERENCES

1) Hicks, L. D.; Dresselhaus, M. S., *Phys. Rev. B* **1993**, *47*, 12727.
2) Hicks, L. D.; Dresselhaus, M. S., *Phys. Rev. B* **1993**, *47*, 16631.
3) Hicks, L. D.; Harman, T. C.; Sun, X.; Dresselhaus, M. S., *Phys. Rev. B* **1996**, *53*, 10493.
4) Harman, T. C.; Spears, D. L.; Walsh, M. P., *J. Electron. Mater.* **1999**, *28*, L1-L4.
5) Harman, T. C.; Spears, D. L.; Manfra, M. J., *J. Electron. Mater.* **1996**, *25*, 1121.
6) Koga, T.; Harman, T. C.; Cronin, S. B.; Dresselhaus, M. S., *Phys. Rev. B* **1999**, *60*, 14286.
7) Harman, T. C.; Taylor, P. J.; Spears, D. L.; Walsh, M. P., *J. Electron. Mater.* **2000**, *29*, L1.
8) Venkatasubramanian, R.; Colpitts, T.; Watko, E.; Hutchby, J., *IEEE 15th International Conference on Thermoelectrics* **1996**, 454.

9) Cho, S.; DiVenere, A.; Wong, G. K.; Ketterson, J. B.; Meyer, J. R., *Physical Review B-Condensed Matter* **1999**, *59*, 10691.
10) Martin, C. R., *Adv. Mater.* **1991**, *3*, 457.
11) Brumlik, C. J.; Martin, C. R., *J.Am.Chem.Soc.* **1991**, *113*, 3174.
12) Shingubara, S.; Okino, O.; Sayama, Y.; Sakaue, H.; Takahagi, T., *Jpn. J. Appl. Phys.* **1997**, *36*, 7791.
13) Fleurial, J.-P.; Borshchevsky, A.; Ryan, M. A.; Phillips, W., *Thermoelectric Microcoolers for Thermal Management Applications*; IEEE: Dresden, Germany, **1997**; *16*, 641.
14) Sapp, S. A.; Lakshmi, B. B.; Martin, C. R., *Advan. Mater.* **1999**, *11*, 402.
15) Takahashi, M.; Katou, Y.; Nagata, K.; Furuta, S., *Thin Solid Films* **1994**, *240*, 70.
16) Zhang, Z.; Gekhtman, D.; Dresselhaus, M. S.; Ying, J. Y., *Chem. Mater.* **1999**, *11*, 1659.
17) Keller, F.; Hunter, M. S.; Robinson, D. L., *J. Electrochem. Soc.* **1953**, *100*, 411.

Thermopower of Bi Nanowire Array Composites.

T.E. Huber[a], M.J. Graf[b], C.A. Foss, Jr.[c], and P. Constant[a].
[a] Laser Research, Howard University, Washington, DC 20059,
[b] Department of Physics, Boston College, Chestnut Hill, MA 02467,
[c] Department of Chemistry, Georgetown University, Washington, DC 20059.

The small effective mass and high mobility of electrons in Bi, make Bi nanowires a promising system for thermoelectric applications. Dense arrays of 20-200 nm diameter Bi nanowires were fabricated by high pressure injection of the melt. Transport properties and Seebeck coefficient were investigated for Bi nanowires with various wire diameters as a function of temperature (1 K < T < 300 K) and magnetic fields (B< 0.6 T). We discuss the problem of the contact resistance of Bi nanowire arrays.

INTRODUCTION

Progress in the study of 1D quantum wire systems has been slow due to the difficulty of fabricating such materials. Since Bi has the smallest electron effective mass among all known materials, quantum confinement effects in Bi are more manifest and can be observed in nanowires of larger diameter than those of any other nanowire material. One technique that is known to be applicable to the problem of fabricating Bi nanowires, and that yields an array of ultrafine nanowires embedded in a porous dielectric template is the high-pressure injection (HPI) of the conducting melt [1] in porous templates. This technique has been applied to fabricate a variety of metal and semiconductor nanowire arrays [2] in insulating porous matrices. The HPI technique was succesfully applied by Gurvitch [3] to the synthesis of single nanowires of Bi in glass pipes.

Recently, the HPI technique has been employed has been employed to fabricate fine Bi nanowire arrays for thermoelectric applications [4,5,6,7]. For such applications, it is important to minimize the template contribution to the thermal conductivity and therefore only samples with very high Bi content, such as wire arrays or networks are of interest. Bulk Bi, a semimetal, and Bi-Sb, a semiconducting alloy, have the highest thermoelectric figure-of-merit Z at 100 K. Quantum size effects are predicted to result in an enhancement of Z for fine wires. Bogachek, Scherbakov and Landman [8] studied theoretically the quantum transport in 3D nanowires. For small diameter wires they predict conductance quantization that is manifest by the step-like behavior of the conductance and the appearance of thermopower peaks as a function of constriction diameter. Hicks and Dresselhaus [9] have predicted an enhancement for 1D conductors. These effects are expected to become very relevant for Bi wire diameters d< 30 nm.

EXPERIMENTS

In this work we use porous anodic aluminium oxide (PAAO) templates. These materials support an array of parallel, largely non-interconnected, cylindrical channels running perpendicular to the plate surface. Two types of porous anodic aluminium oxide templates (PAAO) have been used in this work. One type is a commercial membrane sold for microfiltration under the trade name Anopore [10]. It consists of an alumina plate 25 mm in diameter and about 55 μm thick. The channel pore diameter is about 200 nm. The second type of PAAO template used in this work was prepared in our laboratories by anodyzing a high purity aluminium substrated in acid solutions. PAAO channel diameter and packing density can be systematically controlled by changing the processing parameters. The channel packing density depends mainly on the anodyzing voltage, while the channel diameter depends mainly on the anodizing voltage, the type of electrolyte and the bath temperature. We prepared anodic alumina films of channel diameter down to 20 nm. The details of the film synthesis and detachment from the aluminium substrated are given in Ref. 11.

Like many other conductors, molten Bi does not wet insulators such as alumina or silica at temperatures near their melting points. Therefore, molten Bi will not spontaneously penetrate the channels of an insulating matrix and high-pressure is required to infiltrate the template. The design and operation of the injection apparatus have been described elsewhere [1]. Briefly, the reactor is first heated to a temperature above the melting point of the impregnant.

Scanning electron microscope (SEM) images of a 30 nm diameter Bi nanowire. Top view. Dark is Bi. Clear is alumina.

The temperature is then gradually raised, forcing the molten material into the matrix channels. We have injected Bi in the Anopore nanochannel with a pressure of 1.5 kbar. When the injection is complete, the reactor is cooled and the impregnant solidifies inside the channels, the pressure is then released. The sample is extracted from the ampoule and cleaned of the surrounding excess impregnant by standard mechanical polishing techniques. The composite can be distinguished from pure Bi because it is not as shiny.

The crystalline direction of the Bi wires is relevant to their thermoelectric applications, since the thermal and electrical conductivities of Bi, as well as the Seebeck coefficient are anisotropic. For example, the thermal conductivity λ at 100 K is 18 W/(m.K) along the trigonal direction and 13 W/(m.K) normal to the trigonal direction. Our X-ray studies show that the individual Bi nanowire crystal structure is rhombohedral, with the same lattice parameters as that of bulk Bi; the wires in the array are predominantly oriented with the trigonal axis along the wire length.

It is important to realize that nanowire arrays have a very low intrinsic resistance (typically 10^{-4} Ohm). With a view at the thermoelectric application, we looked for a way to minimize the electrical and thermal contact resistance of contacts to the copper leads. One solution is to inject the template with space for bulk Bi pads on both sides of the template. Copper leads are attached to the bulk Bi pads using low melting point solder. Since bulk Bi is a good electrical conductor this solution is satisfactory for testing. However, there are some problems. One problem is that these assemblies are not very strong and are prone to cleave. Another, more serious problem, is that pure Bi has a large magnetoresistance, much larger that the wire array. This is a serious drawback in applications requiring a magnetic field. Because of these two problems the bulk Bi electrodes have to be as thin. This is difficult to achieve in practice.

Another way to make low resistance contacts is to take advantage of the injection process to create a Bi layer between the metal electrodes and the wire array. Electrical contact was made via copper leads attached to the copper electrodes. The space between the electrodes and the template, as well as the template, was injected with molten Bi. This resulted in a thin (20 micrometers thick) layer of Bi between the copper electrodes and the wire array . The resulting devices are strong, do not cleave on thermal cycling, and are easily handled. Cu has a high solubility of 0.2% in liquid Bi at 325 C [12] whereas there is only slight solubility for bulk Bi . According to Ref. 12, X-rays studies show no evidence of intermetallic compounds. We have measured the Cu concentration in the Bi nanowire array using a Cameca Electron Microprobe, model MBX, equipped with energy dispersive X-ray spectroscopy and three wavelength dispersive spectrometers. Stage positions and X-ray detection electronics are computer-controlled. Software is utilized to correct raw data for background, absorption, fluorescence, and atomic number effects. The resulting spacial resolution is 1 micrometer. The results obtained suggest that there is an average of roughly 0.3% of copper in the nanowire array after processing. This is consistent with the solubility of Cu in liquid Bi at 400 C, the synthesis temperature. It is clear from our experiments that Cu remains, either segregated or truly dissolved, in the nanowires. This can be understood because the diffusion of Cu back to the Cu electrode during solidification is severely impaired by the nanowire geometry.

It would appear that Cu segregation would contribute very little to the nanowire electronic transport properties, resistance, magnetoresistance, and thermopower because there is not enough of it to even form a continuous monolayer on the wire surface, especially for the small diameter nanowires [13]. However, we have found several effects that can be attributed to copper in Bi. One effect is the following: 30 nm diameter nanowires show a superconducting transition at 0.50 ± 0.05 K. It has been shown by Alekseevskii, Bondar, and Polukarov of the Institute of Physical Problems [14] that the intermetallic compound BiCu can be produced by electrolytic thin film co-deposition of copper and Bi. This compound is presumably unstable for temperatures above 120 C.

Another effect that is directly relevant to thermoelectricity is the anomalous thermopower thermopower of 200 nm Bi nanowires. We measured S over the range of temperatures between 77 K and room temperatures . The equipment used is shown in Fig. 2. The disposition of thermal and electrical contacts, as an anvil, maximizes the thermal conductivity to the copper electrodes whole keeping the electrical contacts away from the heat flow.

We measure a Seebeck coefficient that is, within experimental error, temperature independent between 77 K and room temperature, and equal to 15 microvolts per degree This is anomalous for Bi. In comparison, the thermopower of bulk Bi and Bi microwires is between 50 and 100 microvolts per degree at room temperature and decreases at low temperatures [15]. In conclusion, two methods of providing low contact resistance between copper electrodes and Bi nanowire arrays are evaluated. Neither method is completely satisfactory.

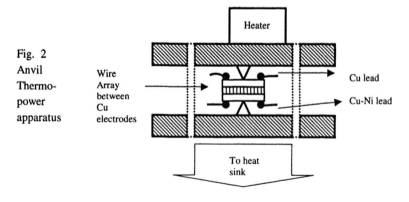

Fig. 2
Anvil Thermo-power apparatus

Wire Array between Cu electrodes

ACKOWLEDGEMENTS

The work of T.E.H. was supported by the Army Research Office through DAA H04-95-1-0117 and by the National Science Foundation through DMR-9632819. M.J.G. was supported in part through Research Corporation grant RA0246. C.F.'s work was supported by the Division of Materials Research of the National Science Foundation through DMR 9625151. We thank Prof. D. Farkas of Virginia Tech. for valuable discussions.

REFERENCES

1. C.A. Huber and T.E. Huber, J. Appl. Phys. 64, 6599 (1988).
2. C.A. Huber, M. Sadoqi, and T.E. Huber, Advanced Materials 7, 316 (1995).
3. M. Gurvitch, J. Low Temp. Phys. 38, 777 (1980).
4. Z. Zhang, J.Y. Ying, and M.S. Dresselhaus, J. Mater. Res. 13, 1745 (1998); Z. Zhang, X. Sun, M.S. Dresselhaus, J.Y. Ying, and J.P. Heremans, Appl. Phys. Lett. 73 1589 (1998).
5. M.S. Dresselhaus, X.Sun, S.B. Cronin, T. Koga, K.L. Wang, and G. Chen, Proc. of the 16th International Thermoelectric Society Conference, Dresden, Germany, edited by A. Heinrich and J. Schumann (IEEE, 1997), p. 12.
6. T.E. Huber and R. Calcao,, Proc. of the 16th International Thermoelectric Society Conference, Dresden, Germany, edited by A. Heinrich and J. Schumann (IEEE, 1997), p. 404.
7. T.E. Huber and C. Foss, Proc. of the 17th International Thermoelectric Society Conference, Nagoya, Japan, edited by K. Koumoto and S. Yamaguchi (IEEE, 1998), p. 244.
8. E.N. Bogachek, A.G. Scherbakov, and U. Landman, in "Nanowires", edited by P.A. Serena and N. Garcia (Kluwer, Dordrecht, 1997).
9. L.D. Hicks and M.S. Dresselhaus, Phys. Rev. B47, 16631 (1993).
10. Whatam Laboratory Division, Clifton, NJ.
11. N.A.F. Al-Rawashdeh, M.L. Sandrock, C.J. Sengling, and C.A. Foss, Jr., J. Phys. Chem. B102, 361 (1998).
12. D.J. Chakrabarti and D.E. Laughlin, Bull. Of Alloy Phase Diagrams 5, 3061 (1972).
13. For a wire of diameter d, assuming a Bi volumetric concentration of c, and that Bi forms a continuous film of thickness t at the periphery of the wire, we obtain that $t = c\, d/4$. Therefore, for a concentration of 0.3% we obtain $t = 0.2$ nm for a wire diameter of 200 nm.
14. N.E. Alekseevskii, V.V. Bondar, and Y.M. Polukarov, J.E.T.P. Lett. 38 294 (1960).
15. T.E. Huber and P. Constant, these Symposium proceedings.

Experimental Investigation of Thin Film InGaAsP Coolers

Christopher J. LaBounty, Ali Shakouri[1], Gerry Robinson, Luis Esparza, Patrick Abraham, and John E. Bowers
Electrical and Computer Engineering, University of California, Santa Barbara, CA 93106
[1]Jack Baskin School of Engineering, University of California, Santa Cruz, CA 95046

ABSTRACT

Most optoelectronic devices for long haul optical communications are based on the InP/InGaAsP family of materials. Thin film coolers based on the same material system can be monolithically integrated with optoelectronic devices such as lasers, switches, and photodetectors to control precisely the device characteristics such as wavelength and optical power. Superlattice structures of InGaAs/InP and InGaAs/InGaAsP are used to optimize the thermionic emission resulting in a cooling behavior beyond what is possible with only the Peltier effect. A careful experimental study of these coolers is undertaken. Mesa sizes, superlattice thickness, and ambient temperature are all varied to determine their effect on cooling performance. A three-dimensional, self-consistent thermal-electric simulation and an effective one-dimensional model are used to understand the experimental observations and to predict what will occur for other untested parameters. The packaging of the coolers is also determined to have consequences in the overall device performance. Cooling on the order of 1 to 2.3 degrees over 1-micron thick barriers is reported.

INTRODUCTION

Thermoelectric (TE) coolers have encountered widespread use in the temperature stabilization of optoelectronic components (lasers, switches, detectors, etc.) in high speed and wavelength division multiplexed (WDM) fiber optic communication systems. This is even more so in dense WDM systems where the spacing between adjacent wavelengths can be from 0.8nm (100GHz) to as small as 0.2nm (25GHz) [1]. Since typical InGaAsP-based DFB lasers operating around 1.55 μm have a wavelength drift of approximately 0.1 nm/°C, the temperature must be controlled to less than a degree of variance to prevent excessive loss in multiplexers / demultiplexers or crosstalk interference. While TE coolers have successfully met this requirement, they have added greatly to the total cost of components since they are not easily integrated with devices [2].

Another disadvantage to the use of TE coolers is the large mismatch in thermal mass between that of the cooler and the device. The smallest TE coolers are a couple of millimeters squared, whereas a typical optoelectronic device is an order of magnitude smaller. Much work is currently underway in thin film thermoelectric refrigeration for other applications, however the same problems of integration with optoelectronics still exist. The InGaAsP/InP family of materials has poor thermoelectric properties due to the inherently small Seebeck coefficient [3]. However, the use of thermionic emission in heterostructures was recently proposed and has been demonstrated in the InGaAsP system to increase the cooling power [4,5] by selectively emitting only the hot electrons over a heterobarrier. An order of magnitude improvement beyond the bulk Peltier properties is possible [6].

In the following we investigate the behavior of several InGaAsP-based thin film thermionic coolers. A qualitative picture of device operation is constructed, and an effort is made to understand the current device limitations by comparing measurements for various device sizes, thickness, and operating temperatures.

MATERIAL STRUCTURE

The material structures investigated were all n-type. Each was grown by metal organic chemical vapor deposition (MOCVD) on n+ InP substrates and was composed of a superlattice barrier layer surrounded by anode and cathode layers of InGaAs. The anode and cathode layers were 0.5μm and 0.3μm thick respectively. Most of the results presented in this work are for a 25 period superlattice of 10nm InGaAs and 30nm InGaAsP (λ_{gap}=1.3μm). All compositions were grown lattice matched to the InP growth substrate.

FABRICATION & MEASUREMENT

Reactive Ion Etching was used to form mesas ranging in area from 3200 μm^2 to 20,000μm^2. In each case the etching depth was through the top cathode and superlattice layers, stopping in the lower anode region. Ohmic metal contacts were formed by electron-beam deposition of 50Å-Ni / 100Å-AuGe / 1000Å-Ni/ 10,000Å-Au. The contacts were then alloyed by rapid thermal annealing at a temperature of 450°C. The specific contact resistivity was measured on separate characterization samples with the transmission line model [7] and determined to be approximately 5×10^{-7} Ω/cm^2. The InP substrate was mechanically lapped to a thickness of 125μm in order to reduce the thermal resistance between the hot side of the cooler and the heat sink. The samples were then cleaved, mounted in packages, and wire bonded for testing.

Micro-thermocouples were used to monitor the temperature of the devices while the current bias was varied. A differential measurement with two thermocouples was used where one is placed on the device and the other on the reference stage. The stage was thermoelectrically controlled to maintain a set heat sink temperature.

EXPERIMENTAL RESULTS & DISCUSSION

It would be useful to begin by discussing qualitatively the device operation. The device can be broken into three regions as shown in figure 1, where the electrical conductivity (σ), thermal conductivity (κ), and area (A) are defined for each region. The equivalent circuit model is shown on the right with the arrows indicating sources or sinks of heat flux. Q_{TI} refers to thermionic heating/cooling, Q_{TE} to thermoelectric cooling (metal-semiconductor interface), and Q_C to heat generation by contact resistance. From circuit analysis, an expression can be found for the temperature at the cold side of the device (between regions 1 & 2). Since the thermal resistance of the wire (R_w) is usually at least an order of magnitude larger than the sum of the cooler and substrate, it is assumed to be zero to simplify the analysis. The resulting expression is,

$$\Delta T = \{(Q_{TI} + Q_{TE})(R_d^{th} + R_{sub}^{th}) - Q_{TI}R_{sub}^{th}\} - \{(\frac{1}{2}R_{Au} + R_C)(R_d^{th} + R_{sub}^{th})\}I^2 \quad (1)$$

where R_d^{th} and R_{sub}^{th} are the barrier and substrate thermal resistances, and R_{Au} and R_C are the gold wire and contact electrical resistances respectively. Q_{TE} and Q_{TI} are both linearly proportional to current [4], hence the first term in equation 1 is linearly proportional to current representing the cooling effects, and the second term is proportional to the square of current representing the heating effects. This equation includes all of the important non-ideal parameters such as contact resistance, wire bond heat load, and substrate thermal resistance.

Cooling Vs. Size

Most of the terms in equation 1 are area dependent, and so studying the cooling dependence on device size provides much information about the behavior of the device. Figure 2(a) shows the measured cooling versus current density for several device areas. At low current densities, all the devices operate nearly identically as expected. As the current is increased, the area dependent non-ideal heating becomes apparent. The smallest size device (5000μm^2) cools best since it requires less current to reach a given current density. Referencing equation 1, the curves in figure 2(a) can be fitted with a second order polynomial and the corresponding linear and quadratic coefficients extracted. Figure 2(b) plots these coefficients versus area. Using equation 1 and the known material properties, the area dependence can be modeled and the device operation understood. In order to develop a more accurate model, the three-dimensional electrical and thermal spreading resistance was simulated for the given geometry. These effective values are used when applying equation one to fit experimental data. In order to replicate the area dependence in figure 2(b), it was necessary to include all of the non-ideal terms

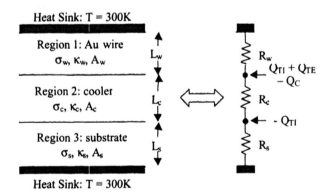

Figure 1. One-dimensional model and boundary conditions. Electrical conductivity (σ), thermal conductivity (κ), and area (A), are defined in each region. The equivalent circuit model is shown on the right with the arrows indicating sources or sinks of heat flux. Q_{TI} refers to thermionic heating/cooling, Q_{TE} to thermoelectric cooling, and Q_C to heat generation by contact resistance.

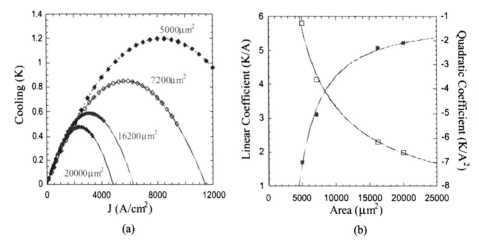

Figure 2. *(a)Cooling versus current density for several sizes and (b) the corresponding linear (cooling) and quadratic (heating) coefficients from a second order polynomial fit as in equation 1. The points are the experimental values, and the curves are simulated from equation 1. Measurements were performed at 300K.*

from equation 1 indicating that all three (contact resistance, wire bond heat load, substrate thermal resistance) have room for improvement.

There does exist a discrepancy between the model and experimental results. With all the non-ideal parameters included, the simulations predict a performance increase for thicker superlattice barriers. Experimentally, little or no improvement has been observed with thicker devices, however the superlattice barriers differed in each case. A more thorough investigation of cooling for thicker identical superlattice barriers is currently underway.

Cooling Versus Temperature

Ideally the height of the heterobarrier and the Fermi level are engineered to be optimum for a given operating range. All of the devices examined were designed for room temperature operation. The cooling behavior was examined for various heat sink temperatures to determine the effects. Figure 3 shows the cooling versus current bias at different temperatures for a 5000 μm^2 cooler. The observed trend is increased cooling at higher operating temperatures. The origin of the temperature dependence stems from the change in material properties (thermal and electrical conductivity) and in the thermionic and thermoelectric cooling mechanisms. Borrowing from the analogous case of conventional thermoelectrics, the maximum cooling and optimum current can be expressed as [8],

$$\Delta T_{max} = \frac{1}{2}\frac{(\Phi_B + 2k_B T/e)^2 \sigma}{\kappa} \quad (2a) \qquad I_{opt} = \frac{\Phi_B + 2k_B T/e}{R} \quad (2b)$$

where $(\Phi_B + 2k_BT/e)$ is the effective cooling for the thermionic effect in the limit of Boltzmann statistics [4], σ and κ are the electrical and thermal conductivity respectively, and R is the electrical resistance. Over the temperature range of interest, the effective cooling is thus approximately $\propto T$ and thermal conductivity $\propto T^{1.4}$ [9]. The temperature dependence of electrical conductivity is determined by the mobility which is approximately constant over these temperature values since it is mostly dominated by impurity scattering. From equations 2a & 2b the maximum cooling and optimum current are hence roughly proportional to $T^{2.4} + T^{3.4}$ and T respectively. The data points in Figure 3(b) fit very well to these corresponding powers, however we are only looking at a small temperature window and a larger spectrum is needed to confirm the validity of these derivations. For an analysis over a wider temperature range, the dependence of the electrical conductivity will have to be estimated, as well as a more accurate model of the effective cooling using Fermi-Dirac statistics.

Cooling Versus Packaging

The packaging has proven to be an important factor to optimize [10]. The addition of a package between the substrate and heat sink adds another thermal barrier for heat to pass through. Improvements in reducing this added thermal resistance by using silicon or copper packages and by optimizing the length of the wire bond have resulted in a maximum cooling increase greater than 100%. It is believed that the package no longer limits device performance.

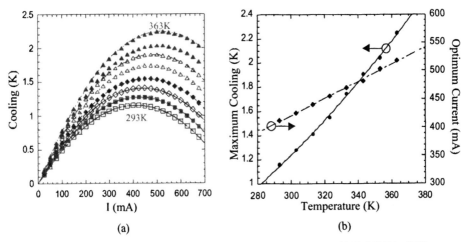

Figure 3. (a) Cooling versus current bias for heat sink temperatures of 293-363K in 10K increments. (b) Corresponding maximum cooling (solid) and optimum current (dashed) versus heat sink temperature. The points are the experimental values and the curves are the theoretical curve fits, $\Delta T_{max} \propto T^{2.4}+T^{3.4}$, $I_{opt} \propto T$. Cooling is measured with respect to the heat sink temperature.

CONCLUSIONS

A thorough experimental investigation of InGaAsP-based thin film thermionic coolers has been presented. Through modeling of the geometry and temperature dependent cooling, an understanding of device performance has been achieved. It was found that smaller area coolers at higher temperatures performed the best, and 2.3K of cooling was demonstrated at 363K for a 5000 μm^2 size device. This amount of cooling over a 1 μm thick barrier corresponds to cooling power densities of several hundred watts per square centimeter. With further improvements in contact resistance, wire bond heat load, and substrate thermal resistance, maximum achievable cooling is expected to reach tens of degrees for single stage devices. These types of thin film coolers should prove to have far reaching applications in optoelectronic devices where a large cooling power, small thermal time constant, and a low cost alternative to thermoelectric coolers are necessary.

ACKNOWLEDGMENTS

This work was supported by the Army Research Office through the DARPA/HERETIC program and by the Office of Naval Research.

REFERENCES

1. Y. Yamada, S.I. Nakagawa, K. Takashina, T. Kawazawa, H. Taga, K. Goto, 25GHz spacing ultra-dense WDM transmission experiment of 1 Tbit/s (100WDM x 10Gbit/s) over 7300km using non pre-chirped RZ format, *Elec. Lett.*, **35**, 2212 (1999).
2. L. Rushing, A. Shakouri, P. Abraham, J.E. Bowers, Micro theroelectric coolers for integrated applications, *Proceedings of 16th International Conference on Thermoelectrics*, Dresden, Germany, August 1997.
3. A. Shakouri, C. LaBounty, Material Optimization for Heterostructure Integrated Thermionic Coolers, *Proceedings of 18th International Conference on Thermoelectrics*, Baltimore, MD, USA August (1999).
4. A. Shakouri, E.Y. Lee, D.L. Smith, V. Narayanamurti, J.E. Bowers, Thermoelectric effects in submicron heterostructure barriers, *Microscale Thermophysical Engineering* **2**, 37 (1998).
5. C. LaBounty, A. Shakouri, P. Abraham, J.E. Bowers, Integrated cooling for optoelectronic devices, *Proceedings of SPIE Photonics West Conference*, San Jose, CA, USA, Jan. 2000.
6. A. Shakouri, C. LaBounty, P. Abraham, J. Piprek, J.E. Bowers, Enahanced Thermionic Emission Cooling in High Barrier Superlattice Heterostructures, *MRS Procedings*, **545**, 449-458, (1998).
7. G.K. Reeves, H.B. Harrison, Obtaining the Specific Contact Resistance from Transmission Line Model Measurments, *Elec. Dev. Lett.*, EDL-3, 5, 111-113, (1982).
8. D.M. Rowe, *CRC Handbook or Thermoelectrics* (CRC Press, New York, 1995).
9. S. T. Huxtable, A. Shakouri, P. Abraham, Y.-J. Chiu, X. Fan, J. E. Bowers, A. Majumdar, Thermal Conductivity of Indium Phosphide Based Superlattices, *to be published in Microscale Thermophysical Engineering*, (2000).
10. C. LaBounty, A. Shakouri, G. Robinson, P. Abraham, J.E. Bowers, Design of integrated thin film coolers, *Proceedings of the 18th International Conference on Thermoelectrics*, Baltimore, MD, USA, August (1999).

AUTHOR INDEX

Abraham, Patrick, Z14.4
Ahn, Channing C., Z11.5
Alboni, Paola N., Z5.1

Badalyan, George, Z8.21
Barkyoumb, John H., Z11.1
Barnes, Peter A., Z5.4, Z9.4
Bergman, David J., Z6.5
Beyer, Harald, Z2.1, Z2.5
Bhattacharya, Sriparna, Z5.2
Bobev, Svilen, Z13.5
Bodiul, Pavel P., Z8.16
Borshchevsky, Alexander, Z1.4, Z11.3, Z11.6
Böttner, Harald, Z2.1, Z2.5
Bouree-Vigneron, F., Z1.2
Bowers, John E., Z11.5, Z14.4
Braun, Thomas P., Z8.2
Brazis, Paul W., Z3.5, Z7.4, Z7.5, Z8.4, Z8.11
Bucher, Ernst, Z3.4

Caillat, Thierry, Z1.4, Z3.3, Z11.6
Chakoumakos, Bryan C., Z7.1, Z13.3
Chapon, L., Z1.2
Chen, G., Z2.4, Z8.3, Z9.1, Z11.4
Chen, Wei, Z8.8, Z10.3
Cho, Sunglae, Z2.4, Z9.1
Choi, Kyoung-Shin, Z8.4, Z8.8
Chung, Duck-Young, Z3.5, Z6.2, Z7.4, Z7.5, Z8.4, Z8.6, Z8.8, Z8.11
Cohn, Joshua L., Z13.1
Constant, Pierre, Z8.5, Z14.2
Croke, Edward, Z11.5
Cronin, Stephen B., Z4.3

Davis, Philip S., Z5.4
DeMattei, Robert C., Z1.4
Demske, David L., Z11.1
de Nardi, Stephan, Z8.6
Derrick, M. Brooks, Z3.1
DiSalvo, Francis J., Z7.6, Z8.2
DiVenere, Antonio, Z2.4
Dong, Jianjun, Z6.1, Z13.2

Dresselhaus, Mildred S., Z4.3
Dyck, Jeffrey S., Z8.8, Z10.3

Esparza, Luis, Z14.4

Fan, Xiaofeng, Z11.5
Feigelson, Robert S., Z1.4
Fel, Leonid G., Z6.5
Feldman, Joseph L., Z6.3
Fleurial, Jean-Pierre, Z1.4, Z3.3, Z11.3, Z11.4, Z11.6
Fornari, Marco, Z6.3
Foss, Jr., Colby A., Z14.2
Freeman, Arthur J., Z2.4
Fritz, Gilbert, Z8.21

Gagnon, Robert, Z5.1
Ghelani, Nishant A., Z3.5, Z8.6
Giordano, Nicholas, Z8.21
Gitsu, Dmitrii V., Z8.16
Graf, Michael J., Z14.2
Gronsky, Ronald, Z10.4, Z14.1
Gryko, Jan, Z6.1, Z13.2
Guardala, Noel A., Z11.1
Gulian, Armen, Z8.21
Gyulamiryan, Ashot, Z8.21

Harper, Jennifer S., Z10.4
Harutyunyan, Sergey, Z8.21
Hecker, Michael, Z8.13
Heinrich, Armin, Z8.13
Herman, Jennifer A., Z11.3
Hoffman, Christopher B., Z8.2
Hoffman, Craig A., Z2.4
Hogan, Timothy P., Z3.5, Z8.4, Z8.6
Horwitz, James, Z8.21
Huang, Chen-Kuo, Z11.3
Huber, Tito E., Z8.5, Z14.2

Inada, Yoichi, Z11.2
Ireland, John R., Z7.4, Z7.5, Z8.4, Z8.11

Jacobs, Todd, Z8.21
Johnson, David C., Z1.1, Z2.3, Z10.1, Z10.2

Kaeser, Michael, Z1.1, Z2.3, Z10.1, Z10.2, Z13.1
Kanatzidis, Mercouri G., Z3.5, Z6.2, Z7.4, Z7.5, Z8.4, Z8.6, Z8.8, Z8.11
Kang, Hyoung Ho, Z11.1
Kannewurf, Carl R., Z3.5, Z7.4, Z7.5, Z8.4, Z8.11
Keppens, Veerle M., Z13.3
Ketchum, Douglas R., Z3.1, Z7.3
Ketterson, John B., Z2.4, Z9.1
Kido, Hiroyasu, Z8.10
Kim, Yunki, Z2.4, Z9.1
Kleint, Christoph A., Z8.13
Kloc, Christian, Z3.4
Koga, Takaaki, Z4.3
Kolis, Joseph W., Z3.1, Z5.1, Z5.4, Z7.3
Kumar, A. Ravi, Z11.4
Kuzanyan, Armen, Z8.21
Kyratsi, Theodora, Z7.5, Z8.8, Z8.11

LaBounty, Christopher J., Z11.5, Z14.4
Laermans, Christiane, Z13.3
Lambrecht, Armin, Z2.1, Z2.5
Lane, Melissa A., Z3.5, Z7.4, Z7.5, Z8.4, Z8.11
Larson, Paul, Z6.2
Legault, Stephan, Z5.1
Lindsey, Norris, Z11.1
Littleton IV, Roy T., Z3.1, Z5.1, Z5.2, Z7.3, Z10.1
Loo, Sim Y., Z8.6

Mahan, Gerald D., Z9.5
Mahanti, Subhendra D., Z6.2
Mandrus, David G., Z7.1, Z13.3
Mazin, Igor I., Z6.3
McMillan, Paul F., Z6.1, Z13.2
Meisner, Gregory P., Z10.3
Meyer, Jerry R., Z2.4
Mrotzek, Antje, Z8.4
Muehl, Thomas, Z8.13
Myles, Charles W., Z6.1

Nelson, Elizabeth, Z1.1, Z2.3, Z10.1, Z10.2
Nikolaeva, Albina A., Z8.16
Nolas, George S., Z1.1, Z2.3, Z8.6, Z10.1, Z10.2, Z13.1
Nomura, Kimitaka, Z11.2
Nurnus, Joachim, Z2.1, Z2.5

Ohnaka, Itsuo, Z11.2
Olafsen, Linda J., Z2.4

Para, Gheorgii, Z8.16
Petrosyan, Silvia, Z8.21
Ponnambalam, Vijayabharathi, Z5.2
Poon, S. Joseph, Z5.2
Pope, Amy L., Z5.1, Z5.2, Z5.4
Price, Jack L., Z11.1
Prieto, Amy L., Z14.1

Qadri, Syed B., Z8.21

Ramachandran, Ganesh K., Z6.1, Z13.2
Ravot, D., Z1.2
Robinson, Gerry, Z11.5, Z14.4
Ryan, Margaret A., Z11.3

Salamanca-Riba, Lourdes, Z11.1
Sales, Brian C., Z7.1, Z13.3
Sander, Melissa S., Z14.1
Sands, Timothy, Z14.1
Sankey, Otto F., Z6.1, Z13.2
Schmitt, Lothar, Z2.5
Schneidmiller, Robert, Z5.1, Z5.4
Schumann, Joachim, Z8.13
Sellinschegg, Heike, Z1.1, Z10.1, Z10.2
Sevov, Slavi C., Z13.5
Shakouri, Ali, Z11.5, Z14.4
Sharma, Jagadish, Z11.1
Simkin, Mikhail V., Z9.5
Singh, David J., Z6.3
Snyder, G. Jeffrey, Z1.4, Z3.3, Z11.3, Z11.6
Song, David W., Z9.1

Sportouch, Sandrine, Z8.6
Stacy, Angelica M., Z14.1
Strelniker, Yakov M., Z6.5
Strom-Olsen, John, Z5.1

Tani, Jun-ichi, Z8.10
Tedenac, Jean-Claude, Z1.2
Tritt, Terry M., Z1.1, Z2.3, Z3.1,
   Z5.1, Z5.2, Z5.4, Z7.3, Z10.1,
   Z10.2, Z13.1

Uher, Ctirad, Z8.8, Z10.3
Ulrich, Marc D., Z9.4

Van Vechten, Deborah, Z8.21
Vartanyan, Violetta, Z8.21
Vining, Cronin B., Z5.4, Z9.4
Völklein, Friedemann, Z2.5
Vurgaftman, Igor, Z2.4

Wang, Ying C., Z7.6

Watcharapasorn, Anucha, Z1.4
Williams, Joshua R., Z1.1, Z2.3
Winkler, Donny, Z5.1
Wölfing, Bernd, Z3.4
Wong, George K.L., Z2.4
Wood, Kent, Z8.21
Wu, Huey-Dau, Z8.21

Xia, Y., Z5.2

Yang, B., Z8.3
Yang, Jihui, Z10.3
Yang, Rong Gui, Z11.4
Yasuda, Hideyuki, Z11.2
Yoon, Gene, Z1.1

Zawilski, Bartosz M., Z3.1, Z5.1,
   Z7.3
Zeng, Gehong, Z11.5

SUBJECT INDEX

alkali metal chalcogenides, Z8.8
alloying, Z8.11
alloys, Z13.1
AlPdMnRe, Z5.1
atomic displacement parameter, Z7.1

band-structures, Z6.3
Bi, Z8.5
BiSb, Z2.4
Bi/Sb superlattice, Z9.1
bismuth, Z14.2
Bi_2Te_3, Z11.3
Bridgman growth, Z8.8

cesium bismuth telluride, Z3.5
chalcogenide(s), Z3.4, Z7.6
clathrates, Z6.1, Z13.1, Z13.2, Z13.3, Z13.5
composites, Z8.5, Z14.2
conductivity, Z8.6, Z13.2
cooler, Z11.5, Z14.4
crystal symmetry, Z7.6
$CsBi_4Te_6$, Z6.2, Z7.5

defects, Z10.4
detector, Z8.21
doping, Z7.4, Z8.10, Z8.11, Z11.1

efficiency, Z11.6
electrochemical, Z11.3
electrodeposition, Z14.1
electronic structure, Z6.2
enhancement, Z6.5

$FeSi_2$, Z11.2
IV-VI compounds, Z2.5

growth, Z2.1

half Heusler, Z5.2
hexaborate, Z8.21

iron disilicide, Z8.10

kinetically stable, Z10.2

lattice dynamics, Z8.3
low-dimensional structures, Z2.5

metastable, Z1.1
molten flux method, Z8.4
multivalley semiconductors, Z8.2

nanowires, Z11.1, Z14.1
  bismuth, Z8.16
new materials, Z8.6
n-type, Z7.5

Pb/Bi/Se, Z7.4
pentatelluride, Z3.1, Z7.3
periodic composite microstructure, Z6.5
phonon confinement, Z8.3
phosphide, Z1.4
porous medium, Z11.2
power generation, Z11.6

quasicrystal, Z5.1
quaternary bismuth selenides, Z8.4

rational synthesis, Z13.5
refrigeration, Z9.4
refrigerator, Z8.21

SiGe, Z8.13, Z11.5
silicon, Z4.3
single crystals, Z13.5
skutterudites, Z1.1, Z1.4, Z2.3, Z6.3, Z7.1, Z10.1, Z10.2, Z10.3, Z10.4
solid solutions, Z3.4, Z3.5
spinel, Z3.3
strain symmetrized, Z8.13
superlattice(s), Z1.1, Z2.1, Z2.3, Z4.3, Z8.3, Z8.13, Z11.5
  alloys, Z2.4
surface scattering, Z8.16

tellurides, Z3.5
TEM, Z10.4
ternary and quaternary, Z8.8

thermal
- conductance, Z7.3
- conductivity, Z5.1, Z5.2, Z5.4, Z6.1, Z9.1, Z10.1
- transport, Z13.1

thermionic(s), Z9.4, Z14.4

thermoelectric(s), Z1.4, Z2.4, Z3.1, Z3.3, Z3.4, Z4.3, Z5.4, Z6.2, Z7.1, Z7.4, Z7.5, Z7.6, Z8.5, Z8.6, Z8.11, Z10.1, Z11.1, Z11.3, Z11.4, Z13.2, Z14.1, Z14.2
- energy conversion, Z11.2
- figure of merit, Z8.16
- material(s), Z2.1, Z2.5, Z6.1, Z8.2
- power factor, Z6.5
- property(ies), Z8.4, Z8.10

thermopower, Z10.3

thin film, Z14.4

3-Omega, Z5.4

3-w method, Z9.1

tin filling, Z10.2

TiNiSn, Z5.2

transient, Z11.4

transport, Z10.3
- properties, Z6.3

ultrasound, Z13.3

unicouple, Z11.6

ZT, Z3.1, Z7.3

CPSIA information can be obtained at www.ICGtesting.com
Printed in the USA
LVOW10s0452300514

387817LV00013B/346/P